Lecture Notes in Control and Information Sciences

Edited by A. V. Balakrishnan and M. Thoma

Vol. 22: Optimization Techniques
Proceedings of the 9th IFIP Conference on Optimization Techniques,
Warsaw, September 4–8, 1979
Part 1
Edited by K. Iracki, K. Malanowski, S. Walukiewicz
XVI, 569 pages. 1980

Vol. 23: Optimization Techniques
Proceedings of the 9th IFIP Conference on Optimization Techniques,
Warsaw, September 4-8, 1979
Part 2
Edited by K. Iracki, K. Malanowski, S. Walukiewicz
XV, 621 pages. 1980

Vol. 24: Methods and Applications in Adaptive Control
Proceedings of an International Symposium
Bochum, 1980
Edited by H. Unbehauen
VI, 309 pages. 1980

Vol. 25: Stochastic Differential Systems – Filtering and Control
Proceedings of the IFIP-WG7/1 Working Conference
Vilnius, Lithuania, USSR, Aug. 28 – Sept. 2, 1978
Edited by B. Grigelionis
X, 362 pages. 1980

Vol. 26: D. L. Iglehart, G. S. Shedler
Regenerative Simulation of Response Times in Networks of Queues
XII, 204 pages. 1980

Vol. 27: D. H. Jacobson, D. H. Martin, M. Pachter, T. Geveci
Extensions of Linear-Quadratic Control Theory
XI, 288 pages. 1980

Vol. 28: Analysis and Optimization of Systems
Proceedings of the Fourth International Conference on Analysis and Optimization of Systems
Versailles, December 16–19, 1980
Edited by A. Bensoussan and J. L. Lions
XIV, 999 pages. 1980

Vol. 29: M. Vidyasagar,
Input-Output Analysis of Large-Scale Interconnected Systems – Decomposition, Well-Posedness and Stability
VI, 221 pages. 1981

Vol. 30: Optimization and Optimal Control
Proceedings of a Conference Held at Oberwolfach, March 16–22, 1980
Edited by A. Auslender, W. Oettli, and J. Stoer
VIII, 254 pages. 1981

Vol. 31: Berc Rustem
Projection Methods in Constrained Optimisation and Applications to Optimal Policy Decisions
XV, 315 pages. 1981

Vol. 32: Tsuyoshi Matsuo,
Realization Theory of Continuous-Time Dynamical Systems
VI, 329 pages, 1981

Vol. 33: Peter Dransfield
Hydraulic Control Systems – Design and Analysis of Their Dynamics
VII, 227 pages, 1981

Vol. 34: H.W. Knobloch
Higher Order Necessary Conditions in Optimal Control Theory
V, 173 pages, 1981

Vol. 35: Global Modelling
Proceedings of the IFIP-WG 7/1 Working Conference Dubrovnik, Yugoslavia, Sept. 1–5, 1980
Edited by S. Krčevinac
VIII, 232 pages, 1981

Vol. 36: Stochastic Differential Systems
Proceedings of the 3rd IFIP-WG 7/1 Working Conference
Visegrád, Hungary, Sept. 15–20, 1980
Edited by M. Arató, D. Vermes, A.V. Balakrishnan
VI, 238 pages, 1981

Vol. 37: Rüdiger Schmidt
Advances in Nonlinear Parameter Optimization
VI, 159 pages, 1982

Vol. 38: System Modeling and Optimization
Proceedings of the 10th IFIP Conference
New York City, USA, Aug. 31 – Sept. 4, 1981
Edited by R.F. Drenick and F. Kozin
XI, 894 pages. 1982

Vol. 39: Feedback Control of Linear and Nonlinear Systems
Proceedings of the Joint Workshop on Feedback and Synthesis of Linear and Nonlinear Systems
Bielefeld/Rom
XIII, 284 pages. 1982

Vol. 40: Y.S. Hung, A.G.J. MacFarlane
Multivariable Feedback: A Quasi-Classical Approach
X, 182 pages. 1982

Vol. 41: M. Gössel
Nonlinear Time-Discrete Systems – A General Approach by Nonlinear Superposition
VIII, 112 pages. 1982

Vol. 42: Advances in Filtering and Optimal Stochastic Control
Proceedings of the IFIP-WG 7/1 Working Conference
Cocoyoc, Mexico, February 1–6, 1982
VIII, 391 pages. 1982

For information about Vols. 1–21 please contact your bookseller or Springer-Verlag.

Lecture Notes in Control and Information Sciences

Edited by M. Thoma and A. Wyner

IFIP

97

I. Lasiecka, R. Triggiani (Eds.)

Control Problems for Systems Described by Partial Differential Equations and Applications

Proceedings of the IFIP-WG 7.2 Working Conference
Gainesville, Florida, February 3-6, 1986

Springer-Verlag
Berlin Heidelberg GmbH

Editor of Conference Proceedings of the series:
Computational Techniques in Distributed Systems IFIP-WG 7.2
Irena Lasiecka
Dept. of Applied Mathematics
Thornton Hall
University of Virginia
Charlottesville, VA 22903
USA

Editors
Irena Lasiecka

Roberto Triggiani
201 Walker Hall
Department of Mathematics
University of Florida
Gainesville, Fl 32611
USA

Library of Congress Cataloging in Publication Data
Control problems for systems described by partial differential equations and applications.
(Lecture notes in control and information sciences ; 97)
1. Control theory – Congresses. 2. Differential equations, Partial – Congresses.
I. Lasiecka, I. (Irena), II. Triggiani, R. (Roberto)
III. IFIP-WG 7.2. IV. Series.
QA402.3.C638 1987 003 87-16549
ISBN 978-3-540-18054-8 ISBN 978-3-540-47722-8 (eBook)
DOI 10.1007/978-3-540-47722-8

Originally published by Springer-Verlag Berlin Heidelberg New York in 1987

2161/3020-543210

PREFACE

This volume comprises the Proceedings at the IFIP TC-7/WG-7.2 Conference on Control Problems for Systems Described by Partial Differential Equations and Applications, held at the University of Florida, Gainesville, Florida, February 3 to 6, 1986.

The Conference was sponsored by the Working Group 7.2 of the Technical Committee 7 of the International Federation for Information Processing (IFIP) and organized through the Center for Applied Mathematics, University of Florida. Financial support was received from the following organizations: IFIP, National Science Foundation, Air Force Office of Scientific Research, and the Center for Applied Mathematics, University of Florida. The support of these organizations is most gratefully acknowledged.

The Conference was devoted to recent advances in the following areas: Linear and Nonlinear Optimal Control Problems, Numerical and Computational Techniques for Control Problems, Stability and Stabilization, Variational Techniques in Control Problems, and Shape Optimization.

It featured six main speakers: A.V. Balakrishnan, V. Barbu, L. Cesari, A. Friedman, R. Glowinski, J.L. Lions; and also 31 invited lecturers. We wish to express our thanks to all the authors for their contributions.

The conference was the first specialized workshop which came to fruition under the auspices and the active involvement of the recently reorganized Working Group 7.2 of the IFIP Technical Committee 7 on System Modelling and Optimization. The TC-7 was founded in 1970 by A.V. Balakrishnan, J.L. Lions, and L.S. Pontryagin. The reorganization of WG-7.2, sponsored by the then chairman of TC-7, Dr. J. Stoer, University of Würzburg, The Federal Republic of Germany, was implemented in the fall of 1984, with the aim of enhancing the cooperation of scientists in the area of Optimal Control of Systems Governed by Partial Differential Equations and related computational aspects.

We acknowledge our appreciation and gratitude to all those who have assisted us at various stages of the Conference; in particular Professors A.R. Bednarek and K. Millsaps, the co-directors of the Center for Applied Mathematics, and also the staff of the Mathematics Department and of the Reitz Union, University of Florida, for their contributions to a successful organization of the Conference. Finally, we wish to express our special thanks to A.B. Aries for invaluable help in preparing these Proceedings for publication.

Irena Lasiecka
Roberto Triggiani

CONFERENCE ON CONTROL PROBLEMS FOR SYSTEMS DESCRIBED BY PARTIAL DIFFERENTIAL EQUATIONS AND APPLICATIONS

February 3-6, 1986
Gainesville, Florida, USA

INTERNATIONAL PROGRAM COMMITTEE

I. Lasiecka (Chairperson)
University of Florida, Gainesville, Florida, USA

A.Bermudez
Universidad de Santiago, Santiago de Compostela, Spain

A. Butkovskij
Institute of Control Sciences, Moscow, USSR

R. Curtain
University of Groningen, Groningen, The Netherlands

G. Da Prato
Scuola Normale Superiore, Pisa, Italy

R. Glowinski
INRIA, Rocquencourt, France

K. Hoffman
University of Augsburg, Augsburg
The Federal Republic of Germany

A. Kurzhanskij
IIASA, Laxenburg, Austria

W. Krabs
Technische Hochschule, Darmstadt
The Federal Republic of Germany

J.L. Lions
College de France and C.N.E.S, Paris, France

U. Mosco
Universita di Roma, Rome, Italy

O. Pironneau
INRIA, Rocquencourt, France

J.P. Zolesio
Universite de Nice, Nice, France

CONFERENCE ON CONTROL PROBLEMS FOR SYSTEMS DESCRIBED BY PARTIAL DIFFERENTIAL EQUATIONS AND APPLICATIONS

LOCAL ORGANIZING COMMITTEE:

A. Bednarek, I. Lasiecka, V. Popov, R. Triggiani Chairmen:

ORGANIZERS:

Professor Irena Lasiecka
Professor Roberto Triggiani

University of Florida, Gainesville, FL, USA

PARTICIPANTS:

1. N.U. Ahmed
2. J.-P. Aubin
3. A.V. Balakrishnan
4. H.T. Banks
5. V. Barbu
6. A. Bermudez
7. V. Capasso
8. L. Cesari
9. C. Corduneanu
10. M.C. Delfour
11. H.O. Fattorini
12. A. Friedman
13. R. Glowinski
14. W.W. Hager
15. K.B. Hannsgen
16. K. Hoffman
17. M. Jacobs
18. F. Kappel
19. J.P. Kernevez
20. K.A. Kime
20. W. Krabs
21. K. Kunisch
22. J. Lagnese
23. E.B. Lee
24. S.J. Lee
26. N. Levan
27. J.-L. Lions
28. W. Littman
29. U. Mackenroth
30. U. Mosco
31. O. Pironneau
32. M. Polisdman
33. T.I. Seidman
34. J. Sokolowski
35. L. Taylor
36. D. Tiba
37. P.K.C. Wang
38. N. Weck
39. J.-P. Zolesio
40. T. Zolezzi

TABLE OF CONTENTS

PART I

PLENARY LECTURES

ON A LARGE SPACE STRUCTURE CONTROL PROBLEM*

A.V. Balakrishnan

Department of Electrical Engineering
University of California, Los Angeles

1. Introduction

This paper concerns a practical problem—in Control *Engineering* rather than Numerical Analysis. The emphasis is *not* on solving partial differential equations. What is needed is the development of good implementable feedback control laws in the absence of total knowledge of the system.

The particular problem is that of slewing an antenna while maintaining the structural stability of the flexible supporting mast—essentially the objective of the NASA SCOLE project using the Space Shuttle under the direction of L.W. Taylor [1], Space Craft Controls Branch, NASA Langley FRC. A laboratory experiment—a test-bed for control laws—is now under way at that center.

Here we shall be concerned with the mast stabilization problem only and the model we use assumes that the angular velocity of the shuttle-antenna system is small enough to be neglected. We model the mast as a thin prismatic beam. There is then the question of whether a finite element model or a continuum (involving partial differential equations) model should be used. Here we deal only with the latter, the basic governing equations being beam bending and torsion equations with control at the boundaries.

With reference to Figure 1, the beam of length L is along the Z axis, z being zero at the shuttle end. $u_\phi(\cdot)$, $u_\theta(\cdot)$ will denote the displacements along the Y-Z, X-Z planes and $u_\psi(\cdot)$ the angular deflection about the Z axis. In addition proof-mass controllers are provided at points s_1 and s_2, on the beam, the locations to be chosen optimally. Control moments are applied at both ends as well as control forces at the reflector center. The various moments of inertia and masses are specified in [1], [2].

* Research supported in part under AFOSR Grant No. 83-0318, USAF.

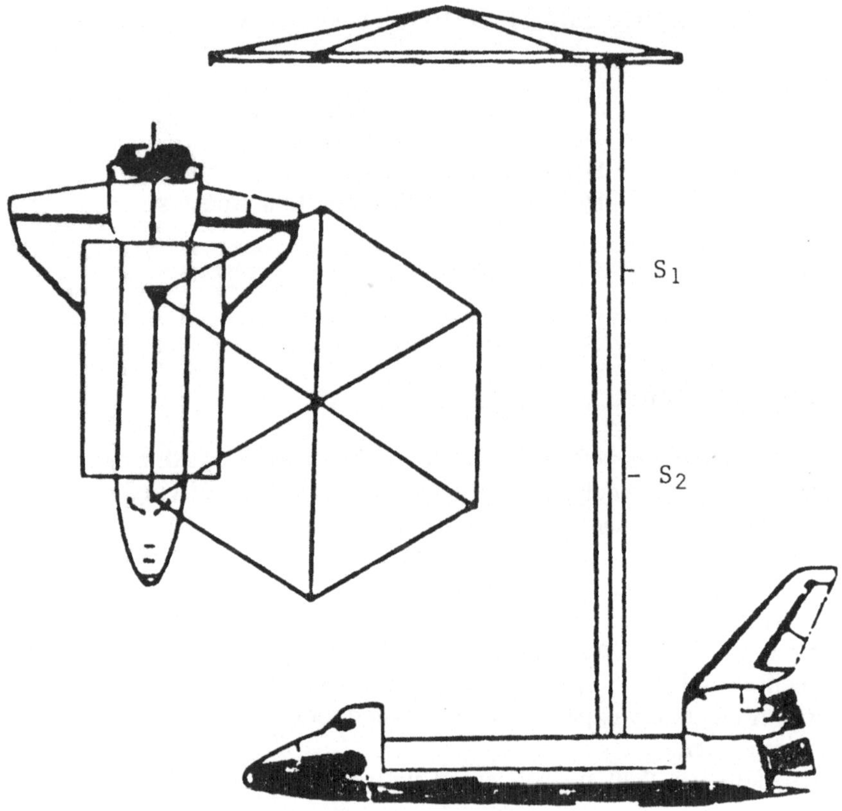

Figure 1

Shuttle/Antenna Configuration

The main result of the paper is the derivation of a robust linear feedback control law which is robust in the sense that it does not depend on quantitative knowledge of system parameter and yet yields global asymptotic stability. This is given in Section 3 following an abstract state-space formulation as a nonlinear wave equation in Section 2, where a controllability result is also established.

2. Abstraction Formulation

We shall confine ourselves to a statement of the problem in the abstract setting if only for reasons of length, referring to [2] for a more detailed description.

We define

$$H = L_2[0,L]^3 \times R^{14} \qquad 0 < L < \infty$$

with the usual inner-product thereon denoted [,]. We fix the points $0 < s^2 < s^3 < L$ and define a linear operator A into H with domain D in H defined as follows. We use $u_\phi(\cdot)$, $u_\theta(\cdot)$, $u_\psi(\cdot)$ to denote the functions in $L_2[0,L]^3$. Thus an element x in H is denoted

$$\begin{array}{c} u_\phi(\cdot) \\ u_\theta(\cdot) \\ u_\psi(\cdot) \\ x_4 \\ \vdots \\ x_{17} \end{array}$$

The domain D consists of elements x such that u_ϕ', u_ϕ'', $u_\phi''' \in L_2[0,L]$ and $u_\phi'''(\cdot)$ has L_2-derivatives in $[0, s_2]$, $[s_2, s_3]$ and $[s_3, L]$; similarly for $u_\theta(\cdot)$; $u_\psi(\cdot)$ such that $u_\psi'(\cdot)$ and $u_\psi''(\cdot) \in L_2[0,L]$; the remaining components of x are specified as

$$\begin{aligned}
x_4 &= u_\phi(0+) & x_{11} &= u_\phi'(L-) \\
x_5 &= u_\theta(0+) & x_{12} &= u_\theta'(L-) \\
x_6 &= u_\phi(L-) & x_{13} &= u_\psi(L-) \\
x_7 &= u_\theta(L-) & x_{14} &= u_\phi(s_2) \\
x_8 &= u_\phi'(0+) & x_{15} &= u_\theta(s_2) \\
x_9 &= u_\theta'(0+) & x_{16} &= u_\phi(s_3) \\
x_{10} &= u_\psi(0+) & x_{17} &= u_\theta(s_3)
\end{aligned}$$

Thus at least for x in D, we may identify the finite-dimensional part as the "boundary." The operator A is then defined by

$$y = Ax$$

where the functional part (in $L_2[0,L]^3$) is given by:

$$EI_\phi u_\phi''''(\cdot)$$

$$EI_\theta u_\theta''''(\cdot)$$

$$-GI_\psi u_\psi''(\cdot)$$

and the boundary part by:

$$
\begin{aligned}
y_4 &= EI_\phi u_\phi'''(0+) & y_{11} &= EI_\phi u_\phi''(L-) \\
y_5 &= EI_\theta u_\theta'''(0+) & y_{12} &= EI_\theta u_\theta''(L-) \\
y_6 &= -EI_\phi u_\phi'''(L+) & y_{13} &= GI_\psi u_\psi'(L-) \\
y_7 &= -EI_\theta u_\theta'''(L+) & y_{14} &= EI_\phi(u_\phi'''(s_2+) - u_\phi'''(s_2-)) \\
y_8 &= -EI_\phi u_\phi''(0+) & y_{15} &= EI_\theta(u_\theta'''(s_2+) - u_\theta'''(s_2-)) \\
y_9 &= -EI_\theta u_\theta''(0+) & y_{16} &= EI_\phi(u_\phi'''(s_3+) - u_\phi'''(s_3-)) \\
y_{10} &= -GI_\psi u_\psi'(0+) & y_{17} &= EI_\theta(u_\theta'''(s_3+) - u_\theta'''(s_3-))
\end{aligned}
$$

It may then be verified that D is dense and A is self-adjoint and nonnegative definite. Moreover A has a compact resolvent with a complete orthonormal set of eigenfunctions (modes). Zero is an eigenvalue.

The control system dynamics can then be characterized as a nonlinear wave-equation:

$$M\ddot{x}(t) + Ax(t) + K(\dot{x}(t)) + Bu(t) = 0 \tag{2.1}$$

where M is a 17×17 nonsingular nonnegative definite matrix, and defines self-adjoint positive definite linear operator H onto H. The control $u(\cdot)$ is in R^{12}, and

$$Bu = x$$

$$x = \text{col. } [0,0,0,0,0,u_1,u_2,u_3,u_4,u_5,u_6,u_7,u_8,u_9,u_{10},u_{11},u_{12}]$$

We have thus only "boundary" control. The nonlinearity is kinematic:

$$K(x) = \left| \begin{matrix} 0 \\ 0 \\ 0 \\ 0 \\ 0 \\ 0 \\ 0 \\ \Omega_1 \otimes I_1 \Omega_1 \\ \Omega_4 \otimes I_4 \Omega_4 \\ 0 \\ 0 \\ 0 \\ 0 \end{matrix} \right|$$

where

$$\Omega_1 = \text{col}\,(x_8, x_9, x_{10})$$

$$\Omega_4 = \text{col}\,(x_{11}, x_{12}, x_{13})$$

I_1, I_4 are symmetric positive definite (moment) matrices and $\otimes$ denotes vector cross-product.

Two relevant properties of the function $K(\cdot)$ are:

(i) $[K(x), x] = 0$

(ii) $\|K(x)\| \leq \text{const.}\ \|x^2\|$

We do allow for "state noise" and let

$$N(t) = \left| \begin{matrix} N_1(t) \\ N_2(t) \\ N_3(t) \end{matrix} \right|$$

$$FN(t) = x(t)$$

where $N(t)$ is white Gaussian with spectral density matrix Λ, and the components of $x(t)$ are defined by

$$
\begin{aligned}
x_i(t) &= 0 & i &= 1,\dots,7 \\
x_8(t) &= N_1(t) \\
x_9(t) &= N_2(t) \\
x_{10}(t) &= N_3(t) \\
x_i(t) &= 0 & i &> 10 .
\end{aligned}
$$

Note that the "boundary" values are part of the state.

State-space Form

With

$$Y = \begin{vmatrix} x(t) \\ \dot{x}(t) \end{vmatrix}$$

we go over to the state-space form:

$$\dot{Y}(t) = \mathcal{A}Y(t) + \mathcal{K}(Y(t)) + \mathcal{B}u(t) + \mathcal{F}(N(t)) \tag{2.2}$$

where

$$\mathcal{A} = \begin{bmatrix} 0 & I \\ -M^{-1}A & 0 \end{bmatrix}$$

$$\mathcal{B}u(t) = \begin{vmatrix} 0 \\ -M^{-1}Bu(t) \end{vmatrix}$$

and in the notation

$$Y = \begin{vmatrix} y_1 \\ y_2 \end{vmatrix}, \qquad Y \in H \times H$$

we have

$$\mathcal{K}(Y) = \begin{vmatrix} 0 \\ -M^{-1}FN(t) \end{vmatrix} \qquad \mathcal{F}N(t) = \begin{vmatrix} 0 \\ -M^{-1}FN(t) \end{vmatrix}$$

As is well known, we can introduce a new inner product, the "energy" inner product

$$[Y, Z]_E = [\sqrt{A}\, y_1, \sqrt{A}\, z_1] + [My_2, z_2]$$

on $\mathcal{R}(A)\times\mathcal{H}$. $\mathcal{R}(A)$ is the orthogonal complement of the null space of A. We denote the completed space $\mathcal{H}_E$. We shall from now on consider on $\mathcal{H}_E$. We have:

$$\mathcal{A} + \mathcal{A}^* = 0$$

and of course $\mathcal{A}$ has a compact resolvent and we have an orthogonal decomposition of $\mathcal{H}_E$ given by

$$Y = \sum_1^\infty P_k Y \tag{2.3}$$

where

$$A\phi_k = \omega_k^2 M\phi_k\,, \quad \omega_k^2 > 0\,,\ \omega_k^2 \to \infty\,; \tag{2.4}$$

$$[M\phi_k, \phi_j] = \delta_j^k\,.$$

Let $S(t)$ denote the semigroup generated by $\mathcal{A}$. Then we have the representation:

$$S(t)Y = \sum_1^\infty S(t)\, P_k\, Y$$

$$P_k S(t)\, P_k = S(t)\, P_k\,.$$

More explicitly, if

$$S(t)Y = \begin{vmatrix} y_1(t) \\ y_2(t) \end{vmatrix}.$$

Then

$$y_2(t) = \dot{y}_1(t)$$

and

$$y_1(t) = \sum_1^\infty [y_1, M\phi_k]\phi_k \cos \omega_k t + \sum_1^\infty [y_2, M\phi_k]\phi_k \frac{\sin\, \omega_k t}{\omega_k}. \tag{2.5}$$

Note that it is required that y_1 satisfy:

$$\sum_1^\infty [y_1, M\phi_k]^2 \, \omega_k^2 \; < \; \infty \, .$$

It is easy to establish existence and uniqueness of solution for the integral version of (2.2):

$$Y(t) = S(t)\,Y(0) + \int_0^t S(t-\sigma)\,\mathcal{B}u(\sigma)\,d\sigma + \int_0^t S(t-\sigma)\,\mathcal{F}N(\sigma)\,d\sigma$$
$$+ \int_0^t S(t-\sigma)\,\mathcal{K}(Y(\sigma))\,d\sigma \, , \tag{2.6}$$

without invoking any nonlinear semigroup theory, by just Picard iteration. See [2].

Controllability

Our first result is that the deterministic system (obtained by taking $F(\cdot)$ in (2.2) to be zero):

$$\dot{Y}(t) = \mathcal{A}Y(t) + \mathcal{K}(Y(t)) + \mathcal{B}u(t) \tag{2.7}$$

is controllable. Because of the special form of $\mathcal{K}(\cdot)$, we can define

$$u(t) = u_1(t) + u_2(t)$$

where

$$\mathcal{B}u_1(t) = -\mathcal{K}(Y(t))$$

so that it is enough to establish controllability for the linear system:

$$\dot{Y}(t) = \mathcal{A}Y(t) + \mathcal{B}u_2(t) \, . \tag{2.8}$$

For any x in $\mathcal{H}$, let

$$b = B^* x$$

and let

$$b_k = B^* \phi_k \, .$$

Suppose for some Y in $\mathcal{H}_E$:

$$[Y, S(t)\mathcal{B}u]_E = 0 \, ;$$

$$Y = \begin{vmatrix} y_1 \\ y_2 \end{vmatrix} \, ,$$

for every t and all u in R^{12}. Using the modal (eigenfunction) expansion (2.5) this will imply that for $k \geq 1$,

$$[y_1, M\phi_k]\,[\phi_k, Bu] = 0$$
$$[y_2, M\phi_k]\,[\phi_k, Bu] = 0\,.$$

If

$$[y_1, M\phi_k] \neq 0\,,$$

we must have

$$[b_k, 0] = 0$$

for every u in R^{12}, where

$$b_k = B^*\phi_k\,.$$

Hence to establish controllability, it is enough to show that

$$b_k \neq 0$$

for any k. Suppose

$$b_k = 0$$

for some k. Then

$$B^*A\phi_k = \omega_k^2 B^*M\phi_k = 0\,.$$

But

$$b_k = 0\,; \qquad B^*(A\phi_k) = 0$$

together will imply that ϕ_k is zero by making the correspnding boundary values:

$$\begin{aligned} u_\phi(L) = u_\theta(L) = u_\theta'(0) &= u_\theta'(0) \\ = u_\psi(0) &= u_\phi'(L) = u_\theta'(L) = u_\psi'(0) \\ = u_\phi'''(L) &= u_\theta'''(L) = u_\phi''(0) \\ &= u_\theta''(0) = u_\psi'(0) = u_\phi''(L) \\ &= u_\theta''(L) = u_\psi'(L) \\ &= 0\,. \end{aligned}$$

It should be noted that the control $u(\cdot)$ involves only 12 of the 14 boundary derivatives.

3. Feedback Control

Our basic result is:

THEOREM 3.1. Let P be any 12×12 symmetric nonnegative definite nonsingular matrix. Then the feedback control

$$u(t) = -P\mathcal{B}^*Y(t) \tag{3.1}$$

is such that the "closed-loop" system

$$\dot{Y}(t) = \mathcal{A}Y(t) - \mathcal{B}P\mathcal{B}^*Y(t) + \mathcal{K}(Y(t)) \tag{3.2}$$

is globally asymptotically stable. That is to say

$$\|Y(t)\|_E \to 0 \qquad \text{as } t \to \infty .$$

PROOF. Let $S_B(t)$ denote the semigroup generated by

$$\mathcal{A} - \mathcal{B}P\mathcal{B}^* = \begin{bmatrix} 0 & I \\ -M^{-1}A & -M^{-1}BPB^* \end{bmatrix} .$$

We shall show first that it is strongly stable. That is to say

$$\|S_B(t)Y\|_E \to 0 \qquad \text{as } t \to \infty$$

This follows from Benchimol [3], since the semigorup $S(t)$ is dissipative, has a compact resolvent and

$$(\mathcal{A}, \sqrt{P}\,\mathcal{B})$$

is controllable. Armed with this result, let us go on to the nonlinear system (3.2). Assume first that $Y(0)$ is in the domain of $\mathcal{A}$. Then (3.2) holds in the strong sense, and hence

$$\frac{d}{dt}\frac{1}{2}\|Y(t)\|_E^2 = -[\mathcal{B}P\mathcal{B}^*Y(t), Y(t)]_E$$

since

$$[\mathcal{K}(Y(t)), Y(t)]_E = 0 .$$

In particular it follows that

$$\int_0^\infty [\mathcal{B}P\mathcal{B}^*Y(t),\, Y(t)]_E \, dt \;<\; \infty \,. \tag{3.3}$$

Since P is nonsingular it follows that

$$\int_0^\infty \|\mathcal{B}^*Y(t)\|^2 \, dt \;<\; \infty \,.$$

But

$$\|\mathcal{K}(Y(\sigma))\|_E \;\leq\; (\text{const.})\; \|\mathcal{B}^*Y(\sigma)\|^2 \,.$$

Hence

$$\int_0^\infty \|\mathcal{K}(Y(t))\|_E \, dt \;<\; \infty \,. \tag{3.4}$$

But by strong stability of $S_B(t)$, this is enough to imply that

$$\|Y(t)\|_E \;\to\; 0 \qquad \text{as } t \to \infty \,.$$

To proceed to the more general case when the differential version may not hold, let us first consider the case where

$$Y(0) \;\in\; \mathcal{D}(\mathcal{A}) \,,$$

Then we define iteratively

$$Y_n(t) \;=\; S_B(t)\, Y(0) \;+\; \int_0^t S_B(t-\sigma)\, \mathcal{K}(y_{n-1}(\sigma))\, d\sigma$$

with

$$Y_0(t) \;=\; S_B(t)\, Y(0) \,.$$

Then because $\mathcal{K}(Y_0(t))$ is strongly differentiable, in any closed interval $[0,\, t]$, it is not difficult to see that so is $Y_1(t)$ and hence by iteration this property holds for every $Y_n(\cdot)$. Moreover we also have the differential version:

$$\dot{Y}_n(t) \;=\; (\mathcal{A} - \mathcal{B}P\mathcal{B}^*)\, Y_n(t) \;+\; \mathcal{K}(Y_{n-1}(t)) \,.$$

$Y_n(t)$ converges uniformly on every closed finite interval to $Y(\cdot)$, because $\mathcal{K}(\cdot)$ is locally Lipschitzian (see [2] if necessary). Moreover

$$\frac{1}{2}[\|Y_n(t)\|_E^2 - \|Y(0)\|_E^2]$$

$$\leq -\int_0^t [\mathcal{B}P\mathcal{B}^*Y_n(\sigma), Y_n(\sigma)]_E\, d\sigma + \int_0^t [\mathcal{K}(Y_{n-1}(t)), Y_n(t)]_E\, 1\sigma$$

$$= \int_0^t \left(-[\mathcal{B}P\mathcal{B}^*Y_n(\sigma), Y_n(\sigma)]_E - \int_0^t [\mathcal{K}[(Y_n(\sigma)) - \mathcal{K}(Y_{n-1}(\sigma))], Y_n(\sigma)]_E\right) d\sigma$$

using the fact that

$$[\mathcal{K}(Y_n(\sigma)), Y_n(\sigma)]_E = 0 .$$

Hence as $n \to \infty$ we obtain

$$\frac{1}{2}[\|Y(t)\|_E^2 - \|Y(0)\|_E^2] = -\int_0^t [\mathcal{B}P\mathcal{B}^*Y(\sigma), Y(\sigma)]\, d\sigma .$$

Since the domain of $\mathcal{A}$ is dense, it is not difficult to see that by the usual arguments that this holds for every $Y(0)$ in $\mathcal{H}_E$. Hence we again get that

$$\int_0^\infty \|\mathcal{K}(Y(\sigma))\|_E\, d\sigma < \infty$$

and hence it follows that

$$\|Y(t)\|_E \to 0 \quad \text{as } t \to \infty .$$

In particular we see that the "boundary values":

$$b(t) = B^*x(t)$$

satisfy:

$$\int_0^\infty [P\dot{b}(t), \dot{b}(t)]\, dt = \frac{1}{2}\|Y(0)\|_E^2 \tag{3.5}$$

This concludes the proof.

We have thus obtained a feedback control law:

$$u(t) = PB^*\dot{x}(t)$$

(rate feedback) which stabilizes the system and at the same time has the robustness property that it does not require quanntitative knowledge of the system parameters. Given the usual model uncertainties, this is a desirable feature. The control law possesses also an optimality property: it is optimal for the quadratic cost functional

$$\int_0^\infty \|\sqrt{P}\,\mathcal{B}^*Y(t)\|^2\,dt \;+\; \int_0^\infty \|u(t)\|^2\,dt$$

for the linear system: $\dot{Y}(t) = \mathcal{A}Y(t) + \mathcal{B}\sqrt{P}\,u(t)$; see [4] for a proof.

There is a catch however; we have not taken account of the inherent damping in the system; and in the absence of knowledge of the damping model our controllability argument may not be valid; unfortunately damping models are in short supply at present, even if one could be used.

Another point worth noting is that the closed-loop semigroup $S_B(t)$ does not any longer have the orthogonal decomposition (2.3). In particular we do not know whether the eigenfunctions are complete. For the SCOLE configuration, the additional damping introduced is small in the sense that

$$\frac{[b_k, b_k]^2}{4\omega_k^2} \ll 1 ,$$

so that the eigenvalues of the closed-loop system are approximately of the form:

$$\sigma_k + i\omega_k$$

where

$$\frac{[Pb_k, b_k]}{2} = \sigma_k = \varepsilon_k \omega_k , \qquad \varepsilon_k \ll 1 .$$

Hence we may continue use of the open-loop eigenfunctions "approximately." Of course a precise "perturbation" theory would be useful. Also we leave open here the choice of P since that would involve the sensor-noise problem.

References

[1] *Scole Workshop Proceedings,* 1984. Compiled by L.W. Taylor. NASA Langley FRC, Hampton, Virginia 23665.

[2] A.V. Balakrishnan. "A Mathematical Formulation of the SCOLE Control Problem." Part 1, NASA CR-172581. Revised July 1985.

[3] C.D. Benchimol. "Feedback Stabilization in Hilbert Spaces." *J. Appl. Math. Optim.*, vol. 4 (1978); 225-48.

[4] A.V. Balakrishnan. "Strong Stabilizability and the Steady State Ricatti Equation." *J. Appl. Math. Optim.*, vol. 7 (1981): 335-45.

THE TIME OPTIMAL PROBLEM FOR A CLASS OF NONLINEAR SYSTEMS

Viorel Barbu

University of Iasi, Romania

1. INTRODUCTION

This work is concerned with the time optimal control problem for nonlinear systems of the form

$$(1.1)\quad \begin{aligned} & y' + Ay + Fy \ni u \\ & y(0) = yo \end{aligned}$$

where A is a linear generator of a sèmigroup of class C_0 in a real Hilbert H, and F is a nonlinear m-accretive operator in H. The plan of the work is the following: In Section 2 we shall derive a maximum principle for the time optimal control problem in the case where F is a smooth operator. In Section 3 we shall study the special case where $H = L^2(\Omega)$, $A = -\Delta$, $D(A) = H_0^1(\Omega) \cap H^2(\Omega)$ and $Fy = \beta(y)$ where β is a maximal monotone graph in R x R. In particular it will be studied in the case where (1.1) reduces to a parabolic variational inequality.

In Section 4 will be presented an algorithm for the time optimal control problem associated with Eq. (1.1). For other literature on time optimal control problems associated with nonlinear systems of the form (1.1), we refer to [1], [2], [3], [5].

2. THE TIME OPTIMAL CONTROL PROBLEM FOR SMOOTH NONLINEARITY F.

We will consider here system (1.1) under the following asumption:

(i) $F:H \to H$ <u>is continuously differentiable and lipschitzian on</u>

bounded subsets; $-A$ is the infinitesimal generator of a C_0-contraction semigroup e^{-At} which is analytic and compact for $t > 0$.

Here H is a real Hilbert space with the scalar product $(.,.)$ and norm $|.|$ We note that for every $u \in L^2(0,T,H)$ and $y_0 \in H$ the control system

$$(2.1) \quad \begin{aligned} & y' + Ay + Fy = u \quad \text{in } [0,T] \\ & y(0) = y_0 \end{aligned}$$

has a unique mild solution $y \in C([0,T];H)$. If $y_0 \in D(A)$ (the domain of A) then y is absolute by continuous, $y' \in L^2(0_1T;H)$ and satisfies Eq. (2.1) a.e. in (0, T). Moreover, the map $u \to y$ is compact from $L^2(0,T;H)$ to $C([0,T];H)$.

Let $U_0 = \{u \in H; |u| \leqslant \rho\}$ and

$$U = \{u \in L^\infty(R+;H);\ u(t) \in U_0 \text{ a.e. } t > 0\}.$$

Let y_0, y_1 be two fixed elements of H. A control $u \in U$ is called admissible if steers y_0 to y_1 on the trajectory of (1.1) (at some time T). The infimum $T(y_0,y_1)$ of all such times for $u \in U$ is called optimal time i.e.,

$$T(y_0,y_1) = \inf \{T;\ y(T,y_0,u) = y_1;\ u \in U\}$$

(Here $y(t,y_0,u)$ is the solution to Eq. (2.1).). A control u^* for which $y(T(y_0,y_1),y_0,u^*) = y_1$ is called time optimal control. The pair $(y(t,y_0,u^*),u^*)$ is called time optimal pair. It follows by a standard device that if there is one admissible control then there exists one time optimal control. This happens for instance if $y_1 \in D(A)$, F. is accretive and

$$(2.2) \quad |Ay_1 + Fy_1| < \rho.$$

Indeed, by the accetivity of $A + F$ we see that the solution $y = y(t)$ to equation

$$y' + (A + F)y + \rho \operatorname{sgn} (y-y_1) \ni 0$$
$$y(0) = y_0$$

steers y_0 to y_1 in finite time.

Now we are ready to formulate the main result of this section. In a few words, Theorem 1 below amounts to saying that every time optimal control for system (2.1) is extremal, i.e., it satisfies the maximum principle.

THEOREM 1 Assume that $y_0, y_1 \in D(A)$ and that assumptions (i) and (2.2) are satisfied. Then, for every time optimal pair (y^*,u^*) there exists a function $p \in C([0,T^*];H)$ which satisfies the system

(2.3) $p^1 - A^*p - (F'(y^*))^*p = 0$ in $[0,T^*]$

(2.4) $u^*(t) = \rho \operatorname{sgn} p(t)\ \forall t \in [0,T]$.

(2.5) $\rho|p(t)| - (Ay^*(t) + Fy(t), p(t)) = 1$ a.e. $t \in [0,T^*]$.

Here $T^* = T(y_0,y_1)$ is the optimal time and $\operatorname{sgn} p = p/|p|$ for $p \neq 0$, $\operatorname{sgn} 0 = \{p; |p| \leq 1\}$.

The idea of the proof consists in aproximating our time optimal control problems by the free time optimal problem.

(2.6) $$\min \{T + \int_0^T (h(u(t)) + \frac{\varepsilon}{2} |u(t)|^2)dt + \frac{1}{2\varepsilon}|e^{-\varepsilon A}(y(T) - y_1)|^2$$
$$+ \frac{1}{2} \int_0^\infty dt \ \Big| \int_0^t (u(s)-u^*(s))ds \ \Big|^2$$
$$;\ y' + Ay + Fy = u \text{ in } R+;\ y(0) = y_0\}$$

where $h(u)$ is the indicator function of U_0, i.e., $h(u) = 0$ if $|u| \leq \rho$, $h(u) = +\infty$ if $|u| > \rho$.

LEMMA 1 Let $(y_\varepsilon,u_\varepsilon,T_\varepsilon)$ be optimal in problem (2.6). Then, for $\varepsilon \to 0$, $T_\varepsilon \to T^* = T(y_0,y_1)$ and

(2.7) $\int_0^t (u_\varepsilon(s)-u^*(s))ds \to 0$ strongly in $L^2(R^+;H)$.

(2.8) $u_\varepsilon \to u^*$ weak star in $L^\infty(0,T^*;H)$.

(2.9) $y_\varepsilon \to y^*$ strongly in $C([0,T^*];H)$ and weakly in $W^{1,2}([0,T^*];H)$.

Proof. We have

$$(2.10)\quad T_\varepsilon + \int_0^{T_\varepsilon} \left(\frac{\varepsilon}{2}|u_\varepsilon|^2 + h(u_\varepsilon)\right)dt + (2\varepsilon)^{-1}\left|e^{-A\varepsilon}(y_\varepsilon(T_\varepsilon)-y_1)\right|^2 + \frac{1}{2}\int_0^\infty dt \left|\int_0^t (u_\varepsilon - u^*)ds\right|^2 \leq T^* + \frac{\varepsilon}{2}\int_0^{T^*} |u^*(t)|^2 dt$$

(we have extended u^* by 0 on $[T^*,+\infty)$).
Hence, $\limsup_{\varepsilon\to 0} T_\varepsilon \leq T^*$ and $\varepsilon \to 0$

$$(2.11)\quad |y_\varepsilon(T_\varepsilon) - y_1| \to 0 \text{ for } \varepsilon \to 0.$$

let $\varepsilon_n \to 0$ be such that $T_{\varepsilon_n} \to \tilde{T}$ and

$$(2.12)\quad u_{\varepsilon_n} \to \tilde{u} \text{ weak star in } L^\infty(R+;H).$$

Since the semigroup e^{-At} is analytic we have

$$\|y'_\varepsilon\|_{L^2(0,\tilde{T};H)} \leq C, \ \forall\, \varepsilon > 0.$$

Moreover, since e^{-At} is compact for $t > 0$, we deduce by the Arzela theorem that $\{y_\varepsilon\}$ is compact in $C([0,\tilde{T}];H)$. Hence, on a subsequence, again denoted ε_n, we have

$$y_{\varepsilon_n} \to \tilde{y} \quad \text{strongly in } C([0,\tilde{T}];H) \text{ and weakly in } W^{1,2}([0,\tilde{T}];H)$$

where $\tilde{y}$ is the solution to Eq. (2.1) with $u = \tilde{u}$. Then, by (2.11) we see that $\tilde{y}(\tilde{T}) = y_1$ and therefore $\tilde{u}$ is admissible. Hence, $\tilde{T} = T^*$ and by (2.10) we see that

$$\int_0^t (u_{\varepsilon_n} - u^*)ds \to 0 \text{ strongly in } L^2(R^+;H),$$

along with (2.12) the latter implies that $u^* = \tilde{u}$ thereby completing the proof.

LEMMA 2 *Let* $y_\varepsilon, u_\varepsilon, T_\varepsilon$ *be optimal in problem* (2.6). *Then there exists* $p_\varepsilon \in W^{1,2}([0,T_\varepsilon);H) \cap C([0,T^*_\varepsilon];H)$ *such that*

(2.13) $y'_\varepsilon + A\, y_\varepsilon + Fy_\varepsilon = u_\varepsilon$ a.e. $t \in (0,T_\varepsilon)$

(2.14) $p'_\varepsilon - A^*p_\varepsilon - (F'(y_\varepsilon))^*\, p_\varepsilon = 0$ a.e. $t \in (0,T_\varepsilon)$

(2.15) $y_\varepsilon(0) = y_0,\ p_\varepsilon(T_\varepsilon) = -\ \varepsilon^{-1} e^{-A^*\varepsilon} e^{-A\varepsilon}(y_\varepsilon(T\varepsilon) - y_1)$

(2.16) $p_\varepsilon(t) \in \partial h(u_\varepsilon(t)) + \varepsilon u_\varepsilon(t) + \int_t^{T_\varepsilon} ds \int_0^s (u_\varepsilon(\tau) - u^*(\tau))d\tau,$

$$\forall\, t \in [0,T_\varepsilon]$$

(2.17) $\int_0^t (u_\varepsilon(s) - u^*(s))ds = 0,\ u_\varepsilon(t) = u^*(t)$ for $t > T_\varepsilon$.

(2.18) $-\ (Ay_\varepsilon(T_\varepsilon) + Fy_\varepsilon(T_\varepsilon), p_\varepsilon(T_\varepsilon)) + \rho|p_\varepsilon(T_\varepsilon) - \varepsilon\, u_\varepsilon(T_\varepsilon)| +$

$$+ \frac{\varepsilon}{2}\, |u_\varepsilon(T_\varepsilon)|^2 = 1$$

Here $\partial h(u) = \{w \in H;\ (w,\ u-v) > 0\ \forall\, u \in \mathcal{U}_0\}$.

Proof The proof is elementary, but we outline it for convenience. If $(y_\varepsilon, u_\varepsilon, T_\varepsilon)$ is optimal then we have

$$\int_0^{T_\varepsilon}(h(u_\varepsilon(t)| + \frac{\varepsilon}{2}\, |u_\varepsilon(t)|^2)dt + \frac{1}{2\varepsilon}|e^{-\varepsilon A}(y_\varepsilon(T_\varepsilon) - y_1)|^2$$

$$+ \frac{1}{2}\int_0^\infty dt\ |\int_0^t (u_\varepsilon - u^*)d\tau|^2 <$$

$$\int_0^{T_\varepsilon}(h(u_\varepsilon(t) + \lambda v(t)) + \frac{\varepsilon}{2}|u_\varepsilon(t) + \lambda v(t)|^2 dt$$

$$+ \frac{1}{2\varepsilon}|\ell^{-\varepsilon A}(y(T_\varepsilon, u_\varepsilon + \lambda v, y_0) - y_1|^2 +$$

$$+ \frac{1}{2}\int_0^\infty dt|\int_0^t (u_\varepsilon + \lambda v - u^*)d\tau|^2\ \forall\, \lambda > 0,\ v \in L^\infty(R+;H).$$

Substracting, dividing by λ and letting $\lambda \to 0$ we get

$$\int_0^{T_\varepsilon}(h'(u_\varepsilon(t), v(t)) + \varepsilon(u_\varepsilon(t), v(t)))dt - \int_0^{T_\varepsilon}(p_\varepsilon(t), v(t))dt +$$

$$+ \int_0^\infty dt(\int_0^t (u_\varepsilon(\tau) - u^*(\tau))d\tau,\ \int_0^t v(\tau)d\tau) > 0$$

where p_ε is the solution to Eqs. (2.14), (2.15) and h' is the directional derivative of h.

Equivalently,

$$\int_0^{T_\varepsilon}(h'(u_\varepsilon(t),v(t)) + \varepsilon(u_\varepsilon(t),v(t))-(p_\varepsilon(t),v(t)))dt + \int_0^\infty (v(\tau), \int_\tau^\infty dt \int_0^t (u_\varepsilon-u^*)ds) \geq 0$$

Since v is arbitrary, the latter inequality implies (2.16) and (2.17).

It remains to prove (2.18). To this end we start with the obvious inequality

$$(2.19)\quad T_\varepsilon + (2\varepsilon)^{-1}\left|e^{-\varepsilon A}(y_\varepsilon(T_\varepsilon)-y_1)\right|^2 + \frac{\varepsilon}{2}\int_0^{T\varepsilon} |u_\varepsilon(t)|^2 dt \leq T_\varepsilon-\lambda +$$

$$+ (2\varepsilon)^{-1}\left|e^{-\varepsilon A}(y_\varepsilon(T\varepsilon-\lambda)-y_1)\right|^2 + \frac{\varepsilon}{2}\int_0^{T\varepsilon-\lambda} |u_\varepsilon(t)|^2 dt.$$

Since $(\varepsilon + \partial h)^{-1}$ is lipschitz and $p_\varepsilon \in W^{1,2}([0,T_\varepsilon];H)$

(because $p_\varepsilon(T_\varepsilon) \in D(A^*)$ and the semigroup e^{-A^*t} is analytic, we see by Eq. (2.16) that u_ε is Hölder continuous on $[0,T_\varepsilon]$.
Hence, $y_\varepsilon \in C^1([0,T_\varepsilon];H)$ (see [8]) and we may pass to limit in [2.19) to get

$$-(y'_\varepsilon(T_\varepsilon), p_\varepsilon(T\varepsilon)) + \frac{\varepsilon}{2}|u_\varepsilon(T_\varepsilon)|^2 \leq -1$$

i.e.,

$$(Ay_\varepsilon(T_\varepsilon) + Fy_\varepsilon(T_\varepsilon) - u_\varepsilon(T_\varepsilon),p_\varepsilon(T_\varepsilon)) + \frac{\varepsilon}{2}|u_\varepsilon(T_\varepsilon)|^2 \leq -1,$$

on the other hand, by (2.16), we have

$$-(u_\varepsilon(T_\varepsilon),p_\varepsilon(T_\varepsilon)) + \frac{\varepsilon}{2}|u_\varepsilon(T_\varepsilon)|^2 = (\partial h(u_\varepsilon(T_\varepsilon)), u_\varepsilon(T_\varepsilon)) =$$

$$= \rho|p_\varepsilon(T_\varepsilon) - u_\varepsilon(T_\varepsilon)|$$

and

$$T_\varepsilon + (2\varepsilon)^{-1}|e^{-A\varepsilon}|y_\varepsilon(T_\varepsilon) - y_1)|^2 + \int_0^{T_\varepsilon} \frac{\varepsilon}{2}|u_\varepsilon(t)|^2 dt \le T_\varepsilon + \lambda +$$

$$(2\varepsilon)^{-1}|e^{-A\varepsilon}(y_\varepsilon(T_\varepsilon + \lambda) - y_1)|^2 + \int_0^{T\varepsilon+\lambda} \frac{\varepsilon}{2}|u_\varepsilon(t)|^2 dt$$

and this yields

$$(-u_\varepsilon(T_\varepsilon) + Ay_\varepsilon(T_\varepsilon) + Fy_\varepsilon(T_\varepsilon), p_\varepsilon(T_\varepsilon)) + \frac{\varepsilon}{2}|u_\varepsilon(T_\varepsilon)|^2 \ge -1$$

as claimed

<u>Proof of Theorem 1.</u> By Eqs. (2.1), (2.18) and the accretivity of A, we have

$$\rho|p_\varepsilon(T_\varepsilon)| \le 1 + |Fy_\varepsilon(T_\varepsilon) + Ay_1 \| p_\varepsilon(T_\varepsilon)| + C\varepsilon.$$

Since $y_\varepsilon(T_\varepsilon) \to y_1$ we infer by assumption (2.2) that $\{p_\varepsilon(T_\varepsilon)\}$ is bounded in H. Then, by the variation of constant formula

$$p_\varepsilon(t) = e^{-A^*(T\varepsilon-t)}p_\varepsilon(T_\varepsilon) + \int_t^T e^{-A^*(s-t)}(F'(y_\varepsilon(s)))^* p_\varepsilon(s)ds$$

and the compactness of the semigroup e^{-A^*t}, we infer that on a subsequence, again denotes ε_n, we have

(2.20) $p_\varepsilon(t) \to p(t)$ strongly in H for all $t \in [0,T^*)$. Where p is a solution to Eq. (2.3). Now letting ε tend to zero in Eq. (2.16) it follows by (2.7), (2.8) and (2.20) that

$$p(t) \in \partial h(u(t)) \quad \text{a.e. } t \in [0,T^*]$$

where $\partial h(u) = \{w \in H;\ (w, u-v) \ge 0,\ \forall u \in U_0\}$. In other words, p satisfies Eq. (2.4). It remains to prove (2.5). To this purpose we note that by Eqs. (2.13) and (2.14) we have

(2.21) $\frac{d}{dt}(Ay_\varepsilon(t) + Fy_\varepsilon(t), p_\varepsilon(t)) = (u_\varepsilon(t), p_\varepsilon'(t))$ a.e. $t \in [0,T_\varepsilon]$.

we set

$$v_\varepsilon(t) = \int_t^{T_\varepsilon} ds \int_0^s (u_\varepsilon(\tau)-u^*(\tau))d\tau.$$

Then, by (2.16) we may write

$$u_\varepsilon(t) = \nabla h_\varepsilon^*(p_\varepsilon(t) - v_\varepsilon(t)) \;\forall\; t \in [0,T_\varepsilon]$$

where

$$(2.22)\quad h_\varepsilon^*(p) = \sup\{(p,u)- \frac{\varepsilon}{2}|u|^2;\quad |u| < \rho\}$$

Since $(\nabla h_\varepsilon^*(p_\varepsilon(t)-v_\varepsilon(t)),\ p_\varepsilon'(t)-v_\varepsilon'(t)) = \frac{d}{dt}h_\varepsilon^*\,(p_\varepsilon(t)-v_\varepsilon(t))$

a.e. $t \in [0,T_\varepsilon]$ it follows by (2.21) that

$$\frac{d}{dt}((Ay_\varepsilon(t) + Fy_\varepsilon(t),p_\varepsilon(t)) - h_\varepsilon^*(p_\varepsilon(t) - v_\varepsilon(t))) = (u_\varepsilon(t),v_\varepsilon'(t)),$$

$$\text{a.e. } t \in [0,T_\varepsilon].$$

integrating from t to T_ε, we get

$$(Ay_\varepsilon(T_\varepsilon) + Fy_\varepsilon(T_\varepsilon),\ p_\varepsilon(T\varepsilon)) - h_\varepsilon^*(p_\varepsilon(T_\varepsilon)) = (Ay_\varepsilon(t) + Fy_\varepsilon(t),p_\varepsilon(t)) -$$

$$-h_\varepsilon^*(p_\varepsilon(t)-v_\varepsilon(t)) + \int_t^{T_\varepsilon}(u_\varepsilon(s),v_\varepsilon'(s))ds,\ \forall t \in [0,T_\varepsilon].$$

Then, letting ε tend to zero it follows by (2.7), (2.8) and (2.18) that

$$(Ay(t) + Fy(t),p(t)) - \rho|p(t)| = \lim_{\varepsilon\downarrow 0}\ (\rho|p_\varepsilon(T_\varepsilon)-\varepsilon u_\varepsilon(T_\varepsilon)|-h_\varepsilon^*(p_\varepsilon(T_\varepsilon)))-$$

$$- 1 = -1,\ \text{a.e. } t \in [0,T^*].$$

To get the latter we have used the fact that

$$h_\varepsilon^*(p) = \rho|p| - \frac{\varepsilon\rho}{2} \text{ if } |p| > \rho\varepsilon,\ h_\varepsilon^*(p) = \frac{|p|^2}{2\varepsilon} \text{ if } |p| < \rho\varepsilon.$$

This completes the proof of Theorem 1.

3. THE TIME OPTIMAL CONTROL OF VARIATIONAL INEQUALITIES

We will study in this section the time optimal problem associated with nonlinear parabolic equation

$$
(3.1)\quad \begin{array}{ll} y_t - \Delta y + \beta(y) \ni u & \text{in } Q = \Omega \times R^+ \\ y(x,0) = y_0(x) & \text{in } \Omega \\ y = 0 & \text{in } \Sigma = \Gamma x R^+ \end{array}
$$

Here Ω is a bounded and open subset of R^N with a sufficiently smooth boundary Γ; $\beta: R \to 2^R$ is a maximal monotone graph such that $0 \in D(\beta)$.

Consider the time optimal problem

$$
(P)\ \inf \{T;\ y(T,y_0,u) = y_1,\ u \in U\}
$$

where $y(t,y_0,u)$ is the solution to Eq. (3.1) and

$$
(3.2)\ U = \{u \in L^\infty(Q);\ |u(x,t)| \leq \rho \text{ a.e. } (x,t) \in Q\}.
$$

As regards y_0 and y_1, throughout this section we will assume that $\exists\ w \in L^2(\Omega)$ such that $w(x) \in \beta(y_0(x))$ a.e. $x \in \Omega$ and

$$
(3.3)\quad y_0, y_1 \in H_0^1(\Omega) \cap H^2(\Omega) \cap L^\infty(\Omega).
$$

$$
(3.4)\quad (-\Delta y_1 + \beta(y_1))^0 \in L^\infty(\Omega);\ \|(-\Delta y_1 + \beta(y_1))^0\|_{L^\infty(\Omega)} < \rho,
$$

where $|(-\Delta y_1 + \beta(y_1))^0| = \inf \{|z|;\ z \in -\Delta z + \beta(t)\}$.

Let us prove first that under these assumptions problem (P) admits at least one admissible control, i.e., it has at least one solution.

LEMMA 3 The solution y to equation

$$(3.5)\quad \begin{array}{ll} y_t-\Delta y + \beta(y) + \rho \operatorname{sgn}(y-y_1) \ni 0 & \text{in } Q \\ y(x,0) = y_0(x) & x \in \Omega \\ y = 0 & \text{in } \Sigma \end{array}$$

steers y_0 y_1 in finite time.

Proof Troughout this section we denote by sgn the map

$$\operatorname{sgn} r = r/|r| \text{ if } r \neq 0\text{: } \operatorname{sgn} 0 = [-1,1].$$

Since the map $y \to \operatorname{sgn}(y-y_1)$ is maximal monotone in RxR, Eq. (3.5) has a unique solution

$$y \in W^{1,2}([0,T]; L^2(\Omega)) \cap L^2(0,T; H_0^1(\Omega) \cap H^2(\Omega)) \text{ for all } T > 0.$$

Consider the function

$$w(x,t) = \|y_0-y_1\|_\infty - (\rho-\mu)t$$

where $\|.\|_\infty$ is the $L^\infty(\Omega)$ norm and $\mu = \|(-\Delta y, + \beta/y,))^0\|_\infty$

We have

$$\begin{array}{l} w_t-\Delta w + \rho \operatorname{sgn} w \ni \mu \text{ in } \Omega x(0,(\rho-\mu)^{-1}\|y_0-y_1\|_\infty) \\ w(x,0) = \|y_0-y_1\|_\infty \text{ in } \Omega;\ w > 0 \text{ in } \Gamma x(0,(\rho-\mu)^{-1}\|y_0-y_1\|_\infty). \end{array}$$

we set $z = y-y_1$ and note that

$$\begin{array}{ll} z_t- \Delta z + \beta(z + y_1) + \rho \operatorname{sgn} z \ni \Delta y_1 & \text{in } Q \\ z(x,0) = y_0(x)-y_1(x) & \text{in } \Omega \\ z = 0 & \text{in } \Sigma \end{array}$$

Subtracting the equations (3.7), (3.6) and multiplying the result by

$(z-w)+$ we find that $(z-w)^+ = 0$ in $\Omega x((\rho-\mu)^{-1}\|y_0-y_1\|_\infty)$

Hence,

$y(x,t)-y_1(x) \leq \|y_0-y_1\|_\infty-(\rho-\mu)t$ for $0 \leq t \leq (\rho-\mu)^{-1}\|y_0-y_1\|_\infty$

Similarly, if we take $w = (\rho-\mu)t-\|y_0-y_1\|_\infty$ we get

$$y(x,t)-y_1(x) \geq -\|y_0-y_1\|_\infty + (\rho-\mu)t \text{ for } 0 \leq t \leq (\rho-\mu)^{-1}\|y_0-y_1\|_\infty$$

Hence, $y(x,T) = y_1(x)$ for $T = \|y_0-y_1\|_\infty/(\rho-\mu)$.

Now let (y^*,u^*) be an optimal pair for problem (p) and $T^* = T(y_0,y_1)$ be the optimal time. In order to obtain a maximum principle theorem we will proceed as in the proof of Theorem 1. Namely, consider the free-time optimal control problem

$$(3.6) \quad \inf \{T + \frac{\varepsilon}{2}\int_0^T |u(t)|_2^2 dt + \int_0^T h(u(t))dt + \eta(\varepsilon)|y(T)-y_1|_2^2 +$$

$$+ \frac{1}{2}\int_0^\infty dt\left|\int_0^t (u(s)-u^*(s))ds\right|_2^2\}$$

where y is the solution to system

$$(3.7) \quad \begin{aligned} &y_t-\Delta y + \beta^\varepsilon(y) = u && \text{in } Q \\ &y(x,0) = y_0(x) \text{ in } \Omega;\ y = 0 && \text{in } \Sigma \end{aligned}$$

Here β^ε is a smooth approximation of β. More precisely, $\beta^\varepsilon = \nabla j^\varepsilon$

where $\lim_{\varepsilon\to 0} j^\varepsilon(r) = j(r)$ and $\liminf_{r_\varepsilon\to r} j^\varepsilon(r_\varepsilon) \geq j(r)$, $\beta = \partial j$ (the differential of j). We have denoted by $|\cdot|_2$ the $L^2(\Omega)$-norm and by $h: L^2(\Omega)\to\overline{R}$ the function

$$h(u) = 0 \text{ if } u \in U_0;\ h(u) = +\infty \text{ if } u \notin U_0$$

where $U_0 = \{u \in L^2(\Omega);\ |u(x)| \leq \rho \text{ a.e. } x \in \Omega\}$.

Finally, $\eta(\varepsilon) \to \infty$ as $\varepsilon \to 0$.

Let $(y_\varepsilon,u_\varepsilon,T_\varepsilon)$ be optimal in problem (3.6). Then as seen in Lemma 2 there is $p_\varepsilon \in C([0,T_\varepsilon];L^2(\Omega))\cap L^2(0,T_\varepsilon;H_0^1(\Omega)\cap H^2(\Omega))$ such that

(3.8) $(y_\varepsilon)_t - \Delta y_\varepsilon + \beta^\varepsilon(y_\varepsilon) = u_\varepsilon \text{ in } Q_\varepsilon = \Omega x(0,T_\varepsilon)$

$y_\varepsilon(x,0) = y_0 \text{ in } \Omega;\ y_\varepsilon = 0 \text{ in } \Sigma_\varepsilon = \Gamma x(0,T_\varepsilon).$

(3.9) $(p_\varepsilon)_t + \Delta p_\varepsilon - (\beta^\varepsilon(y_\varepsilon))' p_\varepsilon = 0 \text{ in } Q_\varepsilon.$

$p_\varepsilon(x,T_\varepsilon) = -2\eta(\varepsilon)(y_\varepsilon(x,T_\varepsilon) - y_1(x) \text{ in } \Omega;\ p_\varepsilon = 0 \text{ in } \Sigma_\varepsilon.$

(3.10) $p_\varepsilon(t) \in \partial h(u_\varepsilon(t)) + \varepsilon u_\varepsilon(t) + \int_t^{T_\varepsilon} ds \int_0^s (u_\varepsilon - u^*) d\tau \text{ for } t \in [0,T_\varepsilon].$

$\int_0^t (u_\varepsilon - u^*) d\tau = 0,\ u_\varepsilon(t) = u^*(t),\ \forall t \in [T_\varepsilon, +\infty)$

(3.11) $\int_\Omega (\nabla y_\varepsilon(x;T_\varepsilon) \nabla p_\varepsilon(x,T_\varepsilon) + \beta^\varepsilon(y_\varepsilon(x,T_\varepsilon))\, p_\varepsilon(x,T_\varepsilon)) dt +$

$+ \rho \int_\Omega |p_\varepsilon(x,T_\varepsilon) - \varepsilon u_\varepsilon(x,T_\varepsilon)| dx + \frac{\varepsilon}{2} |\ u_\varepsilon(T_\varepsilon)|_2^2 = 1.$

Here $\partial h: L^2(\Omega) \to 2^{L^2(\Omega)}$ is the subdifferential of h, i.e.,

$$\partial h(u) = \{w \quad L^2(\Omega);\ (w,u-v) > 0,\ \forall u \in \cup_0\}.$$

Arguing as in the proof of Lemma 1 we find that

LEMMA 3 <u>For</u> $\varepsilon \to 0,\ T_\varepsilon \to T^* = T(y_0,y_1)$

<u>and</u>

(3.12) $u_\varepsilon \to u^*$ weakly in $L^2(0,T^*; L^2(\Omega))$

(3.13) $\int_0^t (u_\varepsilon - u^*) ds \to 0$ strongly in $L^2(R^+; L^2(\Omega))$

(3.14) $y_\varepsilon \to y^*$ strongly in $(C[0,T^*]; L^2(\Omega))$
and weakly in $L^2(0,T^*; H_0^1(\Omega) \cap H^2(\Omega)) \cap$
$W^{1,2}([0,T^*]; L^2(\Omega)).$

Now we need some apriori estimates on p_ε. To this end we multiply Eq. (3.9) by sgn p_ε and integrate on Q_ε to get after some calculation involving (3.4) and (3.9)

$$(3.15)\quad |p_\varepsilon(t)|_1 + \int_{Q\varepsilon} |(\beta^\varepsilon)'(y_\varepsilon)p_\varepsilon|\,dxdt \le |p_\varepsilon(T_\varepsilon)|_1 \le C,\ \forall\varepsilon > 0$$

(We have denoted by $|\cdot|_1$ the $L^1(\Omega)$-norm.)

We note that for $h_i \in L^2(0,T^*;\ L^2(\Omega))$, $i = 1,\dots, N$ and $q > 2$

there is $v \in L^2(0,T^*;\ H_0^1(\Omega))$, $v_t \in L^2(0,T^*;\ H^{-1}(\Omega))$ such that

$$(3.16)\quad \begin{aligned} & v_t - \Delta v = \sum_{i=1}^N (h_i)_{x_i} \ \text{ in } Q^* = \Omega x(0,T^*) \\ & v(x,0) = 0 \text{ in } \Omega;\ v = 0 \text{ in } \Sigma^* = \Gamma x(0,T^*). \end{aligned}$$

Moreover, if $q > N$ then $v \in L^\infty(Q^*)$ and (see [6])

$$(3.17)\quad \|v\|_{L^\infty(Q^*)} \le C \sum_{i=1}^N \|h_i\|_{L^2(0,T^*;\ L^q(\Omega))}.$$

Then if we multiply Eq. (3.9) by v and integrate on Q^* we find after some calculation that

$$(3.18)\quad \sum_{i=1}^N \int_{Q^*} (p_\varepsilon)_{x_i} h_i\,dx\,dt \le C \sum_{i=1}^N \|h_i\|_{L^2(0,T^*;L^q(\Omega))}.$$

Hence

$$(3.19)\quad \|p_\varepsilon\|_{L^2(0,T^*;\ W_0^{1,q'}(\Omega))} \le C,\ \forall\varepsilon > 0$$

where $q^{-1} + (q')^{-1} = 1$. We may infer therefore that $\{(p_\varepsilon)_t\}$ is bounded in $L^1(0,T;\ H^{-s}(\Omega)) + W^{-1,q}(\Omega))$ where $s > N/2$. (We extend u_ε by 0 on $[T_\varepsilon, +\infty)$, and so we may assume that $y_\varepsilon, p_\varepsilon$ are defined on R^+.) Then, by the Helly theorem, there is a function $p \in BV([0,T^*];H^{-s}(\Omega) + W^{-1,q}(\Omega)) \cap L^2(0,T^*;\ W_0^{1,q}(\Omega))$ and a subsequence (again denoted ε_n) such that

(3.20) $p_{\varepsilon_n}(t) \to p(t)$ strongly in $H^{-s}(\Omega) + W^{-1,q}(\Omega)$

for $t \in [0,T]$ and weakly in $L^2_0(0,T; W^{1,q'}(\Omega))$.

On the other hand, for every $\delta > 0$ there is $\eta(\delta) > 0$ such that

$$\|p_{\varepsilon_n}(t) - p(t)\|_{L^{q'}(\Omega)} \leq \delta \|p_{\varepsilon_n}(t) - p(t)\|_{W_0^{1,q'}(\Omega)} +$$

$$+ \eta(\delta) \|p_{\varepsilon_n}(t) - p(t)\|_{H^{-s}(\Omega)+W^{-1,q}(\Omega)}, \quad t \in [0,T^*].$$

Hence

(3.21) $p_{\varepsilon_n} \to p$ strongly in $L^2(0,T^*; L^{q'}(\Omega))$.

Letting ε_n tend to zero in Eqs. (3.9), (3.8), we see that

(3.22) $p_t + \Delta p - \nu = 0$ in $Q^* = \Omega X(0,T^*)$.

(3.23) $u^*(x,t) \in \rho \operatorname{sgn} p(x,t)$ a.e. $(x,t) \in Q$.

Now let us assume that

(3.24) $y_0 \in W_0^{2-\frac{2}{q},q}(\Omega)$; $q > \max(N,2)$

If multiply Eq. (3.8) by $|\beta^\varepsilon(y_\varepsilon)|^{q-2}\beta^\varepsilon(y_\varepsilon)$ and integrate on Q^*

we see that $\{\beta^\varepsilon(y_\varepsilon)\}$ is bounded in $L^q(Q^*)$. Then by Theorem 9.1 in [6] we conclude that $\{y_\varepsilon\}$ is bounded in $W_q^{2,1}(Q^*)$ and therefore for $\varepsilon \to 0$

(3.25) $\Delta y_\varepsilon - \beta^\varepsilon(y_\varepsilon) \to \Delta y^* - \beta(y^*)$ weakly in $L^q(Q^*)$

(3.26) $y_\varepsilon \to y^*$ in $C(\overline{Q})$.

Now, arguing as in the proof of Theorem 1, we get

$$(3.27)\quad \int_\Omega(-\Delta y_\varepsilon(x,T_\varepsilon) + \beta^\varepsilon(y_\varepsilon(x,T_\varepsilon)))p_\varepsilon(x,T_\varepsilon)dx - h^*_\varepsilon(p_\varepsilon(x,T_\varepsilon)) =$$

$$= \int_\Omega(-\Delta y_\varepsilon(x,t) + \beta^\varepsilon(y_\varepsilon(x,t)))p_\varepsilon(x,t)dx - h^*_\varepsilon(p_\varepsilon(t)-v_\varepsilon(t))$$

$$+ \int_t^T \int_\Omega u_\varepsilon(x,t)(v_\varepsilon)_t(x,t)dxdt,\ \forall t \in [0,T_\varepsilon]$$

where

$$v_\varepsilon\ (x,t) = \int_t^T ds \int_0^s (u_\varepsilon(x,\tau) - u^*(x,\tau))d\tau$$

and

$$h^*_\varepsilon(p) = \sup\ \{(p,u)_2 - \frac{\varepsilon}{2}\ |u|_2^2;\ u \in U_0\}$$

Now by (3.21) and (3.25) we see that

$$\lim_{\varepsilon\downarrow 0} \int_\Omega (-\Delta y_\varepsilon(x,t) + \beta^\varepsilon(y_\varepsilon(x,t)))p_\varepsilon(x,t)dx =$$

$$= \int_\Omega (-\Delta y^*(x,t) + \beta(y^*(x,t)))p(x,t)dx \text{ a.e. } t \in [0,T^*].$$

Then taking into account Eq. (3.11) and the fact that

$$|h^*_\varepsilon(p) - \rho|p|_1| \leq C\varepsilon$$

we conclude that

$$(3.28)\quad \int_\Omega (\nabla y^*(x,t)\nabla p(x,t) + \beta|y^*(x,t))p(x,t))dx + \rho|p(t)|_1 = 1$$

$$\text{a.e. } t \in [0,T^*]$$

Summarizing at this point, we have therefore proved

THEOREM 1 _Let_ (y^*,u^*) _be any optimal pair for problem_ (P). _Then under assumptions_ (3.3), (3.4), (3.24) _there exists_ $p \quad L^2(0,T^*;W_0^{1,q'}(\Omega)) \cap Bv([0,T^*];H^{-s}(\Omega) + W^{-1,q}(\Omega))$ _satisfying system_ (3.22), (3.23), (3.28)

where

$$(3.29)\quad \nu = w^*\text{-}\lim_{\varepsilon\to 0} (\beta^\varepsilon(y_\varepsilon))^* p_\varepsilon \text{ in } (L^\infty(Q^*))^*.$$

$((L^\infty(Q^*))^*$ is the dual of $L^\infty(Q^*))$.

As a particular case consider the free boundary problem

$$(3.30)\quad \begin{aligned} & y_t - \Delta y = u && \text{in } \{(x,t); y(x,t) > 0\} \\ & y_t - \Delta y \geq u, y \geq 0 && \text{in } \Omega \times R^+ \\ & y(x,0) = y_0(x) \text{ for } x \in \Omega; && y = 0 \text{ in } \Gamma \times R^+. \end{aligned}$$

In this case, we have

$$(3.31)\quad \beta(r) = 0 \text{ for } r \geq 0;\ \beta(0) = R^-,\ \beta(r) = \emptyset \text{ for } r < 0.$$

and we may take β^ε as (see [7])

$$(3.32)\quad \beta^\varepsilon(r) = \begin{cases} \varepsilon^{-1} r + 2^{-1} & \text{for } r \leq -\varepsilon \\ -(2\varepsilon^2)^{-1} r^2 & \text{for } -\varepsilon \leq r \leq 0 \\ 0 & \text{for } r > 0. \end{cases}$$

We set $X_\varepsilon^1 = \{(x,t);\ y_\varepsilon(x,t) \leq -\varepsilon\}$ and $X_\varepsilon^2 = \{(x,t);\ -\varepsilon \leq y_\varepsilon(x,t) \leq 0\}$. Then we have

$$(3.33)\quad p_\varepsilon \beta^\varepsilon(y_\varepsilon) = p_\varepsilon (\beta^\varepsilon)'(y_\varepsilon) y_\varepsilon + 2^{-1} p_\varepsilon X_\varepsilon^1 + 2^{-1} p_\varepsilon (y_\varepsilon \varepsilon^{-1})^2 X_\varepsilon^2.$$

Using the fact that $\{\beta^\varepsilon(y_\varepsilon)\}$ is bounded in $L^2(Q)$

and $\{p_\varepsilon(\beta^\varepsilon)'(y_\varepsilon)\}$ in $L^1(Q)$, it follows by (3.32), (3.33) that for some $\varepsilon_n \to 0$

$$(3.34)\quad p_{\varepsilon_n} \beta^{\varepsilon_n}(y_{\varepsilon_n}) \to 0 \qquad \text{a.e. in } Q$$

and by (3.21)

$$p_{\varepsilon_n}\beta^{\varepsilon_n}(y_{\varepsilon_n}) \to p\mu \quad \text{weakly in } L^1(Q)$$

where $\mu = -u^* - y_t^* + \Delta y^* \in \beta(y^*)$ a.e. in Q. Hence,

(3.35) $p_{\varepsilon_n}\beta^{\varepsilon_n}(y_{\varepsilon_n}) \to 0$ strongly in $L^1(Q)$

and

(3.36) $p(u^* - y_t^* + \Delta y^*) = 0$ a.e. in Q.

Now using once again (3.33) we see that for $\varepsilon \to 0$

$$p_\varepsilon(\beta^\varepsilon(y_\varepsilon) - (\beta^\varepsilon)'(y_\varepsilon)y_\varepsilon) \to 0 \quad \text{strongly in } L^1(Q).$$

Hence,

(3.37) $p_{\varepsilon_n}(\beta^{\varepsilon_n})'(y_{\varepsilon_n})p_{\varepsilon_n} \to 0$ strongly in $L^1(Q)$.

Since as seen earlier, $y_{\varepsilon_n} \to y^*$ in $C(\overline{Q})$ it follows by (3.22), (3.37) that

$$(p_t + \Delta p)\, y^* = 0 \quad \text{in } Q.$$

We have therefore proved the following theorem

THEOREM 3 <u>Let</u> $y_0 \in H_0^1(\Omega) \cap H^2(\Omega) \cap L^\infty(\Omega) \cap W_0^{2-\frac{2}{q},q}(\Omega)$, $q > \max(N,2)$ <u>be such that</u> $y_0 > 0$ <u>in</u> Ω <u>and let</u> $y_1 \in H_0^1(\Omega) \cap H^2(\Omega) \cap L^\infty(\Omega)$ <u>be such</u> that

$$y_1 > 0 \text{ in } \Omega \text{ and } \|\Delta y_1\|_{L^\infty(\Omega)} < \rho.$$

<u>Let</u> (y_ε^*, u^*) <u>be any optimal pair in time optimal control problem</u> (P). <u>Then there is</u>

$$p \in L^2(0,T^*; W_0^{1,q'}(\Omega)) \cap BV([0,T^*]; H^{-s}(\Omega) + W^{-1,q}(\Omega))$$

<u>such that</u>

$p_t + \Delta p \in (L^\infty(Q^*))^*$ <u>and</u>

(3.38) $p_t + \Delta p = 0$ in $\{(x,t) \in Q^*;\ y^*(x,t) > 0\}$.

(3.39) $p = 0$ in $\{(x,t) \in Q^*;\ y^*(x,t) = 0\}$.

(3.40) $u^*(x,t) \in \rho \operatorname{sgn} p(x,t)$ a.e. $(x,t) \in Q^*$.

(3.41) $\rho|p(t)|_1 + \int_\Omega \Delta y^*(x,t)p(x,t)dx = 1$ a.e. $t \in [0,T^*]$.

Theorem extends in a natural way to time optimal control problems for the variational inequality

$$y_t - \Delta y = u \quad \text{in } \{y > \psi\}$$
$$y_t - \Delta y \geq u \quad \text{in } \Omega X R^+;\ y \geq \psi \text{ in } \Omega X R^+$$
$$y(x,0) = y_0(x) \text{ for } x \in \Omega:\ y = 0 \text{ in } \Gamma\ X R^+$$

where $\psi \in C^2(\overline{\Omega})$ is a given function such that $\psi \leq 0$ in Γ. Replacing $\beta^\varepsilon(y)$ by $\beta^\varepsilon(y-\psi)$ in the previous proof we find that:

_If_ $y_0 \in W^{2-\frac{2}{q},\,q}(\Omega) \cap L^\infty(\Omega)$ _and_ $y_1 \in H_0^1(\Omega) \cap H^2(\Omega) \cap L^\infty(\Omega)$ _are such that_

$$y_0, y_1 \geq \psi \text{ a.e. in } \Omega:\ |\Delta y_1|_\infty < \rho$$

then every time optimal pair (y^*, u^*) _satisfies along with some function_ $p \in L^2(0,T^*;W_0^{1,q'}(\Omega)) \cap BV([0,T^*];\ H^{-s}(\Omega)) + W^{-1,q}(\Omega))$ _the following system_

(3.42) $p_t + \Delta p = 0$ in $\{(x,t) \in Q^*; y^*(x,t) > \psi(x)\}$.

(3.43) $\rho|p| - p\Delta\psi = 0$ a.e. in $\{x,t) \in Q^*; y^*(x,t) = \psi(x)\}$.

(3.44) $u^*(x,t) \in \rho \operatorname{sgn} p(x,t)$ a.e. in Q^*.

(3.45) $\rho|p(t)|_1 + \int_\Omega \Delta y^*(x,t)p(x,t)dx = 1$ a.e. $t \in [0,T^*\}$.

We now consider the special case where $y_1 \equiv 0$ and take in approximating problem $\eta(\varepsilon) = \varepsilon^{-1/2}$. If we multiply Eq. (3.8) where $y = y_\varepsilon$ by $(y_\varepsilon)_t$ and integrate with respect to x and we get

$$\int_\Omega j^\varepsilon(y_\varepsilon(x,t))dx \leq C$$

where C is independent of ε and t. Then by Eq. (3.30) we see that

$$(3.46) \quad \int_{[y_\varepsilon < 0]} y_\varepsilon^2(x,T_\varepsilon)dx \leq M\varepsilon \text{ for all } \varepsilon > 0.$$

Then by Eq. (3.9), we see that

$$\frac{1}{2}\int_\Omega |p_\varepsilon^+ (x,t)|^2 dx + \int_t^T \int_\Omega |\Delta p_\varepsilon^+(x,t)|^2 dx \leq \eta(\varepsilon)\int_\Omega |y_\varepsilon (x,T_\varepsilon)|^2 dx \leq M\varepsilon^{1/2}$$

as we derive from (3.20) that $p^+(x,t) = 0$ a.e. $(x,t) \in \Omega \times (0,T^*)$. (Here p is the function appearing in Theorem 2). We may conclude therefore that the optimal control u* given by Theorem has the form

$$u^*(x,t) = \operatorname{sgn} p(x,t)$$

where

$$p_t + \Delta p = 0 \quad \text{in } \{(x,t) \in Q;\ y^*(x,t) > 0\}$$

and

$$p \leq 0 \quad \text{a.e. in } Q.$$

If the set $\{(x,t) \in Q;\ y^*(x,t) > 0\}$ is connected then by the strong maximum principle we conclude that $p < 0$ in this set and so

$$(3.47) \quad u^* = -1 \text{ in } \{(x,t) \in Q; y^*(x,t) > 0\}.$$

In general we may say that $u^* = -1$ in at least one conexe component of noncoincidence set $[y^* > 0]$.

We have obtained a feedback representation for optimal control u* which, as noted earlier, remains valid for problem (3.42).

If β is locally lipschitz then arguing as in [1], [3] we see that the optimal pair (y^*, μ^*) satisfies the system

$$p_t + \Delta p - \partial\beta(y^*)p \ni 0 \text{ in } Q^*$$

(3.48) $$u^* = \rho p/|p| \qquad \text{in } [p > 0]$$

where $\partial\beta$ is the generalized gradient of β.

Now we will consider as another example the time optimal problem for the controlled diffusion reaction equation

(3.49) $$y_t - \Delta y + a|y|^{s-1}y = u \quad \text{in } \Omega \times R^+$$

$$y(x,0) = y_0(x) \qquad x \in \Omega$$

$$y = 1 \qquad \text{in } \Gamma \times R^+$$

where is a positive constant and $o < s < 1$. It is known that for the solution y to such an equation the "dead core" $\{(x,t);\ y(x,t) \neq 0\}$ is nonempty and has positive measure.

Define β^ε as

$$\beta^\varepsilon(y) = \int \beta(y - \varepsilon\theta)\rho(\theta)d\theta$$

where $\beta(y) = a|y|^{s-1}y$ and ρ is a C_0^∞-mollifier on R.

If $y_\varepsilon, p_\varepsilon$ are the solution to system (3.8), (3.9), (3.6) we note first that

$$|\beta^\varepsilon(y_\varepsilon)p_\varepsilon - (\beta^\varepsilon)'(y_\varepsilon)y_\varepsilon p_\varepsilon| \leq C\varepsilon|p_\varepsilon| +$$

$$+ (1-s)|\beta^\varepsilon(y_\varepsilon)p_\varepsilon| \text{ a.e. in } Q.$$

We will assume that y_0 satisfies (3.24). Since $y_\varepsilon \to y^*$ uniformly in $\overline{Q}$ we infer that

$$\beta^\varepsilon(y_\varepsilon)p_\varepsilon \to 0 \text{ a.e. in } [(x,t) \in Q;\ y^*(x,t) = 0\}$$

and $\nu = 0$ in $[y^* \neq 0]$.

To summarize, we may conclude therefore that <u>for every time optimal pair</u> (y^*,a^*) <u>of problem</u> (3.49) <u>there exists a function</u>

$p \in L^2(0,T^*; W_0^{1,q'}(\Omega)) \cap BV([0,T^*]; H^{-s}(\Omega) + W^{-1,q}(\Omega))$ <u>satisfying the optimality system</u>

$$(3.50) \quad p_t + \Delta p - as|y|^{s-1}p = 0 \qquad \text{in } \{(x,t);\ y^*(x,t) \neq 0\}$$

$$(3.51) \quad p = 0 \qquad \text{in } \{(x,t);\ y^*(x,t) = 0\}$$

$$(3.52) \quad u^*(x,t) \in \rho \operatorname{sgn} p(x,t) \qquad \text{a.e. } (x,t) \in Q$$

$$(3.53) \quad \int_\Omega (\Delta y^* + a|y^*|^{s-1}y^*)p\,dx + \rho \int_\Omega |p(x,t)|\,dx = 1 \qquad \text{a.e. } t \in [0,T^*].$$

4. APROXIMATION OF TIME OPTIMAL CONTROLS

Consider the time optimal problem (P) corresponding to system (3.1). Let $h_\varepsilon : L^2(\Omega) \to R$ be defined by

$$(4.1) \quad h_\varepsilon(u) = \inf \{|u - v|_2^2 / 2\varepsilon;\ v \in U_0\}$$

and consider the free time optimal control problem

$$(4.2) \quad \inf \{ \int_0^T (1 + h_\varepsilon(u))dt + (2\varepsilon)^{-1}|y(T) - y_1|_2^2;\ u \in L^2(0,T;L^2(\Omega))\}$$

where y is the solution to control system (3.7). Arguing as in the proof of Lemma 1, we infer that if $(y_\varepsilon, u_\varepsilon)$ is a family of optimal pairs for problem (4.2) corresponding to transition time T_ε, then

$$(4.3) \quad \lim_{\varepsilon \to 0} T_\varepsilon = T$$

and every weak-star limit point in $L^\infty(0,T^*; L^2(\Omega))$ of $\{u_\varepsilon\}$ is a time optimal control for problem (P).

In other words, u_ε <u>is a suboptimal control for problem</u> (P).

To compute the optimal controls u_ε to problem (4.2) we may use an algorithm of gradient type. To this end we set

$$(4.4)\quad \phi(u,T) = \int_0^T (1 + h_\varepsilon(u(t)))dt + (2\varepsilon)^{-1}|y(T) - y_1|_2^2$$

where y is the solution of Eq. (3.7).

A little calculation shows that

ϕ: $L^2(R^+;H)XR^+ \to R$ is Gateaux differentiable at every u which is absolutely continuous on all [0,T]. Moreover, the Gateaux derivative is given by

$$(4.5)\quad \nabla\phi(u,T) = \{\nabla h_\varepsilon(u)-p, 1-(\Delta y(T)-\beta^\varepsilon(y(T) + u(T), p(T))_2 + h_\varepsilon(u(T))\}$$

where p is the solution to dual system (3.9), i.e.

$$(4.6)\quad \begin{aligned} & p_t + \Delta p - (\beta^\varepsilon)'(y)p = 0 && \text{in } \Omega X(0,T) \\ & p = 0 && \text{in } \Gamma X(0,T) \\ & p(x,T) = -\varepsilon^{-1}(y(x,T) - y_1(x)) && \text{in } \Omega. \end{aligned}$$

Thus, the gradient algorithm for problem (P_ε) can be described as follows

<u>Step 0</u> Select $u_0 \in U$ absolutely contnuous and $T_0 > 0$,

<u>Step 1</u> Set $i = 0$.

<u>Step 2</u> Compute (y_i, p_i) from the system

$$(4.7)\quad \begin{aligned} & (y_i)t - \Delta y_i + \beta^\varepsilon(y_i) = u_i +_0 && \text{in } \Omega\, x(0,T_i) \\ & y_i = 0 && \text{in } \Gamma x(0,T_i) \\ & y_i(x,0) = y_0(x) && x \in \Omega. \end{aligned}$$

$$(4.8)\quad \begin{aligned} & (p_i)_t + \Delta p_i - (\beta^\varepsilon)'(y_i)p_i = 0 && \text{in } \Omega x(0,T_i) \\ & p_i = 0 && \text{in } \Gamma x(0,T_i) \\ & p_i(x,T_i) = -\varepsilon^{-1}(y_i(x,T_i) - y_1(x)), && x \in \Omega \end{aligned}$$

Step 3 Set

$$H_i = (\nabla h_\varepsilon(u_i) - p_i, 1-(\Delta y_i(T_i) - \beta^\varepsilon(y_i(T_i)) , p_i(T_i))_2 -$$

$$- (p_i(T_i), u_i(T_i))_2 + h_\varepsilon(u_i(T_i))).$$

If $H_i = 0$ stop; else go to step 4.

Step 4. Compute the scalar λ_i to be the smallest non-negative number satisfying

$$\phi((u_i, T_i) - \lambda_i H_i) = \min \{\phi((u_i, T_i) - \lambda H_i); \lambda > 0\}.$$

Step 5 Set $u_{i+1} = u_i - \lambda_i(\nabla h_\varepsilon(u_i) - p_i)$ and $T_{i+1} = T_i$
$- \lambda_i(1 - \Delta y_i(T_i) - \beta^\varepsilon(y_i(T_i)) + u(T_i), p_i(T_i))_2 +$

$+ h_\varepsilon(u_i(T_i))$ and go to step 2.

REFERENCES

1. V. BARBU, The time optimal control problem for parabolic variational inequalities, Applied Mathematics and Optimiz., 11: 1-22 (1984).

2. V. BARBU, Optimal Control of Variational Inequalities, Pitman Research Notes in Mathematics 100 London.Boston.Melbourne, 1984.

3. V. BARBU, The time optimal control of variational inequalities; Dynamic programming and the maximum principle, in Recent Mathematical Methods in Dynamic Programming, Capuzzo Dolceta, Fleming and Zolezzi eds. Lecture Notes in Mathematics 1119, Springer Verlag, New York.Heidelberg.Berlin, 1985.

4. H. BREZIS, Problèmes unilateraux. J. Math. Pures Appl., 51, 1-168 (1972).

5. H.O. FATTORINI, The maximum principle for nonlinear nonconvex systems in infinite dimensional spaces, Conference on Control Theory for Distributed Parameter Systems, Lecture Notes in Mathematics, Springer Verlag (to appear).

6. G.A. LADYZENSKAIA, V.A. SOLONNIKOV, N.N. URALTZEVA, Linear and Quasilinear Equations of Parabolic Type, Amer. Math. Soc. Translations, A.M.S. Providence 1968.

7. F. MIGNOT, J. PUEL, Optimal Control in some variational inequalities, SIAM J. Control and Optimiz. 22 (1984), 466-476.

8. A. PAZY, Semigroups of Linear Operators and Applications to Partial Differential Equations. Springer-Verlag, New York.Berlin.Heidelberg.Tokyo 1983.

DISCONTINUOUS SOLUTIONS IN PROBLEMS OF OPTIMIZATION

L. Cesari

University of Michigan, Ann Arbor, Michigan, USA

P. Brandi and A. Salvadori

Università degli Studi, Perugia, Italy

In Section 1 we mention a few points of the theory of functions of bounded variations of $\nu \geqq 1$ independent variables (BV). In Section 2 we state a few results concerning the existence of optimal solutions for problems of the calculus of variations and optimal control with simple integrals. In Section 3 we state analogous results for problems with multiple integrals.

These results are presented in detail in two papers [6], [7], where more extensive bibliographical references are given.

1. BV functions of $\nu \geqq 1$ independent variables

In 1936 ([3]) Cesari introduced a concept of BV real valued functions $z: G \to R$, or $z(t)$, or $z(t^1,\dots,t^\nu)$, from a domain G of R^ν into R. For the case $\nu = 2$, G the rectangle $(a,b;c,d)$ the definition is very simple: we say that z is BV in $G = (a,b;c,d)$ provided $z \in L_1(G)$ and there is a set E of measure zero in G such that the total variation $V_x(y)$ of $z(\cdot,y)$ in (a,b) is of class $L_1(c,d)$, and the total variation $V_y(x)$ of $z(x,\cdot)$ in (c,d) is of classe $L_1(a,b)$, where these total variations are computed completely disregardig the values taken by z in E. The number

$$V_o = V_o(z,G) = \int_a^b V_y(x)\,dx + \int_c^d V_x(y)\,dy$$

may well be taken as a definition of total variation of z in $G = (a,b;c,d)$. Analogous definitions hold for BV functions $z(t^1,\dots,t^\nu)$ in an interval G of R^ν.

We omit the more involved definition of BV function in a general bounded domain G of R^ν.

If z is continuous in G, then no set E need be considered and the concept reduces to Tonelli's concept of BV continuous functions. For discontinuous functions, examples show how essential it is to disregard sets of measure zero. On the other hand, the concept obviously concerns equivalent classes in $L_1(G)$.

We may think of $z(t)$, $t \in G \subset R^\nu$, as defining a nonparametric discontinuous surface $S: z = z(t)$, $t \in G$, in $R^{\nu+1}$, and we may take as generalized Lebesgue area $L(S)$ of S the lower limit of the elementary areas $a(\Sigma)$ of the polyhedral surfaces $\Sigma: z = Z(t)$, $t \in G$, converging to z pointwise a.e. in G (or in $L_1(G)$). More precisely, if (Σ_k) denotes any sequence of polyhedral surfaces $\Sigma_k: z = z_k(t)$, $t \in G$, converging to z pointwise a.e. in G (or in $L_1(G)$), we take for $L(S)$ the number, $0 \leqq L(S) \leqq +\infty$, defined by

$$L(S) = \operatorname*{Inf}_{(\Sigma_k)} \ \varliminf_{k \to +\infty} a(\Sigma_k).$$

Cesari proved ([3]) that $L(S)$ is finite if and only if z is BV in G. This shows that the concept of BV functions is independent of the direction of the axes in R^ν. More than that, the concept of BV functions is actually invariant with respect to 1-1 continuous transformations in R^ν which are Lipschitzian in both directions.

In 1937 Cesari ([4a]) proved that for $\nu = 2$, $G = (0,2\pi;\ 0,2\pi)$ and z BV in G, then the double Fourier series of z converges to z (by rectangles, by lines, and by columns) a.e. in G. Comparable, though weaker, results hold for BV functions of $\nu > 2$ independent variables and their multiple Fourier series [4b].

In 1950 Cafiero ([2]) and later in 1957 Fleming ([11]) proved the relevant compactness theorem: any sequence (z_k) of BV functions with equibounded total variations, say $V_o(z_k,G) \leqq C$, and equibounded mean values in G, possesses a subsequence z_{k_s} which is pointwise convergent a.e. in G as well as strongly convergent in $L_1(G)$ toward a BV function z with $V_o(z,G) \leq \underline{\lim}\, V_o(z_k,G)$, because of the lower semicontinuity of V_o with respect to either such convergences.

In 1967 Conway and Smoller ([8]) used these BV functions in connection with the weak solutions (shock waves) of conservation laws, a class of nonlinear hyperbolic partial differential equations in $R^+ x R^\nu$. Indeed, they proved that, if the Cauchy data on (0) x R^ν are locally BV, then there is a unique solution on $R^+ \times R^\nu$, also locally BV and satisfying an entropy condition. Without any entropy condition there are in general infinitely many weak solutions. Later Dafermos ([9]) and Di Perna ([10]) characterized the properties of the BV weak solutions of conservation laws.

Meanwhile, in the fifties, distribution theory became known, and in 1957 Krickeberg ([13]) proved that the BV functions are exactly those $L_1(G)$ functions whose first order partial derivatives in the sense of distributions are finite measures in G.

Thus a BV function z(t), $t \in G$, G a bounded domain in R^ν, possesses first order partial derivatives in the sense of distributions which are finite measures $\mu_j, j = 1,\dots,\nu$. On the other hand, if we think of the initial definition of z, we see that the set E of measure zero in G has intersection $E \cap \ell$ of linear measure zero on almost all lines ℓ parallel to the axes. Hence, z is BV on almost all such straight lines ℓ when we disregard the values taken by z on E, and has therefore "usual" partial derivatives $D^j z$ a.e. in G, and these derivatives are functions in G of class $L_1(G)$. We call these $D^j z(t)$, $t \in G$, $j = 1,\dots,\nu$, computed by usual incremental quotients disregarding the values taken by z on E, the generalized first order partial derivatives of z in G.

Much work followed on BV functions in terms of the new definition, that is, thought of as those $L_1(G)$ functions whose first order derivatives are finite measures.We mention here Fleming ([11]), Volpert ([14]), Gagliardo ([12]), Anzellotti and Giaquinta ([1]), and also De Giorgi, Da Prato, Giusti, M. Miranda, Ferro, Caligaris, Oliva, Fusco, Temam. However, there are advantages in using both view points.

Great many properties of BV functions have been proved. To begin with, a "total variation" V(z,G) can be defined globally in terms of functional analysis,

$$V(z,G) = \operatorname{Sup}[(\int_G f_1\, d\mu_1)^2+\ldots+(\int_G f_\nu\, d\mu_\nu)^2]^{1/2},$$

where the Sup is taken for all $f_1,\ldots,f_\nu \in C(\bar{G})$ with $f_1^2+\ldots+f_\nu^2 \leq 1$.

One can prove that there are constants c_1, c_2, $0<c_1<c_2<+\infty$, depending only on ν and G such that $c_1 V_o(z) \leq V(z) \leq c_2 V_o(z)$. Moreover, if (z_k) is a sequence of BV functions on G with equibounded total variations, say $V(z,G) \leq C$, and $z_k \to z$ in $L_1(G)$, then z is BV and $V(z,G) \leq \underline{\lim}_{k\to+\infty} V(z_k,G)$.

The question of the existence of traces $\gamma: \partial G \to R$ for BV functions $z: G \to R$ has been discussed under both view points. Note that for a BV function z in an interval G = (a,b;c,d) it is trivial that the generalized limits z(a+,y) and z(b−,y), z(x;c+) and z(x,d−), exist a.e., and are L_1 functions, i.e., the trace $\gamma(z)$ of z on ∂G esists and is $L_1(\partial G)$. For general domains G in R^ν possessing the cone property everywhere on ∂G, a theorem of Gagliardo ([12]) characterizes the properties of ∂G, and one can prove that any BV function z in a bounded domain G with the cone property and $\mathcal{H}^{\nu-1}(\partial G) < \infty$, possesses a trace $\gamma(z)$ on ∂G with $\gamma(z) \in L_1(\partial G)$.

We mention here the following theorem by Gagliardo on bounded domains G with the cone property:

If G is a bounded open domain in R^ν having the cone property, then there is a finite system $\{G_1,\ldots,G_m)$ of open subsets of G with max diam G_s as small as we want, each G_s has the cone property, and has locally Lipschitzian boundary ∂G_s.

From this result, and trace properties for Lipschitzian domains, it is possible to define the traces $\gamma(z)$ of a BV function on ∂G, for G bounded and with the cone property. An equivalent definition of traces of BV functions in terms of the distributional definition is also well known.

We come now to the delicate question of the continuity of the traces $\gamma(z)$ of BV functions z in a domain G, in other words whether $z_k \to z$, say in $L_1(G)$, may actually imply - under assumptions - that $\gamma(z_k) \to \gamma(z)$ in $L_1(\partial G)$. A number of devices have been proposed to this effect. For instance, Anzellotti and Giaquinta ([1]) have recently proved the following statement in terms of the distributional definition of BV functions:

If G has the cone property at every point of ∂G, if $\mathscr{H}^{\nu-1}(\partial G) < \infty$, if the functions z_k are BV with $V(z_k) \leq C$, if $z_k \to z$ in $L_1(G)$ with $V(z_k) \to V(s)$, then $\gamma(z_k) \to \gamma(z)$ in $L_1(\partial G)$. A parallel proof of this statement is available in terms of the original definition of BV functions. We just mention here that it is well known that any BV function $z(t)$, $t \in G \subset R^{\nu}$, can be approximated in $L_1(G)$ by BV smooth functions z_k with $V(z_k) \to V(z)$, hence $V(z_k) \leq C$.

The following example shows a simple situation in which the trace operator is not continuous:

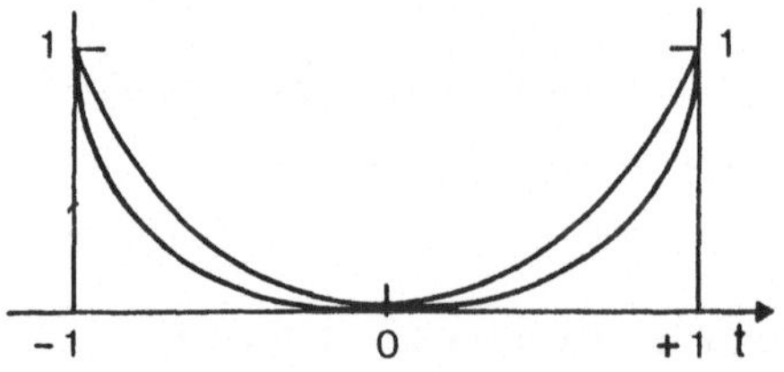

$z_k(t) = t^{2k}$, $-1 \leq t \leq 1$, $k = 1,2,\ldots$,

$z(t) = 0$, $-1<t<1$, $z(-1)=z(1)=1$,

$$V(z_k, [-1,1]) = 2, \quad V(z, [-1,1]) = 2,$$

$$z_k(-1+0) = 1, \; z_k(1-0) = 1, \; z(-1+0) = 0, \; z(1-0) = 0.$$

Of course the condition of Giaquinta and Anzellotti is not satisfied here:

$$V(z_k , (-1,1)) = 2, \quad V(z,(-1,1)) = 0.$$

Another relevant statement concerning the continuity of the trace operator is as follows:

For z , z_k all BV in $G \subset R^\nu$, let μ , μ_k denote the systems of their first order partial derivatives in the sense of distributions, namely, ν-vector valued finite measures in G. Let Z, Z_k denote the mean values of z, z_k in G. If $Z_k \to Z$ and if $\mu_k \to \mu$ weakly, then $\gamma(z_k) \to \gamma(z)$ in $L_1(\partial G)$.

We only mention here that by $\mu_k \to \mu$ weakly we mean, in terms of duality, that $\int_G \langle f,d\mu_k\rangle \to \int_G \langle f,d\mu\rangle$ for all continuous functions f on $\bar{G}$, or $f \in (C(\bar{G}))^\nu$, and in both integrals we mean that G is (bounded) and open in R^ν.

For $\nu = 1$, i.e., for functions $z(t)$, $a \leq t \leq b$, of a real variable t, say of bounded variation in (a,b), then the generalized limits $z(a+)$ and $z(b-)$ (the traces) obviously exist and are finite. If $z_k(t)$, $a \leq t \leq b$, is a sequence of BV functions, say equibounded and with equibounded total variations, then by Helly's theorem, there is a subsequence z_{k_s} which converges pointwise everywhere in [a,b] (as well as in $L_1(G)$) toward a BV function $z(t)$, $a \leq t \leq b$, with $V(z) \leq \underline{\lim}\, V(z_{k_s})$, and in particular, we may require that $z_k(a) \to z(a)$, $z_k(b) \to z(b)$. Thus, for $\nu = 1$, the question of the continuity of the end values is trivially answered in the affirmative by everywhere pointwise convergence and Helly's theorem.

We shall see now how these ideas have been used in questions of optimization, first for simple integrals and then for multiple integrals.

2. Problems of optimization for simple integrals, $\nu = 1$

We may be interested either in problems of the classical calculus of variations involving a vector valued state variable $z(t) = (z^1,\dots,z^n)$, $t_1 \leq t \leq t_2$, or in problems of optimal control involving an analogous state variable $z(t) = (z^1,\dots,z^n)$ and a control variable $u(t) = (u^1,\dots,u^m)$, $t_1 \leq t \leq t_2$, with given control space $U(t,z)$ and constraint $u(t) \in U(t,z(t))$.

It is more general, and more satisfactory, (cfr. [5]), to deparametrize the problems of optimal control, and concern ourselves exclusively with generalized problems of the calculus of variations with constraints on the derivatives, say

$$I(z) = \int_{t_1}^{t_2} f_o(t,z(t),z'(t))dt = \text{minimum},$$

$$(t,z(t)) \in A \subset R^{n+1}, \quad z'(t) \in Q(t,z(t)), \tag{1}$$

where $t \in [t_1,t_2] \subset R$ (a.e.), where A is a subset of R^{n+1} whose projection on the t-axis contains $[t_1,t_2]$, and where, for every $(t,z) \in A$, a set $Q(t,z)$ is given constraining the direction $z'(t)$ of the tangent to the state variable z a.e. in $[t_1,t_2]$.

The process of deparametrization mentioned above can be summarized as follows. Given a problem of optimal control:

$$I(x,u) = \int_{t_1}^{t_2} F(t,x(t), u(t))dt ,$$

with ordinary differential system and constraints

$$x'(t) = f(t,x(t), u(t)) , \quad t \in [t_1,t_2] \text{ (a.e.)},$$

$$(t,x(t)) \in A \subset R^{n+1}, \quad u(t) \in U(t,x(t)) \subset R^m ,$$

let us take

$$f_o(t,x,\xi) = \text{Inf}[\eta \geq F(t,x,u) \,|\, \xi = f(t,x,u), u \in U(t,x)],$$

$$Q(t,x) = [\xi = f(t,x,u) \,|\, u \in U(t,x)].$$

Then, the corresponding problem of the calculus of variations with constraints on the derivatives is as follows:

$$J(x) = \int_{t_1}^{t_2} f_o(t,x(t), x'(t))dt,$$

$$(t,x(t)) \in A \ , \quad x'(t) \in Q(t,x(t)).$$

For what concerns boundary conditions for problem (1), we restrict ourselves here to the Dirichlet type boundary conditions

$$z(t_1) = z_1 \ , \quad z(t_2) = z_2 \ . \tag{2}$$

Above, let M denote the set $M = [(t,z,\xi) \mid (t,z) \in A, \ \xi \in Q(t,z)] \subset R^{1+2n}$, and let $f_o(t,z,\xi)$ be a real valued function on M. Let Ω be a class of admissible functions, i.e., functions $z:[t_1,t_2] \to R^n$, or $z(t) = (z^1,\dots,z^n)$, such that (i) z is BV in $[t_1,t_2]$; (ii) $(t,z(t)) \in A$, $z'(t) \in Q(t,z(t))$ a.e. in $[t_1,t_2]$; (iii) $f_o(\cdot,z(\cdot),z'(\cdot)) \in L_1[t_1,t_2]$.

It is easy to see that the Lebesgue integral definition (1) of the functional I does not yield stable and realistic values for I, and one is forced to use a Serrin type integral. To this effect, for every $z \in \Omega$ we denote by $\Gamma(z)$ the class of all sequences (z_k) of elements $z_k \in \Omega$ with (a) z_k is AC in $[t_1,t_2]$; (b) $z_k \to z$ pointwise a.e. in $[t_1,t_2]$.

If $\Gamma(z)$ is empty we take $\mathcal{J}(z) = +\infty$. If $\Gamma(z)$ is not empty, then we take

$$\begin{aligned} \mathcal{J}(z) &= \inf \underline{\lim} \int_{t_1}^{t_2} f_o(t,z_k(t),z_k'(t))dt \\ &= \inf_{\Gamma(z)} \underline{\lim}_{k\to\infty} I(z_k). \end{aligned} \tag{3}$$

This is a Serrin type definition of the functional which was inspired to the Lebesgue area of nonparametric surfaces.

If problem (1) has assigned boundary conditions, say of the Dirichlet type (2), then let $\Gamma(z)$ denote the class of all sequences (z_k) of elements z_k in Ω with (a) z_k is AC and satisfies the boundary conditions; (b') $z_k \to z$ pointwise a.e. in $[t_1,t_2]$, in particular $z_k(t_i) \to z(t_i)$, $i = 1,2$. Then the analogous integral defined by (3) could be denoted by $\mathscr{I}^*$ and obviously $\mathscr{I} \leq \mathscr{I}^*$.

We can state now a lower semicontinuity theorem and an existence theorem for the integrals I and $\mathscr{I}$ on BV functions. To this purpose we have first to define as usual the "augmented" sets $\tilde{Q}(t,z)$ as follows:

$$\tilde{Q}(t,z) = [(\tau,\xi) \,|\, \tau \geq f_o(t,z,\xi),\ \xi \in Q(t,z)] \subset R^{n+1}.$$

<u>A lower semicontinuity theorem</u>. Let us assume that (i) A is closed; (ii) the sets $\tilde{Q}(t,z)$ are closed, convex, and satisfy property (Q) with respect to (t,z) at every $(t,z) \in A$; (iii) $f_o(t,z,\xi)$ is lower semicontinuous in M, and there exists some function $\lambda \in L_1[t_1,t_2]$ such that $f_o(t,z,\xi) \geq \lambda(t)$ for all $(t,z,\xi) \in M$. Let $z(t)$, $z_k(t)$, $t \in [t_1,t_2]$, $k = 1,2,\dots$, be a sequence of AC functions z_k such that $z_k \to z$ pointwise a.e. in $[t_1,t_2]$, $(t,z_k(t)) \in A$, $z_k'(t) \in Q(t,z_k(t))$ a.e. in $[t_1,t_2]$, and $V(z_k) \leq C$. Then, $(t,z(t)) \in A$, $z'(t) \in Q(t,z(t))$ a.e. in $[t_1,t_2]$, and $I(z) \leq \underline{\lim}_{k\to\infty} I(z_k)$.

A fundamental consequence of this lower semicontinuity theorem is that if (z_k) is any of the sequences of AC elements in $\Gamma(z)$, with $V(z_k) \leq C$, and we take $j = \underline{\lim}_{k\to\infty} I(z_k)$, then

$$I(z) \leq \mathscr{I}(z) \leq j = \underline{\lim}_{k\to\infty} I(z_k).$$

Furthermore, the Serrin integral $\mathscr{I}$ is actually an extension of the integral I. Indeed, if $z \in \Omega \cap AC$, then, by taking $z_k = z$ we conclude that $I(z) \leq \mathscr{I}(z) \leq \underline{\lim}_{k\to\infty} I(z_k) = I(z)$.

Note that, for sequences (z_k) as above with $V(z_k)$ unbounded, it may well occurs that $\mathcal{J}(z) < I(z)$ as it has been proved by examples (cfr. [6]).

We mention here that Kuratowski's property (K) at a point (t_o,z_o) is expressed by the relation

$$\tilde{Q}(t_o,z_o)=\bigcap_{\delta>o} cl[\cup \tilde{Q}(t,z),\ (t-t_o)^2+|z-z_o|^2\leq\delta^2].$$

The analogous condition (Q) at the point (t_o,x_o) is expressed by the relation

$$\tilde{Q}(t_o,z_o) = \bigcap_{\delta>o} cl\ co[\cup \tilde{Q}(t,z),\ (t-t_o)^2 + |z-z_o|^2 \leq \delta^2] .$$

If problem (1) has assigned boundary conditions of the type (2), then in the theorem above we assume that $z_k \to z$ a.e. in $[t_1,t_2]$,in particular $z_k(t_i) \to z(t_i)$, $i = 1,2$, and the same statement holds for $\mathcal{J}^*$.

An existence theorem for the integral $\mathcal{J}$. Let us assume that (i) A is compact and M is closed; (ii) the sets $\tilde{Q}(t,z)$ are closed, convex, and satisfy property (Q) with respect to (t,z) at every point (t,z) of A; (iii) $f_o(t,z,\xi)$ is lower semicontinuous in M. Assume that the class Ω is nonempty and closed, $V(z) \leq C$ for all z Ω, and $\Gamma(z)$ is nonempty for at least one z. Then the functional $\mathcal{J}$ has absolute minimum $z \in BV$ in Ω.

In other words, let i denote the infimum of I(z) for $z \in AC\cap\Omega$, let (z_k) denote a sequence of elements $z_k \in AC\cap\Omega$ with $I(z_k) \to i$. Then, there is an element $z\in\Omega$, $z\in BV$, such that $I(z) \leq \mathcal{J}(z) = i$.

Example 1. Let

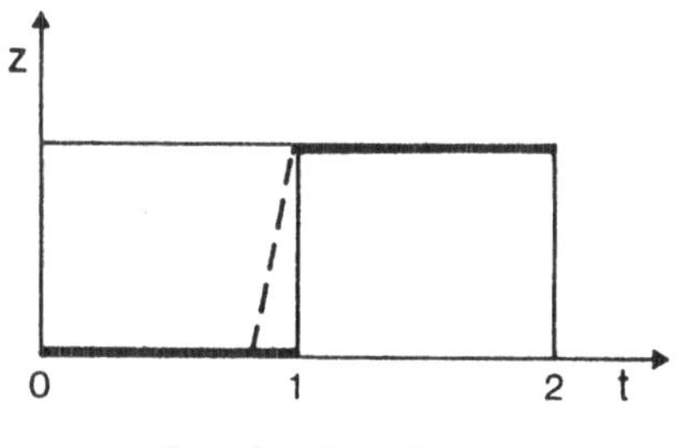

$$I(z) = \int_0^2 |1-t||z'(t)|dt,$$

$$z(0) = 0, \quad z(2) = 1,$$

with $A = [0,2]\times [0,1]$, $n = 1$, $Q(t,z) = [0,+\infty)$, $f_o(t,z,\xi) = |1-t||\xi|\geq 0$.

If we take $z_k(t) = 0$ for $0 \leq t \leq 1-k^{-1}$, $z_k(t) = 1$ for $1 \leq t \leq 2$, $z_k(t) = 1-k+kt$ for $1-k^{-1} \leq t \leq 1$, we have

$$0 \leq \mathscr{I}(z_k) = I(z_k) = \int_{1-k^{-1}}^{1} (1-t)k\,dt = (2k)^{-1} \to 0.$$

Thus $i = 0$, (z_k) is a minimizing sequence. The minimum is attained by the discontinuous function $z(t) = 0$ for $0 \leq t < 1$, $z(t) = 1$ for $1 \leq t \leq 2$, and $I(z) = \mathscr{I}(z) = i = 0$.

Example 2. Let

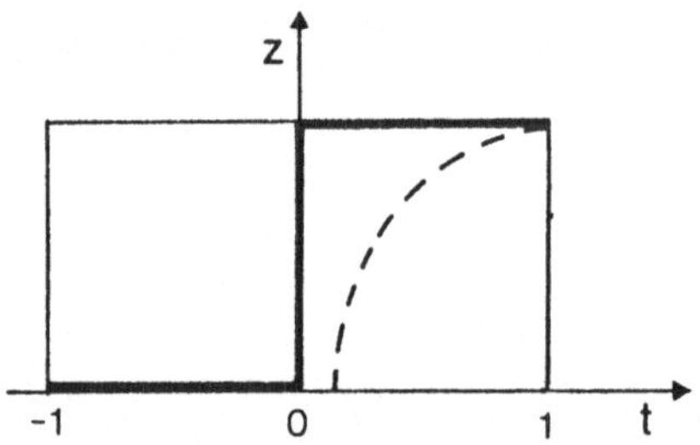

$$I(z) = \int_{-1}^{1} |t| z'^2(t)\,dt,$$

$$z(-1) = 0, \; z(1) = 1,$$

with $A = [-1,1] \times [0,1]$, $n = 1$, $Q(t,z) = [-1,+\infty)$, $f_o(t,z,\xi) = |t|\,\xi^2 \geq 0$, $I(z) \geq 0$, $\mathscr{I}(z) \geq 0$. If we take $z_k(t) = 0$ for $-1 \leq t \leq k^{-1}$, $z_k(t) = (\log k)^{-1} \log t + 1$ for $k^{-1} \leq t \leq 1$, we have $0 \leq \mathscr{I}(z_k) = I(z_k) = \int_{k^{-1}}^{1} (\log k)^{-2} t^{-1}\,dt = (\log k)^{-1} \to 0$. Thus, $i = 0$, (z_k) is a minimizing sequence. The minimum is attained by the discontinuous function $z(t) = 0$ for $-1 \leq t \leq 0$, $z(t) = 1$ for $0 < t \leq 1$, and $I(z) = \mathscr{I}(z) = i = 0$.

Example 3. Let

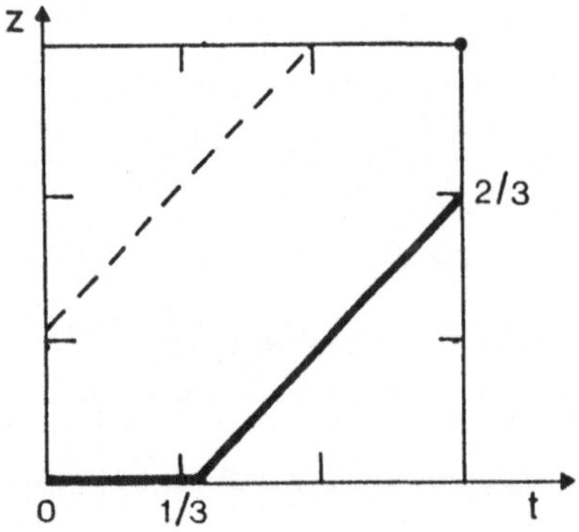

$$I(z) = \int_0^1 |z'(t)|\,dt,$$

$$n = 1, \; Q(t,z) = R,$$

with $A = [(t,z)|0\leq t\leq 1,\ 0\leq z\leq 1,\ t-3^{-1}\leq z\leq t+3^{-1}, f_o(t,z,\xi) = |\xi|$. Let $\phi(t)$, $0\leq t\leq 1$, denote the usual Cantor ternary function. Then $(t,\phi(t))\in A$, $I(\phi) = i_o = 0$, where $i_o = 0$ is the infimum of $I(z)$ in Ω. Let i be the infimum of $I(z)$ in $AC\cap\Omega$. Then $i = 2/3$, and the minimum of $\mathcal{I}$ is attained by $z(t) = 0$ for $0\leq t\leq 1/3$, $z(t) = t-1/3$ for $1/3\leq t\leq 1$. Thus $I(\phi) = i_o = 0 < I(z) = \mathcal{I}(z) = i = 2/3$.

Now consider the same problem with boundary data $z(0)=0$, $z(1)=1$.

The infimum of $I(z) = \Omega$ is still $i_o = 0$. The infimum of $I(z)$ in $AC\cap\Omega$ is now $i = 1$, and $i = 1$ is assumed by the continuous function $\bar{z}(t)$ defined by $\bar{z}(t) = 0$ for $0\leq t\leq 1/3$, $\bar{z}(t) = t - 1/3$ for $1/3\leq t< 1$, $\bar{z}(1) = 1$. Thus $I(\phi) = i_o = 0 < I(\bar{z}) = 2/3 < \mathcal{I}(\bar{z}) = i = 1$.

3. <u>Problems of optimization for multiple integrals and BV discontinuous functions, $\nu > 1$.</u>

Let $\nu > 1$, $n \geq 1$, and let $G\subset R^{\nu}$ be a bounded domain in the t-space R^{ν}, $t = (t^1,\dots,t^{\nu})$, possessing the cone property at every point of its boundary ∂G. Let $A\subset R^{\nu+n}$ be a compact subset of the tz-space $R^{\nu+n}$, whose projection on the t-space contains G.

We shall deal with vector valued functions $z(t) = (z^1,\dots,z^n)$, $z^i\in BV$ in G, therefore possessing first order partial derivatives in the sense of distributions which are measures μ_{ij}, $j = 1,\dots,\nu$, $i = 1,\dots,n$, and in addition also generalized first order derivatives $D^j z^i$ a.e. in G, as functions of class $L_1(G)$, which are obtained as limits of incremental quotients when we disregard the values taken by the functions in suitable sets E of measure zero in G. We may need only a subset of such derivatives $D^j z^i$ as follows.

For every $i = 1,\dots,n$, let $\{j\}_i$ be a system of indices $1\leq j_1< \dots < j_s\leq \nu$, let $D^j z^i$, $j\in\{j\}_i$, denote the corresponding system of first order partial derivatives of the function z^i, and let N be their total number. Then by Dz we denote the N-vector function $Dz(t)= (D^j z^i,\ j\in\{j\}_i,\ i = 1,\dots,n)$, $t\in G$ (a.e.).

For every $(t,z) \in A$ let $Q(t,z)$ be a given subset of R^N. Let $M \subset R^{\nu+n+N}$ denote the set $M = [(t,z,\xi) \mid (t,z) \in A,\ \xi \in Q(t,z)]$, and let $f_o(t,z,\xi)$ be a given real-valued function in M. We are interested in the multiple integral problem of the calculus of variations with constraints on the derivatives

$$
\begin{aligned}
& I(z) = \int_G f_o(t,z(t),\ Dz(t))\,dt = \text{minimum}, \\
& (t,z(t)) \in A, \quad Dz(t) \in Q(t,z(t)),\ t \in G \ (\text{a.e.}),
\end{aligned} \tag{1}
$$

and possible Dirichlet type boundary conditions of the form $\gamma z(t) = \phi(t), t \in \partial G(\mathscr{H}^{\nu-1}$-a.e.) on ∂G. Again we introduce a Serrin type integral.

Let Ω be a class of admissible functions $z(t) = (z^1,\ldots,z^n)$, $t \in G$, such that (i) z is BV in G; (ii) $(t,z(t)) \in A$, $Dz(t) \in Q(t,z(t))$, $t \in G$ (a.e.); (iii) $f_o(\cdot,z(\cdot),Dz(\cdot)) \in L_1(G)$.

To simplify notations, let AC, or AC(G), denote the class of functions $z(t) = (z^1,\ldots,z^n)$, $t \in G$, whose components z^i are of Sobolev class $W^{1,1}(G)$, or, briefly, Beppo Levi functions.

For any element $z \in \Omega$ let $\Gamma(z)$ denote the class of all sequences (z_k) of elements z_k in Ω with (a) z_k is AC in G; (b) $z_k \to z$ strongly in $L_1(G)$.

If $\Gamma(z)$ is empty we take $\mathscr{I}(z) = +\infty$. If $\Gamma(z)$ is not empty then we take

$$
\begin{aligned}
\mathscr{I}(z) &= \inf \underline{\lim} \int_G f_o(t,z_k(t),\ Dz_k(t))\,dt = \\
&= \inf_{\Gamma(z)} \ \underline{\lim}_{k\to\infty} \ I(z_k).
\end{aligned} \tag{2}
$$

If the given problem has assigned Dirichlet type boundary conditions, say $\gamma z(t) = \phi(t)$, $t \in \partial G$ ($\mathcal{H}^{\nu-1}$-a.e.) on G, then let $\Gamma(z)$ denote the class of all sequences (z_k) of elements z_k in Ω with (a') z_k is AC in G and $\gamma(z_k) = \phi$ on ∂G; (b') $z_k \to z$ strongly in $L_1(G)$; and $V(z_k) \to V(z)$. Then the Serrin integral defined by (2) could be denoted by $\mathcal{I}^*$, and obviously $\mathcal{I}(z) \leq \mathcal{I}^*(z)$. By Anzellotti and Giaquinta's theorem then $\gamma(z) = \phi$ on ∂G ($\mathcal{H}^{\nu-1}$-a.e.).

To state an existence theorem we introduce, as usual, the augmented sets $\tilde{Q}(t,z) \subset R^{N+1}$ as follows:

$$\tilde{Q}(t,z) = [(\tau,\xi) \mid \tau \geq f_o(t,z,\xi), \xi \in Q(t,z)].$$

Beside property (Q) we shall require on the sets $Q(t,z)$ another property, or property F_1.

We say that the sets $Q(t,z)$, $(t,z) \in A$, have property F_1 with respect to z at the point $(t_o,z_o) \in A$ provided, given any number $\sigma > 0$, there are constants $C > 0$, $\delta > 0$ which depend on t_o, z_o, σ, such that for any two measurable vector functions $z(t)$, $\xi(t)$, $t \in H$, on a measurable subset H of points t of G with

$$(t,z(t)) \in A, \ |z(t) - z_o| \geq \sigma, \xi(t) \in Q(t,z(t))$$

$$\text{for } t \in H, \ |t-t_o| \leq \delta,$$

there are other measurable vector functions $\bar{z}(t)$, $\bar{\xi}(t)$, $t \in H$, such that

$$(t,\bar{z}(t)) \in A, \ |\bar{z}(t) - z_o| \leq \sigma, \ \bar{\xi}(t) \in Q(t,\bar{z}(t)),$$

$$|\xi(t) - \bar{\xi}(t)| \leq C[|z(t)-\bar{z}(t)| + |t-t_o|] \text{ for } t \in H.$$

We denote by F_2 the same condition with $\bar{z}(t) = z_o$. These conditions are inspired to analogous ones proposed by Rothe, Berkovitz, Browder (cfr. Cesari [5], sect. 13).

An existence theorem. Let us assume that (i) A is compact and M is closed; (ii) the sets $\tilde{Q}(t,z)$ are closed, convex, and satisfy properties (Q) and (F_1) at every point $(t,z) \in A$; (iii) $f_o(t,z,\xi)$ is bounded below and lower semicontinuous in (t,z,ξ). Also assume that the class Ω is nonempty and closed, and $\Gamma(z)$ is nonempty for at least one $z \in \Omega$. Then the functional $\mathcal{J}$ has an absolute minimum z in Ω, $z \in BV$ in G.

In other words, let i denote the infimum of I(z) for $z \in AC \cap \Omega$, let (z_k) denote any sequence of elements $z_k \in AC \cap \Omega$ with $I(z_k) \to i$. Then there is at least one element $z \in \Omega$, $z \in BV$, such that $I(z) \leq \mathcal{J}(z) = i$.

REFERENCES

[1] G. Anzellotti and M. Giaquinta, Funzioni BV e tracce, Rend. Sem. Mat. Padova, 60 (1978), 1-21.

[2] F. Cafiero, Criteri di compattezza per le successioni di funzioni generalmente a variazione limitata, Atti Accad. Naz. Lincei 8 (1950), 305-310.

[3] L. Cesari, Sulle funzioni a variazione limitata, Ann. Scuola Norm. Sup. Pisa, 5 (1936), 299-313.

[4] L. Cesari, (a) Sulle funzioni di due variabili a variazione limitata e sulla convergenza delle serie doppie di Fourier, Rend. Sem. Mat. Univ. Roma, 1 (1937), 277-294; (b) Sulle funzioni di più variabili a variazione limitata e sulla convergenza delle relative serie multiple di Fourier, Pont. Accad. Scienze, Commentationes, 3 (1939), 171-197.

[5] L. Cesari, Optimization-Theory and Applications, Problems with Ordinary Differential Equations, Springer Verlag (1983).

[6] L. Cesari, P. Brandi and A. Salvadori, Existence theorems concerning simple integrals of the calculus of variations for discontinuous solutions, Arch. Rat. Mech. Anal. to appear.

[7] L. Cesari, P. Brandi and A. Salvadori, Existence theorems for multiple integrals of the calculus of variations for discontinuous solutions, to appear.

[8] E. Conway and J. Smoller, Global solutions of the Cauchy problem for quasi-linear first-order equations in several space variables, Comm. Pure App. Math., 19 (1966), 95-105.

[9] C.M. Dafermos, Generalized characteristics and the structure of solutions of hyperbolic conservation laws, Indiana Univ. Math. J. 26 (1977), 1097-1119.

[10] P.J. Di Perna, Singularities of solutions of nonlinear hyperbolic systems of conservation laws, Arch. Rat. Mech. Anal., 60 (1974), 75-100.

[11] W.H. Fleming, Functions with generalized gradient and generalized surfaces, Ann. Math. Pure Appl., 44 (1957), 93-103.

[12] E. Gagliardo, Proprietà di alcune classi di funzioni di più variabili, Ricerche Mat., 7 (1959), 24-51.

[13] K. Krickeberg, Distributionen, Funktionen beschränkter Variation und Lebesguescher Inhalt nichtparametrischer Flächen, Ann. Mat. Pura Appl., 44 (1957), 105-133.

[14] A.L. Volpert, The spaces BV and quasilinear equations, Mat. Sb. 73 (1967), 225-267.

OPTIMAL CONTROL FOR FREE BOUNDARY PROBLEMS

Avner Friedman

Purdue University, Indiana, USA

1. INTRODUCTION

In this talk we review recent results on the bang-bang principle for free boundary problems. The functional $J(k)$ to be optimized is generally non-differentiable. In order to overcome this difficulty we correspond to an optimizer k_0 a family of ε - approximating smooth problems and denote a corresponding optimizer by k_ε. We prove that k_ε satisfies an approximate bang-bang principle and that $k_\varepsilon \to k_0$ strongly as $\varepsilon \to 0$. Using these facts and some additional analysis, a bang-bang principle is then established for k_0.

2. PARABOLIC VARIATIONAL INEQUALITIES

Consider a parabolic variational inequality

$$\begin{aligned} & u_t - \Delta u + \beta(u) \ni -(f+k) \quad \text{in } D_T, \\ & u = g \quad \text{on } \partial_p D_T \end{aligned} \tag{1}$$

where $\partial_p D$ is the parabolic boundary of D_T, f and g are given functions, $g \geq 0$, $\beta(u)$ is the graph $\beta(u) = 0$ if $u > 0$, $\beta(0) = (-\infty, 0]$, $\beta(u) = \phi$ if $u < 0$, $D_T = D \times (0<t<T)$ and D is a bounded domain in $\mathbb{R}^N$ with smooth boundary. Here k is a control variable belonging to a set

$$\mathcal{A} = \{k \in L^\infty(D_T)\,,\ 0 \leq k \leq N\,,\ \int_{D_T} k = M\}.$$

This work is partially supported by National Science Foundation Grant DMS-8420896.

Introduce a functional

$$J(k) = \int_{D_T} F(x,t,u)\,dx\,dt$$

and consider the problem:

$$\text{maximize } J(k) \quad \text{over} \quad k \in \mathcal{A} \tag{2}$$

It is easily proved that (2) has a solution k_0 ; denote the corresponding solution of (1) by u_0. In order to study the structure of k_0, we consider
the ε-approximating problems

$$\begin{aligned} u_t - \Delta u + \beta_\varepsilon(u) &= -f - k \quad \text{in } D_T, \\ u &= g \quad \text{on } \partial_p D_T \end{aligned} \tag{3}$$

where $\beta_\varepsilon(u) \in C^\infty$, $\beta'_\varepsilon(u) \geq 0$, $\beta_\varepsilon(u) \to -\infty$ if $u < 0$, $\varepsilon \to 0$ and $\beta_\varepsilon(u) \to 0$ if $u > 0$, $\varepsilon \to 0$, and denote by K_ε the class of pairs (k,u) where u is the solution of (3). Introduce the functional

$$J_\varepsilon(k) = \int_{D_T} F(x,t,u)\,dx\,dt - \frac{1}{2}\int_{D_T}(k - k_o)^2$$

and denote by $(k_\varepsilon,u_\varepsilon)$ a maximizer of $J_\varepsilon(k)$ over $(k,u) \in K_\varepsilon$. One can prove [2] [7] that

$$k_\varepsilon \to k_0 \quad \text{in} \quad L^2(D_T) \quad \text{as} \quad \varepsilon \to 0 . \tag{4}$$

We proceed to study the structure of k_ε. Let ℓ be a function such that $k_\varepsilon + \delta\ell \in \mathcal{A}$ for any small $\delta > 0$. One can show that the solution $u_{\varepsilon,\delta}$ of (3) corresponding to $k_\varepsilon + \delta\ell$ satisfies:

$$\frac{u_{\varepsilon,\delta} - u_\varepsilon}{\delta} \to z_\varepsilon \quad \text{as} \quad \delta \to 0 \tag{5}$$

where z_ε is the solution of

$$\begin{aligned} & z_t - \Delta z + \beta_\varepsilon'(u_\varepsilon) = \ell \quad \text{in} \quad D_T , \\ & z = 0 \quad \text{on} \quad \partial_p D_T . \end{aligned} \tag{6}$$

Denote by Q_ε the solution of

$$\begin{aligned} & Q_t + \Delta Q - \beta_\varepsilon'(u_\varepsilon)\, Q = -F_u(x,t,u_\varepsilon) \quad \text{in} \quad D_T , \\ & Q = 0 \quad \text{on} \quad \partial_{-p} D_T \end{aligned} \tag{7}$$

where $\partial_{-p} D_T$ consists of $\partial D \times (0,T) \cup D \times \{T\}$.

Taking $\delta \to 0$ in $(J_\varepsilon(u_{\varepsilon,\delta}) - J_\varepsilon(u_\varepsilon))/\delta \le 0$ and using (5)-(7), we arrive at the optimality condition

$$\int_{D_T} (Q_\varepsilon + k_\varepsilon - k_0)\ell \ge 0 . \tag{8}$$

From (4) and (3) one can deduce that for any $\eta > 0$ there is a set Σ_η with measure $< \eta$ such that in $\{u_0>0\}\setminus\Sigma_\eta$ there holds:

$$k_\varepsilon = \begin{cases} N & \text{if} \quad Q_\varepsilon < \lambda_\varepsilon - \eta \\ 0 & \text{if} \quad Q_\varepsilon > \lambda_\varepsilon + \eta \end{cases}$$

for all ε sufficiently small and some constant λ_ε. One can also establish that, for a sequence $\varepsilon \to 0$,

$$Q_\varepsilon \to Q \quad \text{uniformly in compact subsets of} \quad \{u_0>0\}. \tag{9}$$

Hence:

<u>THEOREM 1</u>. For any optimal control k_0 there is a λ such that

$$k_0 = \begin{cases} N & \text{in} \quad \{Q<\lambda\}\cap\{u_0>0\} \\ 0 & \text{in} \quad \{Q>\lambda\}\cap\{u_0>0\}; \end{cases}$$

if further $F_u > 0$, then $\text{meas}[\{Q=\lambda\}\cap\{u_0>0\}] = 0$.

This result, proved in [8], can be extended to the case where the control k appears in the boundary condition, and to more general functionals J(k). We mention only one result concerning the one-phase Stefan problem in which heat flux k is applied on the fixed part Γ_1 of the boundary of D in contact with water. We take

$$\mathscr{A} = \{N_1 \le k(x,t) \le N_2 \; , \; x \in \Gamma_1, \; 0 \le t \le T, \; \int_0^T \int_{\Gamma_1} k = M\}$$

and J(k) to be the volume of ice that has melted by time t=T; we wish to maximize J(k). Taking a slightly mollified form of J(k) we have [8]:

<u>THEOREM 2</u>. For any maximizer k_0 there exists a smooth function $\phi(x)$ defined on Γ_1 such that

$$k_0(x,t) = \begin{cases} N_2 & \text{if} \quad 0 < t < \phi(x), \; x \in \Gamma_1 \\ N_1 & \text{if} \quad \phi(x) < t < T, \; x \in \Gamma_2. \end{cases}$$

3. <u>THE DAM PROBLEM</u>

Consider the general dam problem [1] [4] [5] and assume that at the bottom part S we withdraw fluid at a rate k (k is the flux); k is a control variable belonging to a set

$$\mathscr{A} = \{0 \le k(x) \le N \text{ if } x \in S, \int_S k = M\}$$

with N small enough. Take, for definiteness, a flat bottom S, and N any number <1. We wish to minimize the "total pressure" of the fluid in the dam Ω:

$$J(k) = \int_\Omega g(y)p(x,y)dxdy$$

where p is the pressure corresponding to k, and g is a given function satisfying: $g > 0$, $g' \ge 0$. It is proved in [9] that for any k there is a unique solution p, and thus J(k) is well defined; further, there exists a minimizer k_0 of $J(k)$. Finally:

THEOREM 3. Any minimizer k_0 is a bang-bang, that is, $k_0 = N\chi_A$ where A is some subset of S.

THEOREM 4. For the rectangular dam the minimizer k_0 is unique, and A consists of at most two intervals.

Recall that the pressure p satisfies, formally,

$$\Delta p + \frac{\partial}{\partial y} H(p) = 0;$$

in proving Theorem 3 we choose for the ε-approximating problem the equation:

$$\Delta p + \frac{\partial}{\partial y} H_\varepsilon(p) = 0$$

where $H_\varepsilon(p)$ is a smooth approximation of $H(p)$.

4. PERIODIC CONTROL

In a large body of ice we insert an object D_1 and heat its boundary Γ_1 with heat flux k, thereby causing the ice to melt. Let Γ_2 be a surface in the ice region which contains D_1 in its interior and denote by D the domain bounded by Γ_1 and Γ_2. Since the temperature g on the ice boundary Γ_2 (of D) fluctuates with time in a nearby periodic manner, it is natural to assume that g is periodic (say of period 1) and to choose k also periodic, say in the class:

$$\mathcal{A} = \{N_1 \le k(x,t) \le N_2 \quad \text{if} \quad x \in \Gamma_1,\ 0 < t < \infty,\ \int_0^1 \int_{\Gamma_1} k(x,t) = M,$$

$$k(x,t+1) = k(x,t)\}$$

In this case (see [6]) there exists a unique periodic solution $\hat{u}(x,t)$ such that the solution $u(x,t)$ of the two-phase Stefan problem with data k, g and any initial data $u_0(x)$ satisfies:

$$\sup_D |u(x,t) - \hat{u}(x,t)| \to 0 \quad \text{if} \quad t \to \infty.$$

We introduce the functional

$$J(k) = \int_0^1 \int_D p(x)\hat{u}(x,t)$$

where $p(x) \geq 0$, p piecewise continuous, $p(x) > 0$ if $x \in \Gamma_1$, and consider the problem of minimizing $J(k)$ for $k \in \mathcal{A}$.

THEOREM 5. For any minimizer k_0 there exists a smooth function $\phi(x)$ defined on Γ_1 such that

$$k_0 = \begin{cases} N_2 & \text{if} \quad m < t < m + \phi(x) \\ N_1 & \text{if} \quad m + \phi(x) < t < m + 1 \end{cases}$$

for $m = 0,1,2,\ldots$ and $x \in \Gamma_1$.

This result is proved in [10]. Recall that the two-phase Stefan problem consists of solving the formal equation

$$\frac{\partial}{\partial t}(u+H(u)) - \Delta u = 0$$

together with the appropriate boundary conditions; for the ε-approximating problem we take the differential equation to be

$$\frac{\partial}{\partial t}(u+H_\varepsilon(u)) - \Delta u = 0.$$

5. ELECTROCHEMICAL MACHINING

Consider for simplicity a two dimensional domain $\mathcal{A}$ bounded by $\{x=0\}$, $\{x=1\}$, $\{y=H\}$ and a curve $\{y=k(x), 0 < x < 1$ with $k(x) < H$. Consider, in Ω, the variational inequality

$$-\Delta u + \beta(u) \ni f \tag{9}$$

with boundary conditions:

$$\begin{aligned} u_\nu &= 0 \quad \text{on} \quad \{x=0\},\ \{x=1\} \\ u &= 0 \quad \text{on} \quad \{y=H\}, \\ u &= 1 \quad \text{on} \quad \{y=k(x)\}. \end{aligned} \tag{10}$$

Here k is a control variable belonging to the set

$$\mathcal{A} = \{a \leq k(x) \leq b, \quad |k'(x)| \leq 1 \quad \text{a.e.}, \quad k(0) = \alpha, \quad k(1) = \beta\}$$

and $0 \leq a < b < H$. Introduce the functional

$$J(k) = \int_{\{u>0\}} (u(x,y) - g(x,y))^2 \, dxdy \tag{11}$$

where g is a "target" function, and consider the problem of minimizing $J(k)$ over the set $\mathcal{A}$. In the actual electrochemical machining problem one wishes to choose k so that the free boundary of u will be "as close as possible" to a given curve $\{y=\psi(x)\}$. Thus in the less precise formulation in terms of $J(k)$, we try to choose $g(x,y)$ to "concentrate" on this given curve. Assuming some properties on g, it was proved in [3] that

there is a unique minimizer k_0, and

$k_0'(x) = -1$ if $0 < x < x_0$, $k'(x) = 0$ if $x_0 \leq x \leq y_0$,

$k_0'(x) = 1$ if $y_0 < x < 1$, for some $0 \leq x_0 \leq y_0 \leq 1$.

In constructing the ε-approximate problem one replaces, in (9), $\beta(u)$ by $\beta_\varepsilon(u)$ and the controls k by mollifications k_ε.

REFERENCES

[1] H.W. Alt, Stromungen durch inhomogene poruse median mit freien rand, J. Reine Angew. Math., 305(1979), 89-115.

[2] V. Barbu, Optimal control of variational inequalities, Pittman, London, 1984.

[3] V. Barbu and A. Friedman, Optimal control of a free boundary for variable domains, to appear.

[4] H. Brezis, D. Kinderlehrer and G. Stampacchia, Sur une nouvelle formulation de probleme de l'ecoulement a travers une digue, C.R. Acad. Sci. Paris, 287(1978), 711-714.

[5] J. Carrillo-Menendez and M. Chipot, On the dam problem, J. Diff. Eqs., 45(1982), 234-271.

[6] E. DiBenedetto and A. Friedman, Periodic behavior for the evolutionary dam problem and related free boundary problems, Comm. P.D.E., to appear.

[7] A. Friedman, Variational Principles and Free Boundary Problems, Wiley, New York, 1982.

[8] A. Friedman, Optimal control for parabolic variational inequalities, SIAM J. Control and Optim., to appear.

[9] A. Friedman, S. Huang and J. Yong, Bang-bang optimal control for the dam problem, to appear.

[10] A. Friedman, S. Huang and J. Yong, Optimal periodic control for the two-phase Stefan problem, to appear.

CONTROL METHODS FOR THE NUMERICAL COMPUTATION OF PERIODIC SOLUTIONS OF AUTONOMOUS DIFFERENTIAL EQUATIONS

Giles Auchmuty,* Edward Dean, Roland Glowinski, and S.C. Zhang*

University of Houston, Texas, USA

1. INTRODUCTION

Minimization methods have been used for a long time, and by many mathematicians, to find approximate solutions of nonlinear differential equations. For references see [1] or [2] amongst others.

When the minimization is done in an appropriate L^2-norm, one has a non-linear least squares method. In practice, it is often useful to view these least squares problems as optimal control problems. Such methods have been described by several authors and applied to the simulation of various non-linear problems including

(i) potential transonic flows (see [1]-[4]),

(ii) the compressible and incompressible Navier-Stokes equations (see [1],[2],[5]-[7]),

(iii) the Von Karman equations [8], and

(iv) problems with multiple solutions, using continuation methods (see [2],[4],[7] and [8]).

The paper of O. Pironneau in these proceedings also describes an application of these methods to the numerical solution of some nonlinear fluid dynamical problems.

In this paper, we will apply these nonlinear least squares/optimal control methods to the numerical computation of the <u>periodic solutions</u>, or <u>free oscillations</u> of nonlinear autonomous differential equations.

---

*The research of these authors was partially supported by the Welch Foundation.

These problems can be formulated as follows:

$$(1.1)\qquad \begin{cases} \textit{Find } \{u,T\} \textit{ such that} \\ \dfrac{du}{dt} + A(u)=0, \\ u(0)=u(T). \end{cases}$$

Here $u(t)$ may be vector-valued, with values in R^n, so that (1.1) is an autonomous system of ordinary differential equations, or, more generally, $A(\cdot)$ may be a nonlinear elliptic operator and (1.1) would be a (system of) nonlinear evolutionary partial differential equations. We are particularly interested in the case where (1.1) is a system of reaction-diffusion equations of the form

$$\frac{\partial u_i}{\partial t} = D_i \Delta u_i + f_i(u_1,\dots,u_m) \qquad \text{in } \Omega x(0,T),$$

for $1 \le i \le m$, subject to some boundary conditions for the u_i on $\partial\Omega x(0,T)$. The construction of time-periodic solutions for these equations is described in [9], [15], [16] and many more recent papers.

There is an extensive literature on (1.1). Many methods have been developed to treat this problem, including

(i) methods based on solving the initial value problem for (1.1) and findings a limit cycle. This method often is efficient, but will only find stable solutions. More sophisticated methods have been developed based on using the Poincare' map; see Urabe [17] for some early work on this.

(ii) Normalize the period T to be 1 and consider (1.1) as a one-parameter family of boundary value problems or as a nonlinear eigenvalue problem. These may be solved using continuation and/or Newton's method and is described in [10],[11]. These methods can be very effective if the dimension of u is not too large. (When (1.1) is obtained from the discretization of a partial differential equation, these methods may require considerable storage).

(iii) Use a shooting method in which the initial value u(0) and the period T are adjusted systematically until a solution of (1.1) is found.

Here, we shall combine the philosophy of each of these approaches to develop a least squares/continuation shooting method. The continuation parameter here is the (unknown) period.

In Section 2, several formulations of the problem (1.1) in a Hilbert space setting will be described and in Section 3, an optimal control/least squares formulation will be given. The calculation of the gradient of the cost function will be discussed in Section 4 and a conjugate gradient algorithm for solving the least squares problem is treated in Section 5. The time discretization via a Crank-Nicholson method is developed in Section 6 together with a quasi-Newton method which is an alternative to the conjugate gradient method for solving the discrete problem.

Finally, in Section 7, we present results of some numerical computations of the periodic solutions of some well-known chemical systems including the Brusselator and the Oregonator model of the Belousov-Zhabotinski reaction.

2. Formulation of the periodic solution problem

For the sake of generality, we consider the following fairly abstract situation:

V and H are two *real Hilbert spaces,* such that $V \subset H$ and that the injection from V into H is continuous; we suppose also that $\overline{V}=H$, which implies then that $H \subset V'$ (V': dual space of V) and $\overline{H}=V'$, the injection from H into V' being also continuous. We shall denote by $(\cdot,\cdot)$ and $\|\cdot\|$ (resp. $\langle\cdot,\cdot\rangle$ and $|\cdot|$) the *scalar product* and the *norm* of V (resp. H). From the above density and continuity properties, it is natural (and classical) to denote by $\langle\cdot,\cdot\rangle$ the *duality pairing* between V' and V coinciding with the H-scalar product when both arguments are smooth enough. The norm $\|\cdot\|_*$ over V' is defined by

$$(2.1) \quad \|f\|_* = \underset{v \in V-\{0\}}{\mathrm{Max}} \frac{|\langle f,v\rangle|}{\|v\|} ;$$

finally, we denote by S the *duality isomorphism* from V onto V' satis-

$$(2.2) \quad \langle Sv,w\rangle=(v,w), \ \forall\, v,w \in V.$$

Let A be a *nonlinear operator* from V into V', the problem under consideration is defined by

$$(2.3)\begin{cases} \textit{Find } \{u,T\},\ T>0,\ u(t)\in V \textit{ a.e. on } [0,T], \textit{ such that} \\ u\in C^0([0,T];\ H) \textit{ and} \\ \frac{du}{dt}+A(u)=0, \\ u(0)=u(T). \end{cases}$$

We do not discuss here the problem of the existence and uniqueness of the solutions of (2.3), being mainly concerned with the numerical calculation of such solutions (assuming that they exist).

Remark 2.1: If u solves

$$(2.4)\quad A(u)=0,$$

then it solves also (2.3), $\forall$ T>0.

Remark 2.2: If {u,T} solves (2.3) then, $\forall\tau$, $\{u_\tau,T\}$ solves (2.3), assuming that u_τ is defined by $u_\tau(t)=u(t+\tau)$.

Remark 2.3: If {u,T} solves (2.3), then {u,nT} solves (2.3), $\forall$ n>0, n an integer.□

Reduction to a fixed interval. Further comments: Using t/T as a new time variable yields to consider the one parameter family of periodic problems

$$(2.5)\begin{cases} \textit{Find } \{u,T\} \textit{ such that} \\ \frac{du}{dt}+TA(u)=0, \\ u(0)=u(1). \end{cases}$$

We observe that {u,0}, with u arbitrary in V, solves (2.5); also, if A is *linear*, {u,T} solution of (2.5) implies that {λu,T} is also a solution of (2.5), $\forall\ \lambda\in\mathbb{R}$.

3. Least Squares Formulations of Problem (2.5)

The method to be described is essentially a *shooting method* whose principle is given just below

(1) *Denote by* {μ,θ} *a guess of* {u(0),T};

(2) *Solve by initial value problem methods (i.e. marching techniques) the initial value problem*

$$(3.1)\begin{cases} \frac{dv}{dt}+\theta A(v)=0, \\ v(0)=\mu; \end{cases}$$

(3) *If* $v(1)=\mu$, *problem* (2.5) *is solved; if not "improve"* $\{\mu,\theta\}$ *and go back to* (1).□

The above program can be implemented via a *least squares/continuation method* to be described below.

<u>The basic least squares formulation</u>: The basic least squares formulation is given by

$$(3.2)\qquad \operatorname*{Min}_{\{\mu,\theta\}} \{\tfrac{1}{2}|\mu-v(1)|^2\};\ \mu\in H,\ \theta>0,$$

with v a function of μ and θ through the solution of the *initial value problem*

$$(3.3)\qquad \begin{cases} \frac{dv}{dt}+\theta A(v)=0, \\ v(0)=\mu. \end{cases}$$

Problem (3.2), (3.3) has clearly the structure of an *Optimal Control* problem.

If one uses the above least squares formulation to solve (2.5), any "reasonable" optimization algorithm (like one of those described in, e.g., [12]) will converge fairly quickly to T=0 and u=constant over [0,1], which is indeed a (trivial) solution of (2.5). In order to avoid convergence to solutions such as the one above, or to solutions of the steady equation (2.4), we shall employ *penalty* techniques:

To avoid the steady state solutions we introduce a function H with the following properties:

(i) $H:R_+\to R_+$, $H\in C^2$,

(ii) $H(0)$ *is "large"*, $H(t)>0$, $\forall t$, $0\le t<1$,

(iii) $H(t)=0$, $\forall\ t\ge 1$,

(iv) H *is decreasing over* R_+;

the graph of H looks like the one in Figure 3.1.

Now, with $\varepsilon>0$, we define H_ε: $R_+\to R_+$ by

$$H_\varepsilon(t)=H(t/\varepsilon).$$

If, on the other hand, one wants to restrict the period T to the closed interval $[T_1,T_2]$ we should use a penalty function K_ε, satisfying - like the one in Figure 3.2 - $\forall\ \varepsilon'>0$,

(i) $K_{\varepsilon'}$ *is* C^2 *and convex over* R,

(ii) $K_{\varepsilon'}(t)=0$, $\forall\ t\in[T_1,T_2]$,

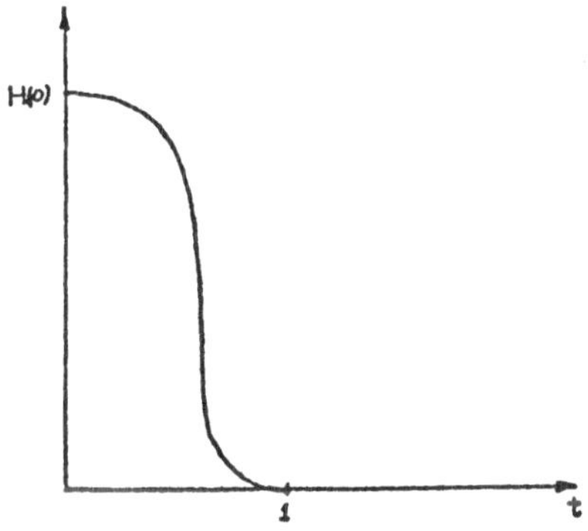

Figure 3.1

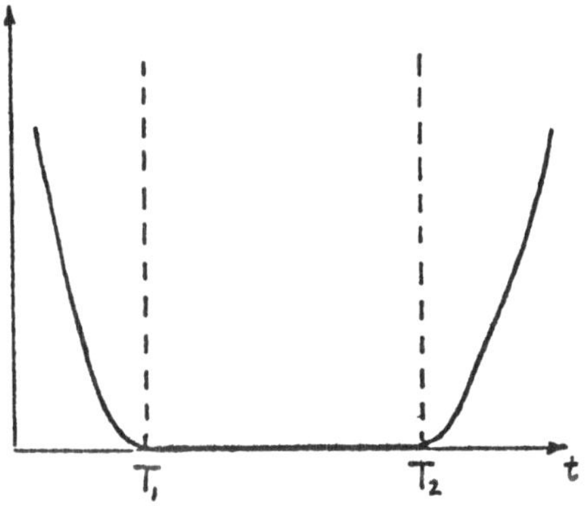

Figure 3.2

(iii) $K_{\varepsilon'}(t)>0$, $\forall t\in R-[T_1,T_2]$,

(iv) $\lim_{\varepsilon'\to 0} K_{\varepsilon'}=I_{[T_1,T_2]}$,

where $I_{[T_1,T_2]}$ is the *indicator functional* of $[T_1,T_2]$, i.e. the function from R into $R\cup\{+\infty\}$ defined by

$$I_{[T_1,T_2]}(t)=0,\ \forall t\in[T_1,T_2];\ I_{[T_1,T_2]}(t)=+\infty \text{ if } t\notin[T_1,T_2].$$

Introducing H_ε and $K_{\varepsilon'}$ in the least squares formulation (3.2), (3.3) we obtain as eventual least squares formulation of problem (2.5), the minimization problem below (where ε and ε' have been dropped from H_ε and $K_{\varepsilon'}$):

$$(3.5)\quad \begin{cases} \textit{Find } \{u,T\} \textit{ such that} \\ J(\lambda,T)\le J(\mu,\theta),\ \forall\{\mu,\theta\}, \end{cases}$$

with

$$(3.6)\quad J(\mu,\theta)=\tfrac{1}{2}|\mu-v(1)|^2+H(y(1))+K(\theta),$$

and where v and y are solutions of the initial value problems

$$(3.7)\quad \begin{cases} \dfrac{dv}{dt}+\theta A(v)=0, \\ v(0)=\mu, \end{cases}$$

$$(3.8)\quad \begin{cases} \dfrac{dy}{dt}=\dfrac{\theta^2}{2}\|A(v)\|_*^2 \\ y(0)=0, \end{cases}$$

respectively. It is clear that the penalty term associated to H will make difficult convergence to T=0 and to *constant* solutions u (either solutions of the steady state problem (2.4), or not).

To solve the least squares/optimal control problem by either *conjugate gradient* or *quasi-Newton* methods we shall have to introduce first a time discretization, like the specific one discussed in Section 6; a next step will be to compute the gradient of the approximate cost function. Owing to the importance of this last step, we shall consider first (in Section 4) the calculation of the gradient of the cost function in (3.5) - (3.8) and then go back to the discrete case in Section 6.

4. Derivative of the Cost Function J

In (3.5) - (3.8) the *independant variables* are μ and θ. To compute $\frac{\partial J}{\partial \mu}$ and $\frac{\partial J}{\partial \theta}$ we proceed by a *perturbation analysis;* therefore, with obvious notation we have:

$$(4.1)\quad \begin{cases} \delta J=<\frac{\partial J}{\partial \mu},\delta\mu>+\frac{\partial J}{\partial \theta}\,\delta\theta = <\mu-v(1),\delta\mu>+<v(1)-\mu,\delta v(1)> \\ \qquad +H'(y(1))\delta y(1)+K'(\theta)\delta\theta, \end{cases}$$

$$(4.2)\quad \begin{cases} \frac{d}{dt}\,\delta v+\theta A'(v)\delta v+\delta\theta A(v)=0, \\ \delta v(0)=\delta\mu, \end{cases}$$

$$(4.3)\quad \begin{cases} \frac{d}{dt}\delta y=\theta\delta\theta\,\|A(v)\|_*^2+\theta^2(A'(v)\delta v,A(v))_*, \\ \delta y(0)=0, \end{cases}$$

where in (4.1) - (4.3) the symbol ' denotes differentiation. Now, let p be a function defined over [0,1] and taking its values in V; multiplying by p the differential equation in (4.2) we obtain

$$(4.4)\quad <\frac{d}{dt}\delta v,p>+\theta<A'(v)\delta v,p>+\delta\theta<A(v),p>=0.$$

Integrating by parts over [0,1] it follows from (4.4) that

$$(4.5)\quad \begin{cases} <p(1),\delta v(1)>-<p(0),\delta v(0)>-\int_0^1<\frac{dp}{dt},\delta v>\,dt \\ \quad +\theta\int_0^1<A'(v)^*p,\delta v>\,dt+\delta\theta\int_0^1<A(v),p>\,dt=0, \end{cases}$$

where $A'(v)^*$ is the *adjoint* operator of $A'(v)$. The V'-scalar product $(\cdot,\cdot)_*$ satisfies

$$(4.6) \qquad (f,g)_* = \langle f, S^{-1} g\rangle, \ \forall\ f,g \in V';$$

integrating (4.3) over [0,1], and using (4.6) we obtain

$$(4.7) \qquad \delta y(1) = \theta\left(\int_0^1 \| A(v) \|_*^2\, dt\right)\delta\theta + \theta^2 \int_0^1 \langle A'(v)^* S^{-1} A(v), \delta v\rangle\, dt.$$

Suppose now that p is the solution of the *(linear) adjoint state equation*

$$(4.8) \begin{cases} -\dfrac{dp}{dt} + \theta A'(v)^* p = \theta^2 H'(y(1)) A'(v)^* S^{-1} A(v), \\ p(1) = v(1) - \mu. \end{cases}$$

It follows then from (4.1), (4.2), (4.5), (4.7), (4.8) that

$$\begin{cases} \delta J = \langle p(0) - p(1), \delta\mu\rangle \\ +\{K'(\theta) + \theta H'(y(1)) \displaystyle\int_0^1 \| A(v) \|_*^2\, dt - \int_0^1 \langle A(v), p\rangle dt\}\delta\theta, \end{cases}$$

i.e.

$$(4.9)_1 \qquad \frac{\partial J}{\partial \theta}(\mu,\theta) = p(0) - p(1),$$

$$(4.9)_2 \qquad \frac{\partial J}{\partial \theta}(\mu,\theta) = K'(\theta) + \theta H'(y(1)) \int_0^1 \| A(v) \|_*^2\, dt - \int_0^1 \langle A(v), p\rangle\, dt.$$

5. A Conjugate Gradient Algorithm for Solving the least squares/optimal control problem (3.5)

Conjugate gradient methods have always been popular tools for solving optimal control problems (see e.g. [13] for more details on the implementation of conjugate gradient methods for solving control problems); on the other hand, conjugate gradient methods have provided effective algorithms for solving large scale least squares problems (see [1], [2], [7] for more details and other references). From these observations, it is therefore quite natural to apply a conjugate gradient algorithm for solving the minimization problem (3.5). The independent variables being μ and θ, problem (3.5) is indeed a minimization problem in $H \times R$; we suppose that $H \times R$ is equipped with the scalar product

(5.1) $\langle\mu_1,\mu_2\rangle + a\theta_1\theta_2$

(a>0) and the corresponding norm.

1 Initialization:

(5.2) $\{\lambda^0,T^0\}\in H \times R$ *is given;*

solve then the linear variational problem

set

$$(5.3)\quad \begin{cases} \{g_1^0,g_2^0\}\in H \times R;\ \forall\{\mu,\theta\}\in H \times R, \\ \langle g_1^0,\mu\rangle + a g_2^0\theta = \langle\frac{\partial J}{\partial\mu}(\lambda^0,T^0)\mu\rangle + \frac{\partial J}{\partial\theta}(\lambda^0,T^0)\theta. \end{cases}$$

If $g_1^0=0$, $g_2^0=0$ *(or at least are "small") then take* $\lambda=\lambda^0$, $T=T^0$*; if not*

(5.4) $w_1^0=g_1^0$, $w_2^0=g_2^0$. □

Then for $n\geq 0$, $\{\lambda^n,T^n\}$, $\{g_1^n,g_2^n\}$, $\{w_1^n,w_2^n\}$ *being known we compute* $\{\lambda^{n+1},T^{n+1}\}$, $\{g_1^{n+1},g_2^{n+1}\}$, $\{w_1^{n+1},w_2^{n+1}\}$ *according to the following steps:*

2 Descent:

$$(5.5)\quad \begin{cases} \text{Find } \rho_n\in R \text{ such that, } \forall\rho\in R \\ J(\lambda^n-\rho_n w_1^n,T^n-\rho_n w_2^n)\leq J(\lambda^n-\rho w_1^n,T^n-\rho w_2^n), \end{cases}$$

$(5.6)_1$ $\lambda^{n+1}=\lambda^n-\rho_n w_1^n$,

$(5.6)_2$ $T^{n+1}=T^n-\rho_n w_2^n$. □

3 Test of the convergence; new descent direction:

Solve

$$(5.7)\quad \begin{cases} \{g_1^{n+1},g_2^{n+1}\}\in H\times R;\ \forall\{\mu,\theta\}\in H\times R \\ \langle g_1^{n+1},\mu\rangle + a g_2^{n+1}\theta = \langle\frac{\partial J}{\partial\mu}(\lambda^{n+1},T^{n+1}),\mu\rangle\ \frac{\partial J}{\partial\theta}(\lambda^{n+1},T^{n+1})\theta. \end{cases}$$

If $\{g_1^{n+1},g_2^{n+1}\}=0$, *or is small we have converged (locally at least);*

if it is not the case proceed as follows:

Define (Polak-Ribiere strategy)

$$(5.8)\quad \gamma_n = \frac{\langle g_1^{n+1},g_1^{n+1}-g_1^n\rangle + a g_2^{n+1}(g_2^{n+1}-g_2^n)}{|g_1^n|^2+a|g_2^n|^2},$$

and set

(5.9) $w^{n+1}=g^{n+1}+\gamma_n w^n$. □

Do $n=n+1$ *and go to (5.5).*

Remark 5.1: A good test of the convergence is clearly

$$(5.10)\quad \frac{|u^n(1)-\lambda^n|}{|\lambda^0|} \le \varepsilon,$$

since $\lambda^n=u^n(0)$.

Remark 5.2: When applying the above conjugate gradient algorithm, we have to solve *at each iteration* the initial value problems

$$(5.11)\begin{cases} \frac{du^n}{dt} + T^nA(u^n)=0, \\ u^n(0)=\lambda^n, \end{cases}$$

and

$$(5.12)\begin{cases} \frac{dy^n}{dt} = \frac{(T^n)^2}{2}\| A(u^n) \|_*^2 \\ y^n(0)=0; \end{cases}$$

and then the *final value problem*

$$(5.13)\begin{cases} -\frac{dp^n}{dt}+T^nA'(u^n)^*p^n=(T^n)^2H'(y^n(1))A'(u^n)^*S^{-1}A(u^n), \\ p^n(1)=u^n(1)-\lambda^n, \end{cases}$$

to compute $\frac{\partial J}{\partial \mu}(\lambda^n,T^n)$ and $\frac{\partial}{\partial \theta}(\lambda^n,T^n)$, via (4.9).

Remark 5.3: The solution of (5.5) is in fact a *line search* which can be achieved by those methods discussed in, e.g., [12], [13]; we shall go back to this problem in Section 7.

Remark 5.4: For *genuine finite dimensional problems*, it is likely that we can take V=H and S=I, making the above calculations much simpler. If the problem arises from the *space discretization* of partial differential equations, it is likely that we shall have to use $S\neq I$.

Remark 5.5: One of the main advantages of the above minimization approach is that it is convergent despite the fact that to any solution, we can associate an infinity of solutions of the same period obtained by translation along the time axis (see Remark 2.2); the above conjugate gradient (or the quasi-Newton methods, below) will pick a solution which, in some sense, is the "closer" to the initial data $\{\lambda^0,T^0\}$.

Remark 5.6: Suppose that V is *finite dimensional* (therefore H=V) with dim V=N. To study the *stability* of a periodic orbit, we should compute the *eigenvalues* of the *Floquet Matrix*

$$M(u) = \frac{Du(T)}{Du(0)}.$$

If N is not too large, we can compute M(u) first and then the eigenvalues of M(u) (if N is large, alternative methods not requiring the calculation of M(u) have to be used). Let us discuss here the construction of M(u) via the time integration of N *final value problems*, associated to the *adjoint operator*

$$-\frac{d}{dt} + A'(u)^*,$$

and which, indeed, *can be integrated in parallel:*
Let u be a periodic orbit of period T; we have

$$\frac{du}{dt} + A(u) = 0,$$

and then by *perturbation*

$$\frac{d}{dt}\delta u + A'(u)\delta u = 0. \tag{5.14}$$

Multiplying (5.14) by q and integrating by parts we obtain

$$\begin{cases} \langle q(T),\delta u(T)\rangle - \langle q(0),\delta u(0)\rangle \\ +\displaystyle\int_0^T \langle -\frac{dq}{dt} + A'(u)^* q, \delta u\rangle\, dt = 0. \end{cases} \tag{5.15}$$

Take for q the solution p_i of the *linear final value problem*

$$\begin{cases} -\dfrac{dp_i}{dt} + A'(u)^* p_i = 0, \\ p_i(T) = e_i, \end{cases} \tag{5.16}$$

where e_i is the ith vector of the canonical Euclidean vector basis of R^N (i.e. $e_i = \{\delta_{ij}\}_{j=1}^N$). The ith component $(\delta u(T))_i$ of $\delta u(T)$, in the above vector basis satisfies then, from (5.15), and $\forall i=1,\dots,N$,

$$(\delta u(T))_i = \langle e_i, \delta u(T)\rangle = \langle p_i(0), \delta u(0)\rangle. \tag{5.17}$$

Relation (5.17) shows that

$$M(u) = \begin{bmatrix} p^t(0) \\ \cdot \\ \cdot \\ \cdot \\ p_N^t(0) \end{bmatrix} .$$

The interested reader should easily check that the calculation of M(u) is just a by product of the calculation of $\frac{\partial J}{\partial \mu}$ and $\frac{\partial J}{\partial \theta}$ through (3.7), (3.8), (4.8). We observe that 1 is always an eigenvalue of M(u) (it corresponds to those solutions obtained from u by translation along the time axis).

6. Time Discretization of the Least Squares/Optimal Control Problem (3.5). Description of Quasi-Newton Solution Methods.

The *computer implementation* of either the conjugate gradient algorithm (5.2) - (5.9), or of the quasi-Newton methods to be described later on, requires the *time discretization* of the cost function (3.6), and of the differential equations (3.7), (3.8). To discretize the differential equations, we shall use an *implicit scheme* of *Crank-Nicholson* type (like the one already used in [14] for computing the periodic solutions of functional-differential equations); for the reader more familiar with the ordinary differential equation terminology, let us mention that Crank-Nicholson scheme is nothing more than an *implicit Runge-Kutta method of order 2*. Such a scheme has the justified reputation of not being very robust and not very appropriate to the time integration of stiff differential systems; however, monitored by a least squares method, it can capture periodic orbits that it will not be able to reach through a pure initial value approach.

Let N be a positive integer; we define the *time discretization* $\Delta t=k$ by $k=1/N$. We approximate then the least squares problem (3.5) by

$$\min_{\{\mu,\theta\}} J_k(\mu,\theta), \tag{6.1}$$

where

$$J_k(\mu,\theta)=\tfrac{1}{2}|\mu-v^N|^2+H(y^N)+K(\theta) \tag{6.2}$$

with

$$\begin{cases} v^0=\mu, \\ \dfrac{v^{n+1}-v^n}{k} + \theta A\left(\dfrac{v^{n+1}+v^n}{2}\right) =0, \ \forall n=0,1,\dots,N-1, \end{cases} \tag{6.3}$$

$$\begin{cases} y^0=0, \\ \dfrac{y^{n+1}-y^n}{k} = \dfrac{\theta^2}{2}\left\| A\left(\dfrac{v^n+v^{n+1}}{2}\right)\right\|_*^2, \ \forall n=0,1,\dots,N-1. \end{cases} \tag{6.4}$$

Using an approach similar to the one used for the continuous problem, we should prove (not too easily) the discrete equivalent of (4.9), namely

$$(6.5)_1 \quad \begin{cases} <\frac{\partial J_k}{\partial \mu}(\mu,\theta),z>=<\mu-v^N+p^o-\frac{k\theta}{2}A'(v^{\frac{1}{2}})^* \\ \qquad\qquad + \frac{k\theta^2}{2}H'(y^N)A'(v^{\frac{1}{2}})^*S^{-1}A(v^{\frac{1}{2}}),z>, \end{cases}$$

$\forall z\in H$, and

$$(6.5)_2 \quad \begin{cases} \frac{\partial J_k}{\partial \theta}(\mu,\theta)=K'(\theta)-k\sum_{n=0}^{N-1} <A(v^{n+\frac{1}{2}}),p^n> \\ \qquad\qquad +H'(y^n)\theta k\sum_{n=0}^{N-1} \| A(v^{n+\frac{1}{2}}) \|_*^2 \quad , \end{cases}$$

where, in $(6.5)_1$, $(6.5)_2$, we have, $\forall n=0,1,\dots,N-1$,

$$v^{n+\frac{1}{2}}=\tfrac{1}{2}(v^n+v^{n+1}),$$

and where $\{p^n\}_{n=0}^{N-1}$ is obtained through the *sequential solution* of the *discrete adjoint equations*

$$(6.6)_1 \qquad p^N=v^N-\mu,$$

$$(6.6)_2 \qquad \frac{p^{N-1}-p^N}{k} + \frac{\theta}{2}(A'(v^{n-\frac{1}{2}})^*p^{N-1}=H'(y^N)\frac{\theta^2}{2}A'(v^{N-\frac{1}{2}})^*S^{-1}A(v^{N-\frac{1}{2}}),$$

and then for $n=N-1,N-2,\dots,1$,

$$(6.6)_3 \quad \begin{cases} \frac{p^{n-1}-p^n}{k} + \frac{\theta}{2}(A'(v^{n-\frac{1}{2}})^*p^{n-1}+A'(v^{n+\frac{1}{2}})^*p^n) \\ =H'(y^N)\frac{\theta^2}{2}(A'(v^{n-\frac{1}{2}})^*S^{-1}A(v^{n-\frac{1}{2}})+A'(v^{n+\frac{1}{2}})^*S^{-1}A(v^{n+\frac{1}{2}})); \end{cases}$$

the above scheme (6.6) is clearly of Crank-Nicholson type, but it is quite different from the scheme that we should obtain by a *direct* Crank-Nicholson discretization of the continuous adjoint equation (4.8). Using the above material, we should derive a discrete variant of the conjugate gradient algorithm (5.2) - (5.9), and also of the construction-described in Remark 5.6 - of the Floquet matrix, in order to study the stability of the discrete orbit.

On Quasi-Newton Methods for Solving Problem (3.5):

We consider now the simple case where $V=H=R^m$; we suppose that R^m is equipped with the usual norm $|v|$ $(=\| v \|)$, defined by

$$|v|=\left(\sum_{j=1}^{m} v_j^2\right)^{\frac{1}{2}}, \quad \forall v\in R^m, \quad v=\{v_j\}_{j=1}^{m}.$$

Problem (3.5) is first discretized in time, providing the fully discrete problem (6.1), which here takes the following form

$$\text{(6.7)} \qquad \underset{\{\mu,\theta\}\in R^{m+1}}{\text{Min}} J_k(\mu,\theta)$$

where (cf. (6.2))

$$\text{(6.8)} \qquad J_k(\mu,\theta)=\tfrac{1}{2}|\mu-v^N|^2+H(y^N)+K(\theta).$$

Let us drop index k, for clarity. If m is not too large, it is tempting to solve (6.7) by *Newton type methods*. We have already seen that $J(\mu,\theta)$ and $\nabla J(\mu,\theta)=\{\frac{\partial J}{\partial \mu},\frac{\partial J}{\partial \theta}\}(\mu,\theta)$ can be computed fairly easily (even for large m); if one wishes to solve (6.7) by Newton type methods, it is necessary to compute the *Hessian matrix* $\nabla^2 J(\mu,\theta)$. Such calculation is possible, but complicated. In order to avoid it, we have concentrated on *Quasi-Newton methods* like those discussed in e.g. Dennis and Schnabel [12]. These are *locally superlinearly convergent* algorithms for finding local solutions to problem (6.7). Globalization modifications have been incorporated to enable the algorithms to converge from arbitrary starting points $\{\mu^0,\theta^0\}$. The two algorithms that we consider below are Quasi-Newton methods in which we use the BFGS least change secant update formula for approximating the Hessian matrix $\nabla^2 J(\mu,\theta)$.

1st Method - A linear search algorithm.

Let $x=\{\mu,\theta\}$ and

$$S=\{H|\,H\in R^{(m+1)^2},\ H=H^t,\ \text{and}\ x^tHx>0,\ \forall x\in R^{m+1}-\{0\}\};$$

the *linear search algorithm* proceeds as follows:

$$\text{(6.9)} \qquad \textit{Take } x_0\in R^{m+1},\ H_0\in S,\ \beta>0;$$

for $n\geq 0$, *until convergence, do*

$$\text{(6.10)} \qquad \textit{Solve } H_n s_n= -\nabla J(x_n),$$

$$\text{(6.11)} \qquad \begin{cases} \textit{Find } \rho_n\in R_+, \textit{ such that} \\ J(x_n+\rho_n s_n)\leq J(x_n)+\beta\rho_n\nabla J(x_n)^t s_n, \end{cases}$$

$$\text{(6.12)} \qquad \textit{Let } x_{n+1}=x_n+\rho_n s_n.$$

Update then H_n *by*

$$(6.13)\quad H_{n+1}=\begin{cases} H_n+\dfrac{y_ny_n^t}{y_n^t\sigma_n}-\dfrac{H_n\sigma_n\sigma_n^tH_n}{\sigma_n^tH_n\sigma_n}, & \text{if } \sigma_n^ty_n>0,\\ H_n, & \text{otherwise,}\end{cases}$$

with $\sigma_n=x_{n+1}-x_n$, $y_n=\nabla J(x_{n+1})-\nabla J(x_n)$. □

We observe that H_{n+1} satisfies the *secant condition*

$$H_{n+1}\sigma_n=y_n.$$

It can also be shown that $H_n\in S$ and $\sigma_n^t\, y_n>0$ imply that $H_{n+1}\in S$.

The initial Hessian approximation is chosen to be

$$H_o=|J(x_o)|\ I,$$

where I is the *identity matrix* in $R^{(m+1)^2}$.

In (6.11), ρ_n is computed by using the backtracking safeguarded cubic line search described in Dennis and Schnabel [12, Chapter 6].

2nd Method - A trust region algorithm.

This alternate algorithm is also described in [12]. Basically, we consider the *quadratic approximation* to $J(x_n+s)$ given by

$$(6.14)\qquad Q_n(s)=\tfrac{1}{2}s^tH_ns+\nabla J(x_n)^ts+J(x_n).$$

Since s_n in (6.10) minimizes Q_n, it can be a poor descent direction when Q_n is a poor approximation to $J(x_n+s)$. From this observation, instead of solving (6.10), we consider the following minimization problem

$$(6.15)\qquad \textit{Minimize } Q_n(s), \textit{ subject to } |s|\le\delta_n.$$

Here, δ_n is the radius of the ball around x_n in which, we "trust" the quadratic functional Q_n to be a good approximation to $J(x_n+s)$. Problem (6.15) can be solved by

$$(6.16)\qquad s_n(\mu)=-(H_n+\mu I)^{-1}\nabla J(x_n),$$

for the *unique* $\mu\ge0$ such that

$$(6.17)\qquad \begin{cases} |s_n(\mu)|=\delta_n, & \textit{if } |s_n(0)|>\delta_n;\\ \mu=0 & \textit{if } |s_n(0)|\le\delta_n\ (\Rightarrow s_n=s_n(0));\end{cases}$$

in the nontrivial case (where $\mu>0$), problem (6.17) can be solved by the Moré–Hebden algorithm described in [12].

7. Numerical Experiments.

7.1. Generalities. Synopsis.

In this section, we shall report the results of numerical experiments in which the methods described in the previous sections have been applied to the solution of three test problems, two of them associated with *chemical reactions*. From a methodological point of view, nonlinear differential systems such as (3.3) have been approximated by the Crank-Nicholson scheme (6.3), using Newton's method to compute v^{n+1} from v^n.

The results reported here have been obtained with the Quasi-Newton methods of Section 6 using

$$(7.1) \qquad |\nabla J(x_n)|/|\nabla J(x_0)|<10^{-6}$$

as *stopping criteria*.

7.2. A first test problem.

The first test problem that we consider is defined by $V=R^2$, $v=\{v_1,v_2\}$, with

$$(7.2) \quad A(v)= -\{-v_2+v_1(1-v_1^2-v_2^2), v_1+v_2(1-v_1^2-v_2^2)\}.$$

An obvious solution of problem (1.1) is defined by

$$u(t)=\{\cos t, \sin t\},\ T=2\pi$$

Taking $k=\Delta t=10^{-2}$ and using the first quasi-Newton method described in Section 6 (i.e. the line search/BFGS algorithm) we have obtained the results summarized in Table 7.1 (the computed period 6.2852 has to be compared to $2\pi=6.2831\ldots$).

7.3. A second test problem: The Brusselator problem.

This problem originates from chemical reactions (see [18],[19] for related references). It is defined by $V=R^2$ and

$$(7.3) \quad A(v)=-\{a-(b+1)v_1+v_1^2v_2, bv_1-v_1^2v_2\}.$$

If $b>1+a^2$ the Brusselator problem has a unique asymptotically stable periodic orbit and for $a=1$, $b=3$ an approximate value of the period is 7.16 as shown in [19] by analytical methods.

The corresponding numerical results are summarized in Table 7.2 and have been obtained with $k=\Delta t=10^{-2}$.

Table 7.1

u^0	T^0	# iterations	computed T	computed $\|u(0)-u(1)\|$	# J evaluations	# ΔJ evaluations
{1,0}	6.28	2	6.2852	1×10^{-5}	3	3
{.1,.1}	2	12	6.2852	6×10^{-5}	14	13
{1,1}	3	21	6.2852	2×10^{-5}	32	22

Table 7.2

u^0	T^0	# iterations	computed T	computed $\|u(0)-u(1)\|$	# J evaluations	# ∇J evaluations
{3,.5}	6	10	7.1529	1×10^{-5}	12	11
{3,.5}	5	16*	7.1529	1×10^{-4}	18	17
{3,.5}	14	32	7.1529	1.4×10^{-5}	35	33
{3.1,1.1}	6	17	7.1529	1.9×10^{-5}	22	18
{4,2}	6	28	7.1529	1.3×10^{-5}	32	29
{.5,.5}	6	31*	7.1529	3.3×10^{-5}	41	32

*: Solved by the trust region algorithm of Section 6 (stopping criteria $|\nabla J(x_n)|<10^{-5}$).

Computed eigenvalues of the Floquet matrix: .994 and $.228\times10^{-5}$.

We have represented, on Figure 7.1, the computed periodic orbit in the phase plane $\{u_1,u_2\}$; on Figures 7.2 and 7.3 we have shown the variations of the computed u_1 and u_2 as functions of the scaled time t/T.

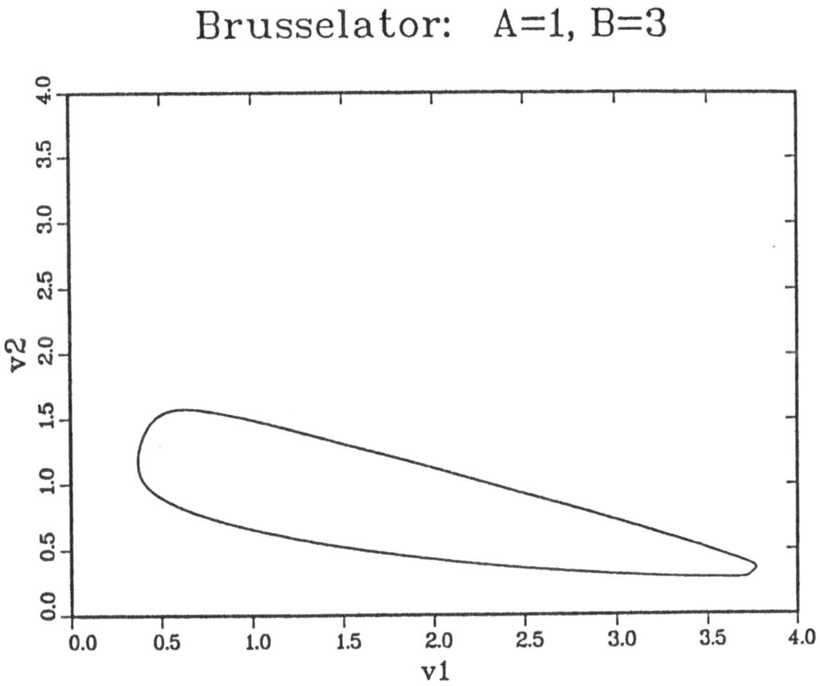

Figure 7.1

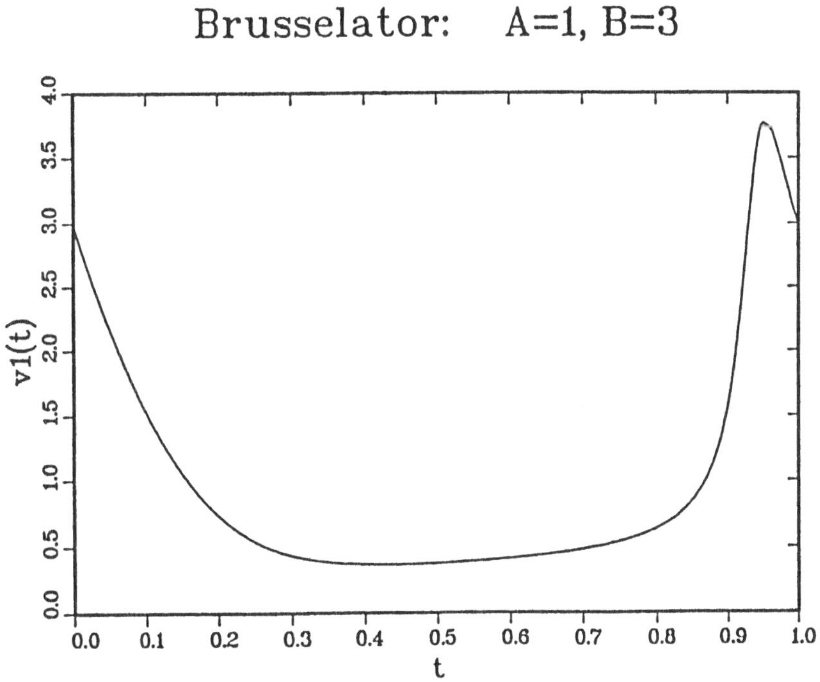

Figure 7.2

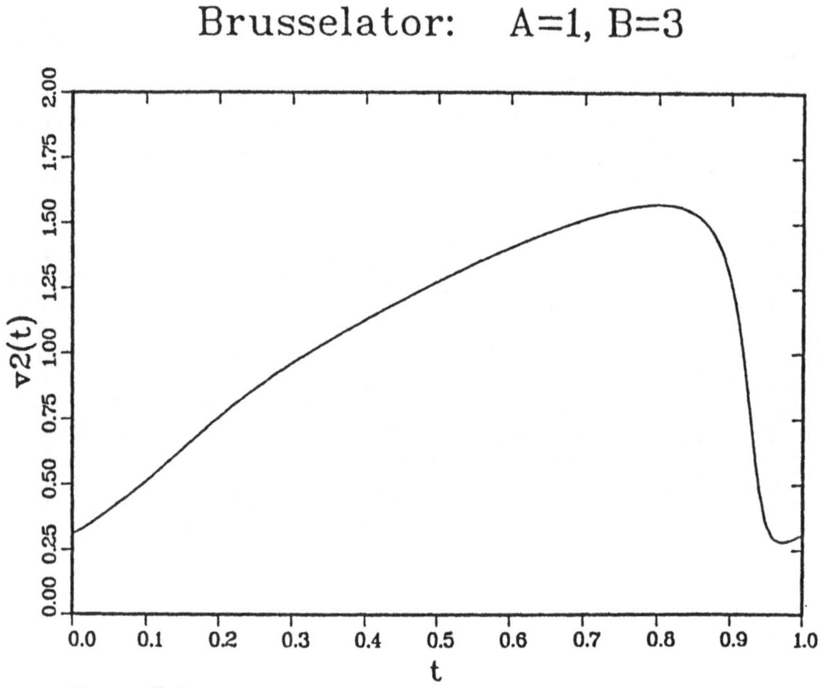

Figure 7.3

7.4. A third test problem: The Field-Noyes model of the Belousov-Zhabotinskii reaction.

This last problem is far more difficult than the previous two. It concerns the search of the periodic solutions of the nonlinear differential system below

$$(7.4)\quad \begin{cases} \dfrac{du_1}{dt} = s(u_2-u_1u_2+u_1-qu_1^2), \\ \dfrac{du_2}{dt} = \dfrac{1}{s}(fu_3-u_2-u_1u_2), \\ \dfrac{du_3}{dt} = w(u_1-u_3). \end{cases}$$

This system has motivated a large number of publications; we shall mention among them [20], [21] (see also the references therein). According to the literature, typical values of the parameters s, q, f, w are given by: $s=77.27$, $q=8.375\times10^{-6}$, $w=1$, $f=1$. For such values, problem (7.4) is highly *stiff* and in fact, our Crank-Nicholson discretization scheme could not handle such a system for the above value of the parameter q. However, in order to test the possibilities of our least squares approach, coupled to the Crank-Nicholson scheme, we have been solving system (7.4) for q successively equal to 2^{-8}, 2^{-10}, 2^{-11} with s=64, f=1, w=1. The time step Δt was taken equal to 2×10^{-3} for $q=2^{-8}$ and 2^{-10} , and to 10^{-3} for $s=2^{-11}$. In fact, we have been using a very elementary *continuation* method in the sense that we used the results at $q=2^{-8}$ (resp. 2^{-10}) to initialize the calculation with $q=2^{-10}$ (resp. 2^{-11}). The corresponding results, obtained by the first Quasi-Newton method, have been reported in Table 7.3.

Table 7.3

q	u^0	T^0	# its.	computed T	# J evaluations	# ∇J evaluations
2^{-8}	{2.5,1.65,2.59}	25	19	22.71	22	20
2^{-10}	*continuation*	*continuation*	20	57.59	25	21
2^{-11}	*continuation*	*continuation*	23	75.32	27	24

Computed eigenvalues of the Floquet matrix ($q=2^{-11}$):

1.005, 10^{-18}, 10^{-18}.

We have shown on Figures 7.4, 7.5, and 7.6 computed periodic orbit in the phase planes (u_1,u_2), (u_2,u_3), (u_3,u_1) and on Figures 7.7, 7.8, and 7.9 the variations of u_1, u_2, u_3 as functions of t/T. We observe that these functions are quite nonsmooth, particularly u_1 and u_3, but nevertheless, we have been able to "capture" them through an elementary Crank-Nicholson scheme with a constant time discretization step. However, for smaller values of q, a more sophisticated time discretization scheme should be required.

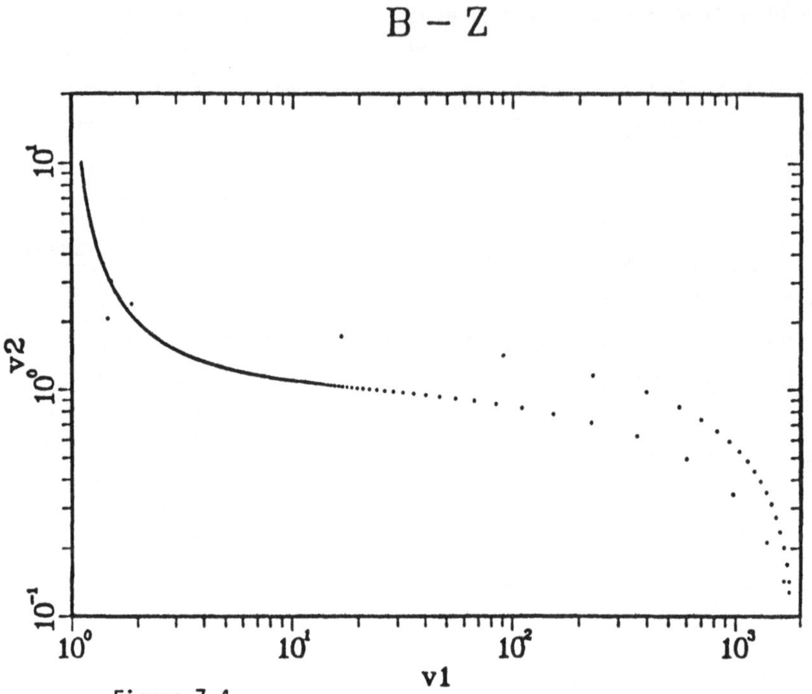

Figure 7.4

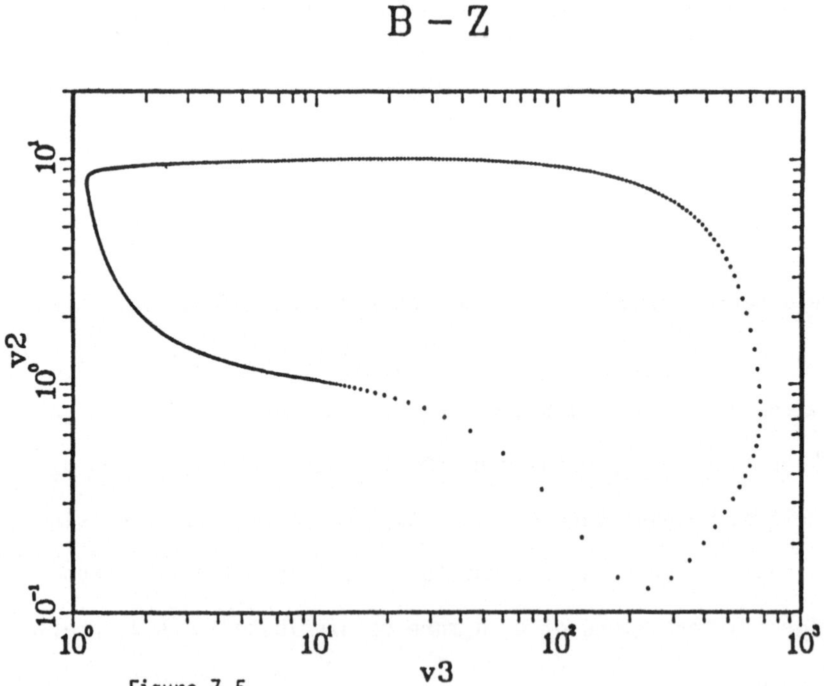

Figure 7.5

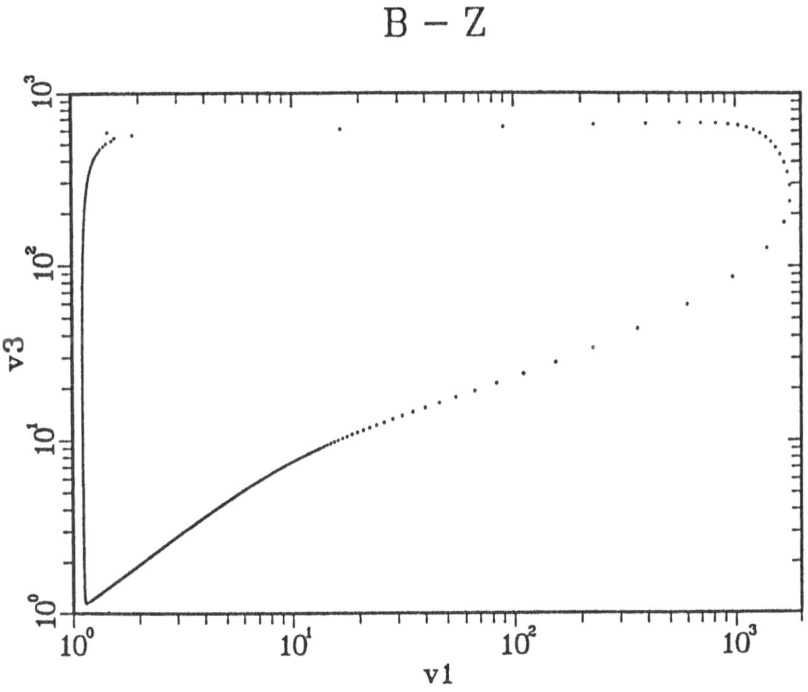

Figure 7.6

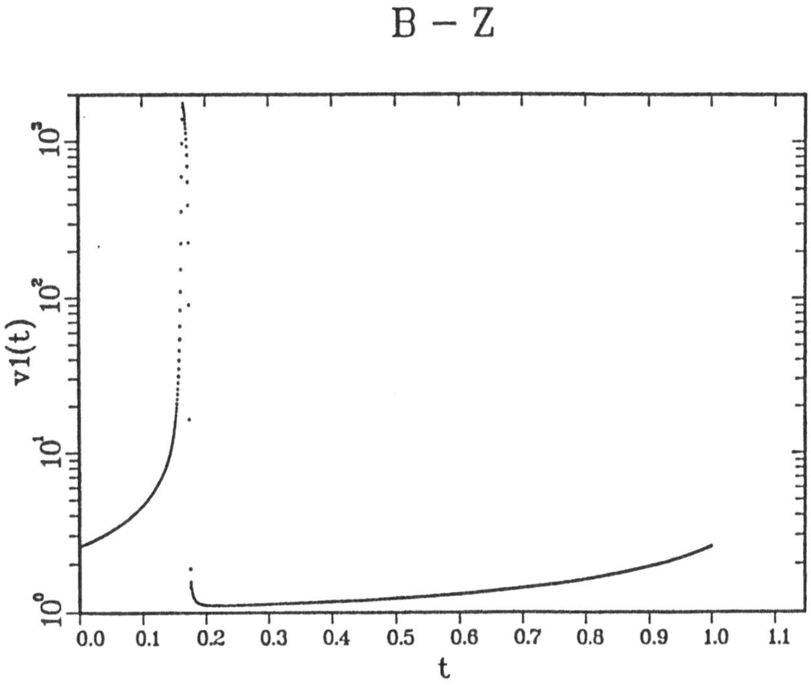

Figure 7.7

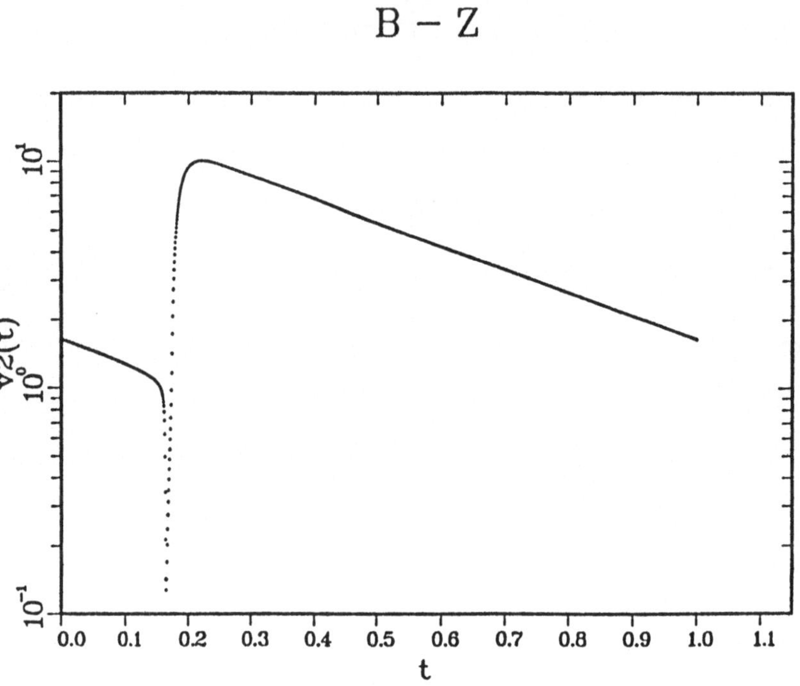

Figure 7.8

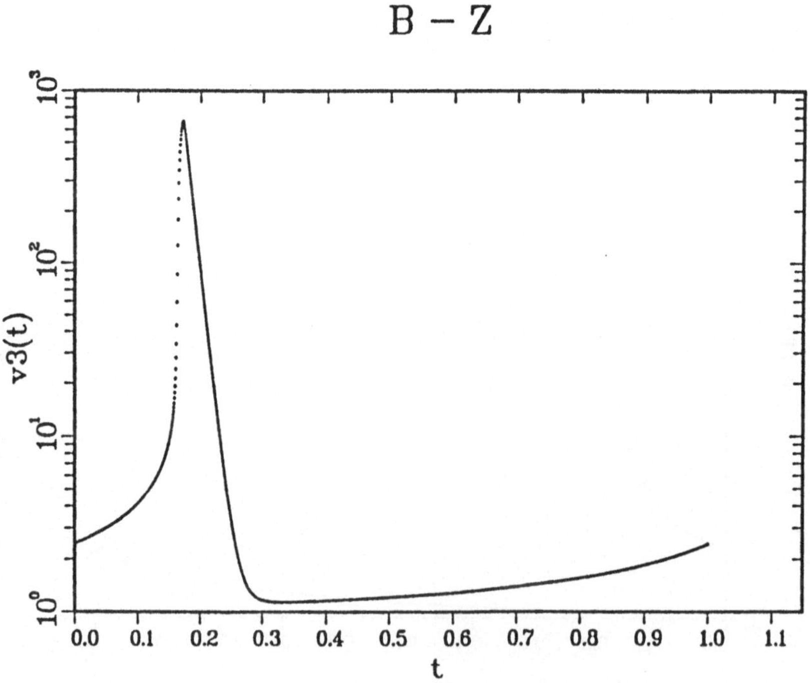

Figure 7.9

8. Conclusion

From the above numerical results, it appears that nonlinear least squares techniques--associated to an optimal control formulation--can be very effective tools for the numerical solution of nontrivial nonlinear differential problems. In the near future, we shall apply these methods to the solution of nonlinear parabolic equations of reaction diffusion type.

We have to observe that with the approach used in this paper, the least squares problem that we are solving has always an infinity of solutions, since, as already mentioned in Remark 2.2, $\{u,T\}$, the solution of (1.1), implies that $\{t \rightarrow u(t+\tau),T\}$ is also a solution, $\forall \tau \in R$. However, we have good convergence properties without requiring additional constants, such as the anchor condition used in [11].

Acknowledgement

We would like to thank Professor R. Buchler, Department of Physics, University of Florida, for the helpful discussions during the meeting and also Yvette Fisk for typing the manuscipt.

REFERENCES

[1] M.O. Bristeau, R. Glowinski, J. Periaux, P. Perrier and O. Pironneau, On the numerical solution of nonlinear problems in fluid dynamics by least squares and finite element methods (I). Least squares formulation and conjugate gradient solutions of the continuous problems, *Comp. Meths. Appl. Mech. Engrg.* 17/18 (1979) 619-657.

[2] R. Glowinski, *Numerical Methods for Nonlinear Variational Problems*, Springer-Verlag, New York, 1984.

[3] M.O. Bristeau, R. Glowinski, J. Periaux, P. Perrier, O. Pironneau and G. Poirier, Transonic flow simulations by finite element and least squares methods in: R.H. Gallagher, D.H. Norrie, J.T. Oden and O.C. Zienkiewicz, eds., *Finite Element in Fluids, Vol. 4*, Wiley, Chichester, 1982, 453-482.

[4] M.O. Bristeau, R. Glowinski, J. Periaux, P. Perrier, O. Pironneau and G. Poirier, On the numerical solution of non-

linear problems in fluid dynamics by least squares and finite element methods (II). Application to transonic flow simulations, *Comp. Meths. Appl. Mech. Engrg.* 51 (1985) 363-394.

[5] R. Glowinski, Viscous flow simulation by finite element methods and related numerical techniques, in: E.M. Murman and S.S. Abarbanel, eds., *Progress and Supercomputing in Computational Fluid Dynamics,* Birkhauser, Boston, 1985, 173-210.

[6] R. Glowinski and J. Periaux, Finite element, least squares and domain decomposition methods for the numerical solution of nonlinear problems in fluid dynamics, in: F. Brezzi, ed., *Numerical Methods in Fluid Dynamics,* Lecture Notes in Mathematics, Vol. 1127, Springer-Verlag, Berlin, (1985) 1-114.

[7] R. Glowinski, H.B. Keller and L. Reinhart, Continuation-Conjugate gradient methods for the least squares solution of nonlinear boundary value problems, *SIAM J. Sci. Stat. Comput.* 6 (1985) 793-832.

[8] L. Reinhart, On the numerical analysis of the Von Karman equations: Mixed finite element approximation and continuation techniques, *Numer. Math.* 39 (1982) 371-404.

[9] G. Meurant and J.C. Saut, Bifurcation and Stability in a chemical system, *J. Math Anal. Appl.* 59 (1977) 69-92.

[10] E.J. Doedel, A.D. Jepson and H.B. Keller, Numerical Methods for Hopf bifurcation and continuation of periodic solution paths, in: R. Glowinski and J.L. Lions, eds., *Computing Methods in Applied Sciences and Engineering VI,* North-Holland, Amsterdam, 1984, 127-138.

[11] E.J. Doedel and J.P. Kernevez, Software for Continuation problems in ordinary differential equations with applications (to appear).

[12] J.E. Dennis and R.B. Schnabel, *Numerical Methods for Unconstrained Optimization and Nonlinear Equations,* Prentice-Hall, Englewood Cliffs, N.J., 1983.

[13] E. Polak, *Computational Methods in Optimization,* Academic Press, New York, 1971.

[14] R. Glowinski, Approximation numerique des solutions periodiques de l equation integro-differentielle $\frac{du}{dt} + \phi(u) + \int_0^1 A(t,\tau)u(\tau)d\tau=f$ *J. Math. Anal. Appl.* 41 (1973) 67-96.

[15] G. Auchmuty and G. Nicolis, Bifurcation analysis of reaction-diffusion equations III, *Bull. Math. Biol.* 38 (1975) 325-350.

[16] G. Auchmuty, Bifurcating Waves, *Ann. N.Y. Acad. Sciences* 316 (1979) 263-278.

[17] M. Urabe, *Nonlinear Autonomous Oscillations*, Academic Press, New York, 1967.

[18] G. Nicolis and I. Prigogine, *Self-Organization in Nonequilibrium Systems*, Wiley, New York, 1977.

[19] S. Zhang, Approximate solution of the Brusselator limit cycle in Biochemistry, *Kexue Tongbao* 27 (1982) 433-437.

[20] J.J. Tyson, *The Belousov-Zhabotinskii reaction*, Lecture Notes in Biomathematics, Vol. 10, Springer-Verlag, Berlin, 1976.

[21] J.J. Tyson, Analytic representation of oscillations, excitability, and traveling waves in a realistic model of the Belousov-Zhabotinskii reaction, *J. Chem. Physics* 66 (1977) 905-915.

PARETO CONTROL OF DISTRIBUTED SYSTEMS. AN INTRODUCTION

J.L. Lions*

1. GENERAL REMARKS

Let us consider a linear (unbounded) operator, say A; the operator A will be a Partial Differential operator in the applications. In this introductory section we consider the *stationary* case, but the whole theory extends to *evolution problems*.

In order to define more precisely the setting, we introduce two real Hilbert spaces V and H (*all* Hilbert spaces we consider here are on R); we assume that $V \subset H$, V dense in H and $V \to H$ continuous. We identify H with its dual and we define V' dual of V; we have $V \subset H \subset V'$. We assume that

A defines an isomorphism from V onto V'. (1.1)

We consider now two (real) Hilbert spaces $\mathcal{U}$ and F; $\mathcal{U}$ is the space of the *control variables* and F is the space of *uncertainties*. We are given

$$B \in \mathcal{L}(\mathcal{U}; V'), \tag{1.2}$$

$$\beta \in \mathcal{L}(F; V'). \tag{1.3}$$

We consider a *closed subspace* G *of* F:

$$G \subset F; \tag{1.4}$$

loosely speaking, G corresponds to the information that we have relative to the uncertainties.

The state of the system is $y(v,g)$, $v \in \mathcal{U}$, $g \in G$; $y(v,g)$ corresponds to a control v and to a possible choice of g in the space G; it is given be

$$Ay(v,g) = Bv + \beta g, \quad y(v,g) \in V. \tag{1.5}$$

---

* Collège de France, Chaire d'Analyse des Systèmes et de leur Contrôle, Paris and C.N.E.S. (Centre National d'Etudes Spatiales), 2, Place M. Quentin 75039 PARIS CEDEX 01, France.

Remark 1.1

If $G = \{0\}$, (1.5) is a classical state equation.

The case where $G = F$ corresponds to the case where we do not have information relative to the uncertainties in the system. □

We consider now the *observation operator*

$$C \in \mathscr{L}(V;\mathscr{H}), \text{ where } \mathscr{H} \text{ is a (real) Hilbert space} \tag{1.6}$$

and we consider

$$J(v,g) = \|Cy(v,g) - z_d\|^2_{\mathscr{H}} + N\|v\|^2_{\mathscr{U}} \tag{1.7}$$

where z_d is given in $\mathscr{H}$, N is given > 0 and where $\| \ \|_X$ denotes the norm in X.

We now introduce *the Pareto controls for the family of functionals*

$$J_g(v) = J(v,g), \ g \in G. \tag{1.8}$$

By a natural extension of the notion of a Pareto point or a Pareto equilibrium, we shall say that u *is a Pareto control* if *there is no* v *in* $\mathscr{U}$ such that

$$\begin{aligned} &J(v,g) \leq J(u,g) \quad \forall g \in G, \\ &J(v,g_0) < J(u,g_0) \text{ for at least one } g_0 \in G. \end{aligned} \tag{1.9}$$

We shall see below that there exists *infinitely many Pareto controls*. We shall say that u *is a Pareto control relative to* u_0 if

$$J(v,g) \leq J(u_0,g) \ \forall g \in G \tag{1.10}$$

(u_0 is here given in $\mathscr{U}$)

We shall see now that

there exists a unique Pareto control relative to u_0. (1.11)

Proof.

Let us see first when one has

$$J(v,g) \leq J(w,g) \quad \forall g \in G. \tag{1.12}$$

We introduce $y(v)$ and $\phi(g)$ by

$$Ay(v) = Bv, \ A\phi(g) = \beta g. \tag{1.13}$$

Then $y(v,g) = y(v) + \phi(g)$ and

$$J(v,g)-J(w,g)=2(C(y(v)-y(w)),C\phi(g))_{\mathscr{H}}+J(v,o)-J(w,o). \tag{1.14}$$

Since G is a *vector space*, one easily verifies that (1.12) is equivalent to

$$(C(y(v) - y(w)), C\phi(g))_{\mathscr{H}} = 0 \ \forall g \in G, \tag{1.15}$$

and

$$J_0(v) \leq J_0(w), \tag{1.16}$$

where

$$J_o(v) = J(v,o).$$

We introduce then

$$\mathcal{M} = \{v \mid v \in \mathcal{U}, (Cy(v), C\phi(g))_{\mathcal{H}} = 0 \ \forall\, g \in G\}; \tag{1.17}$$

we define in this way a closed subspace of $\mathcal{U}$. With this definition we see that (1.12) is equivalent to

$$v - w \in \mathcal{M} \text{ and } (1.16).$$

In that situation (1.14) reduces to

$$J(v,g) - J(w,g) = J_o(v) - J_o(w),$$

and it is then a simple matter to verify that u is a Pareto control relative to u_o if and only if

$$\begin{aligned} & u \in \mathcal{M} + u_o, \\ & J_o(v) \leq J_o(v) \ \forall\, v \in \mathcal{M} + u_o. \end{aligned} \tag{1.18}$$

In other words u *minimizes* $J_o(v)$ *on* $\mathcal{M} + u_o$. This proves (1.11).□

Remark 1.2

Conditions (1.18) prove that finding u, Pareto control relative to u_o, is equivalent to a problem of optimal control *with constraints on the state*. These constraints are expressed by

$$(C(y(v) - y(u_o)), C\phi(g))_{\mathcal{H}} = 0 \ \forall\, g \in G. \square \tag{1.19}$$

Remark 1.3

Let $\mathcal{U}_{ad}$ be a closed convex subset of $\mathcal{U}$-the set of admissible controls-and let us add the contraints

$$v \in \mathcal{U}_{ad}.$$

We have similar definitions for u, Pareto control relative to u_o, where now $u \in \mathcal{U}_{ad}$ and where (to fix ideas) u_o is given in $\mathcal{U}_{ad}$. Again u exists and is unique; it is given by

$$\inf. J_o(v), \ v \in \{\mathcal{M} + u_o\} \cap \mathcal{U}_{ad}. \square \tag{1.20}$$

Remark 1.4

As we already said, if $G = \{0\}$, conditions (1.19) are automatically verified and one deals with a standard optimal control problem.□

Remark 1.5

What we have said extends to other settings, so as to apply to nonhomogeneous boundary value problems and it also extends to evolution operators. We shall give examples below.□

The objectives that we want to pursue are the following:

(i) to obtain *a constructive optimality system*, in particular in order to take into account in an explicit manner the implicit constraints (1.15);

(ii) the optimality system contains, as we shall see below, an element λ-*a Lagrange multiplier*-which in the finite-dimensional case is in general an element of G, but which is generally in the infinite dimensional case *an element of a space* $\hat{G}$ *larger than* G;

(iii) one can define-by extension by continuity-

$J(v,\lambda) - J(u_0,\lambda)$ for $\lambda \in \hat{G}$

(this will be made precise in the examples below) and the Pareto *control relative to* u_0 *is the unique element* u *which minimizes*

$J(v,\lambda) - J(u_0,\lambda)$ *over* $\mathcal{U}$.

But the *choice* of λ is nontrivial and depends on (i).□

This is *not* a set of theorems but *a program* which is verified in all the situations that we have considered so far (the main difficulty lies in the construction of $\hat{G}$, which can be the completion in a suitable norm of a quotient of G). This program has been initiated in Lions [1][2]. We pursue this program here.

In Section 2 below we introduce the optimality system for (1.18).

In Section 3 we give an example where the uncertainties lie on the boundary and in Section 4 we briefly consider a parabolic system. (Hyperbolic systems, or systems of Petrowsky's type, are considered in Lions [1].)

2. ABSTRACT OPTIMALITY SYSTEM

Using the notations of Section 1, let us consider the *augmented state* $\{y(v),\zeta(v)\}$ given by

$$\begin{aligned} &Ay(v) = Bv, \quad A^*\zeta(v) = C^*Cy(v), \\ &y(v),\ \zeta(v) \in V. \end{aligned} \tag{2.1}$$

Then $v \in \mathcal{M}$ is equivalent to $(A^*\zeta(v),\phi(g)) = 0 \ \forall\, g \in G$, i.e., $(\zeta(v),\beta g) = 0$ or $(\beta^*\zeta(v),g)_F = 0 \ \forall\, g \in G$.

Let us introduce:

$$\pi = \text{operator of orthogonal projection } F \to G. \tag{2.2}$$

Then "$v \in \mathcal{M}$" iff $\pi\beta^*\zeta(v) = 0$ and (1.19) is equivalent to

$$\pi(\beta^*(\zeta(v) - \zeta(u_o))) = 0. \tag{2.3}$$

Finding u, Pareto control relative to u_o, is equivalent to

$$\inf. J_o(v),\ v \text{ subject to } (2.3).\ \square \tag{2.4}$$

We approach this problem by *a penalty argument*. We define

$$K_\varepsilon(v) = J_o(v) + \frac{1}{\varepsilon}\|\pi\beta^*(\zeta(v) - \zeta(u_o))\|_F^2,\ \varepsilon > 0, \tag{2.5}$$

and we consider

$$\inf. K_\varepsilon(v),\ v \in \mathcal{U}. \tag{2.6}$$

This problem admits a unique solution u_ε and one has

$$u_\varepsilon \to u \text{ (as } \varepsilon \to 0) \text{ in the space } \mathcal{U}. \tag{2.7}$$

Let us set:

$$y(u_\varepsilon) = y_\varepsilon,\ \zeta(u_\varepsilon) = \zeta_\varepsilon,\ \lambda_\varepsilon = \frac{1}{\varepsilon}\pi\beta^*(\zeta_\varepsilon - \zeta(u_o)).$$

The control u_ε is characterized by

$$\begin{aligned}&(Cy_\varepsilon - z_d, Cy(v))_{\mathcal{H}} + N(u_\varepsilon, v)_{\mathcal{U}} + (\lambda_\varepsilon, \pi\beta^*\zeta(v))_F = 0\\ &\forall\, v \in \mathcal{U}.\end{aligned} \tag{2.8}$$

We have

$$Ay_\varepsilon = Bu_\varepsilon,\ A^*\zeta_\varepsilon = C^*Cy_\varepsilon \tag{2.9}$$

and we introduce p_ε and ρ_ε by

$$A^*p_\varepsilon = C^*(Cy_\varepsilon - z_d) + CC\rho_\varepsilon,\ A\rho_\varepsilon = \beta\lambda_\varepsilon. \tag{2.10}$$

Then

$$\begin{aligned}&(A^*p_\varepsilon, y(v)) + (A\rho_\varepsilon, \zeta(v))\\ &= (Cy_\varepsilon - z_d, Cy(v))_{\mathcal{H}} + (C\rho_\varepsilon, Cy(v))_{\mathcal{H}} + (\beta\lambda_\varepsilon, \zeta(v))\\ &= (p_\varepsilon, Bv) + (\rho_\varepsilon, A^*\zeta(v)) = (p_\varepsilon, Bv) + (C\rho_\varepsilon, Cy(v))_{\mathcal{H}}\end{aligned}$$

so that (2.8) reduces to

$$(B^*p_\varepsilon, v)_{\mathcal{U}} + N(u_\varepsilon, v)_{\mathcal{U}} = 0\ \forall\, v \in \mathcal{U},$$

i.e.,

$$B^*p_\varepsilon + Nu_\varepsilon = 0. \tag{2.11}$$

The difficulty lies in the obtainment of a priori estimate on λ_ε.

Let us introduce $\hat{p}_\varepsilon$ by

$$A^*\hat{p}_\varepsilon = C^*(Cy_\varepsilon - z_d),\ p_\varepsilon \in V \tag{2.12}$$

and σ_ε by

$$A^*\sigma_\varepsilon = C^*C\rho_\varepsilon,\ \sigma_\varepsilon \in V. \tag{2.13}$$

Then

$$p_\varepsilon = \hat{p}_\varepsilon + \sigma_\varepsilon. \tag{2.14}$$

When $\varepsilon \to 0$, since $u_\varepsilon \to u$ in $\mathcal{U}$, we know that $y_\varepsilon \to y$ in V, $\zeta_\varepsilon \to \zeta$ in V, $\hat{p}_\varepsilon \to \hat{p}$ in V, where

$$Ay = Bu, \ A^*\zeta = C^*Cy, \ A^*\hat{p} = C^*(Cy - z_d).$$

Then (2.11) gives

$$B^*\sigma_\varepsilon + B^*\hat{p}_\varepsilon + Nu_\varepsilon = 0$$

so that

$$B^*\sigma_\varepsilon \to (B^*\hat{p} + Nu) \text{ in } \mathcal{U}. \tag{2.15}$$

This remark leads to the following: we consider

$$A\rho = \beta g, \ A^*\sigma = C^*C\rho \tag{2.16}$$

and we set

$$|||g||| = \|B^*\sigma\|_{\mathcal{U}}. \tag{2.17}$$

We define in this way a seminorm on $\overset{\circ}{G}$ (of course it can be a norm!); we define the quotient space $\overset{\circ}{G}$ associated with $|||g|||$ and we define

$$\hat{G} = \text{completion of } G \text{ for the norm } |||g||| \quad \text{(on the quotient space)}. \tag{2.18}$$

Then (we have made exactly what needed to obtain that!)

$$\lambda_\varepsilon \text{ remains in a bounded set of } \hat{G} \tag{2.19}$$

(actually λ_ε denotes here the class equivalent to λ_ε in $\overset{\circ}{G}$).

We obtain *the optimality system*:

$$\begin{aligned} &Ay = Bu, \ A^*\zeta = C^*Cy, \\ &A^*p = C^*(Cy - z_d) + C^*C\rho, \ A\rho = \beta\lambda, \\ &B^*p + Nu = 0, \\ &\lambda \in \hat{G}, \\ &\pi(\zeta - \zeta(u_o)) = 0. \end{aligned} \tag{2.20}$$

Remark 2.1

This is rather abstract! We do not pursue any further. We just wanted to indicate the general ideas we are going to follow in the applications and to show the introduction of the "generalized space" $\hat{G}$ of Lagrange multipliers. □

Remark 2.2

We already met (Lions [3]) in a different framework-situations where the Lagrange multipliers are in generalized spaces-which can be spaces of ultra distributions. □

3. AN ELLIPTIC EXAMPLE

Let Ω be a bounded open set in $\mathbf{R}^n$, with smooth boundary Γ. Let A be second-order elliptic operator in Ω, with smooth coefficients. We suppose that the Neumann problem in Ω is well set for A.

Let the state y(v,g) be given by

$$Ay(v,g) = v \text{ in } \Omega, \tag{3.1}$$

$$\frac{\partial y}{\partial \nu}(v,g)=g \text{ on } \Gamma. \tag{3.2}$$

In (2.2), $\frac{\partial}{\partial \nu}$ denotes the conormal derivative relative to A. The control variable lies in $\mathcal{U} = L^2(\Omega)$.

We assume that

$$F = L^2(\Gamma),\ G \subset F. \tag{3.3}$$

The cost function is given by

$$J(v,g) = \int_\Gamma |y(v,g) - z_d|^2 d\Gamma + N\int_\Omega v^2 dx. \tag{3.4}$$

We are looking for the *Pareto control* u *relative to* u_o, u_o given in $L^2(\Omega)$.□

The Pareto control u relative to u_o exists and is unique (as in Sections 1 and 2). It is characterized as the solution of

$$\inf.\ J_o(v),\ v \in \mathcal{M} + u_o, \tag{3.5}$$

where J_o and $\mathcal{M}$ are defined as follows. One introduces y(v) and $\phi(g)$ by

$$Ay(v) = v \text{ in } \Omega,\ \frac{\partial y(v)}{\partial \nu} = 0 \text{ on } \Gamma, \tag{3.6}$$

$$A\phi(g) = 0 \text{ in } \Omega,\ \frac{\partial \phi(g)}{\partial \nu} = g \text{ on } \Gamma. \tag{3.7}$$

Then

$$J_o(v) = \int_\Gamma |y(v) - z_d|^2 d\Gamma + N \int_\Omega v^2 dx, \tag{3.8}$$

$$v \in \mathcal{M} \text{ iff } \int_\Gamma y(v)\ \phi(g) d\Gamma = 0\ \forall\ g \in G.\square \tag{3.9}$$

We are going to show the following:

<u>Theorem 3.1</u>

The Pareto control u *relative to* u_o *is characterized by the unique solution* $\{u,y,\zeta,p,\rho,\lambda\}$ *of*

$$
\begin{aligned}
&Ay = u \text{ in } \Omega, \ \frac{\partial y}{\partial \nu} = 0 \text{ on } \Gamma, \\
&A^*\zeta = 0 \text{ in } \Omega, \ \frac{\partial \zeta}{\partial \nu_*} = y \text{ on } \Gamma \\
&A^*p = 0 \text{ in } \Omega, \ \frac{\partial p}{\partial \nu_*} = y - z_d + \rho \text{ on } \Gamma, \\
&A\rho = 0 \text{ in } \Omega, \ \frac{\partial \rho}{\partial \nu} = \lambda \text{ on } \Gamma,
\end{aligned}
\tag{3.10}
$$

with

$$p + Nu = 0 \text{ in } \Omega, \tag{3.11}$$

$$\lambda \in \hat{G}, \ \hat{G} = \text{completion of } G \text{ for the norm of } H^{-5/2}(\Gamma), ^{(1)} \tag{3.12}$$

$$\pi(\zeta - \zeta(u_o)) = 0, \text{ where } \pi \text{ denotes the orthogonal projection from } L^2(\Gamma) \text{ onto } G. \square \tag{3.13}$$

Before we prove Theorem 3.1 let us give some examples.

Example 3.1

Let us suppose that G is finite-dimensional, generated by $g^1,\ldots,g^m$, $g^i \in L^2(\Gamma)$. Then $\hat{G} = G$ and one can simplify the optimality system as follows:

$$
\begin{aligned}
&Ay = u, \ A^*p = 0 \text{ in } \Omega, \\
&\frac{\partial y}{\partial \nu} = 0, \ \frac{\partial p}{\partial \nu_*} = y - z_d + \sum_{i=1}^{m} \xi_i \rho^i, \\
&\int_\Gamma (y - y(u_o))\rho^i d\Gamma = 0, \ i = 1,\ldots,m,
\end{aligned}
\tag{3.14}
$$

where we have set $\rho^i = \phi(g^i)$. □

Example 3.2 $G = L^2(\Gamma)$.

Then $\hat{G} = H^{-5/2}(\Gamma)$.

According to (3.13), $\zeta = \zeta(u_o)$ on Γ so that $\zeta = \zeta(u_o)$ in Ω and therefore

$$y = \frac{\partial \zeta(u_o)}{\partial \nu_*} \text{ on } \Gamma.$$

It follows that y is *characterized* by

$$
\left\{
\begin{aligned}
&A^*Ay = 0, \\
&y = \frac{\partial \zeta(u_o)}{\partial \nu_*}, \ \frac{\partial y}{\partial \nu} = 0 \text{ on } \Gamma.
\end{aligned}
\right.
\tag{3.15}
$$

(1) We use the notation of Lions and Magenes [1]; $H^\alpha(\Gamma)$ denotes the space of functions with derivatives in $L^2(\Gamma)$ up to the order α, α not necessarily an integer and $H^{-\alpha}(\Gamma)$ = dual space of $H^\alpha(\Gamma)$.

Then

$$u = Ay \in L^2(\Omega), \text{ and } A^*u = 0 \tag{3.16}$$

We can then *compute*

$$\rho = \frac{\partial p}{\partial \nu_*} - (y - z_d) = -\frac{1}{N}\frac{\partial u}{\partial \nu_*} - (y - z_d) \in H^{-3/2}(\Gamma),$$

and one computes next λ (since $A\rho = 0$ in Ω) by

$$\lambda = \frac{\partial \rho}{\partial \nu} \in H^{-5/2}(\Gamma).$$

This example shows that $H^{-5/2}(\Gamma)$ *cannot* be avoided. □

Example 3.3

Let us suppose that $\Gamma = \Gamma_o \cup \Gamma_1$, Γ_o and Γ_1 being smooth with $\Gamma_o \cap \Gamma_1 = \emptyset$. We assume that

$$G = L^2(\Gamma_o) \times \{o\}. \tag{3.17}$$

Then

$$\hat{G} = H^{-5/2}(\Gamma_o) \times \{o\}. \tag{3.18}$$

In that case (3.13) gives:

$$\zeta = \zeta(u_o) \text{ on } \Gamma_o.$$

And one can write the optimality system as follows:

$$\begin{aligned} &Ay = u,\ A^*\zeta = 0,\ A^*p = 0,\ A\rho = 0 \text{ in } \Omega, \\ &\frac{\partial y}{\partial \nu} = 0,\ \frac{\partial \zeta}{\partial \nu_*} = y,\ \frac{\partial p}{\partial \nu_*} = y - z_d + \rho \text{ on } \Gamma, \\ &\zeta = \zeta(u_o) \text{ on } \Gamma_o,\ \frac{\partial \rho}{\partial \nu} = 0 \text{ on } \Gamma_1, \\ &p + Nu = 0. \end{aligned} \tag{3.19}$$

This system in $\{y,\zeta,p,\rho\}$ *admits a unique solution*. Then λ can be computed as $\lambda = \frac{\partial \rho}{\partial \nu}$ on Γ_o ($\in H^{-5/2}(\Gamma_o)$). □

Sketch of the proof of Theorem 3.1

1. We introduce first the augmented state $\{y(v),\zeta(v)\}$ by

$$\begin{aligned} &Ay(v) = v,\ A^*\zeta(v) = 0 \text{ in } \Omega, \\ &\frac{\partial y(v)}{\partial \nu} = 0,\ \frac{\partial \zeta(v)}{\partial \nu_*} = y(v) \text{ on } \Gamma. \end{aligned} \tag{3.20}$$

We observe then that $v \in \mathcal{M}$ iff $\int_\Gamma \zeta(v) g\, d\Gamma = 0\ \forall\, g \in G$ i.e., iff $\pi(\zeta(v)) = 0$ using the notation (3.13). The problem is then equivalent to minimizing $J_o(v)$ where v is subject to

$$\pi(\zeta(v) - \zeta(u_o)) = 0 \tag{3.21}$$

2. We use a penalty argument as in Section 2. We introduce

$$K_\varepsilon(v) = J_o(v) + \frac{1}{\varepsilon}\|\pi(\zeta(v) - \zeta(u_o))\|^2_{L^2(\Gamma)} . \tag{3.22}$$

Let u_ε be the solution of

$$K_\varepsilon(u_\varepsilon) = \inf. K_\varepsilon(v), \ v \in L^2(\Gamma). \tag{3.23}$$

If we set

$$y(u_\varepsilon) = y_\varepsilon, \ \zeta(u_\varepsilon) = \zeta_\varepsilon, \ \lambda_\varepsilon = \frac{1}{\varepsilon}\pi(\zeta_\varepsilon - \zeta(u_o)),$$

we have

$$\begin{aligned} &\int_\Gamma (y_\varepsilon - z_d) y(v) d\Gamma + N \int_\Omega u_\varepsilon v \, dx + \int_\Gamma \lambda_\varepsilon \pi\zeta(v) d\Gamma = 0 \\ &\forall \, v \in L^2(\Gamma). \end{aligned} \tag{3.24}$$

We introduce p_ε and ρ_ε (compare to Section 2) by

$$\begin{aligned} &Ay_\varepsilon = u_\varepsilon, \ A^*\zeta_\varepsilon = 0, \ A^*p_\varepsilon = 0, \ A\rho_\varepsilon = 0 \text{ in } \Omega, \\ &\frac{\partial y_\varepsilon}{\partial \nu} = 0, \ \frac{\partial \zeta_\varepsilon}{\partial \nu_*} = y_\varepsilon, \\ &\frac{\partial p_\varepsilon}{\partial \nu_*} = y_\varepsilon - z_d + \rho_\varepsilon, \ \frac{\partial \rho_\varepsilon}{\partial \nu} = \lambda_\varepsilon \text{ on } \Gamma. \end{aligned} \tag{3.25}$$

Then (3.24) reduces to

$$p_\varepsilon + Nu_\varepsilon = 0 \text{ in } \Omega. \tag{3.26}$$

Therefore p_ε *remains in a bounded set of* $L^2(\Omega)$ (actually $p_\varepsilon \to p = -Nu$ in $L^2(\Omega)$). We construct now $\hat{G}$ using this information.

3. Given $g \in L^2(\Gamma)$, we define ρ,σ by

$$\begin{aligned} &A\rho = 0, \ A^*\sigma \text{ in } \Omega, \\ &\frac{\partial \rho}{\partial \nu} = g, \ \frac{\partial \sigma}{\partial \nu_*} = \rho \text{ on } \Gamma, \end{aligned} \tag{3.27}$$

and we set (compare to Section 2)

$$|||g||| = \|\sigma\|_{L^2(\Omega)} . \tag{3.28}$$

But $\|\sigma\|_{L^2(\Omega)}$ defines a norm equivalent to $\|\rho\|_{H^{-3/2}(\Gamma)}$ which shows that

$$|||g||| \textit{ is a norm equivalent to the } H^{-5/2}(\Gamma) \textit{ norm}. \tag{3.29}$$

It follows then from (3.25), (3.26) that

$$\lambda_\varepsilon \text{ remains in a bounded subset of } \hat{G}, \tag{3.30}$$

$\hat{G}$ defined as in (3.12), and the Theorem follows. ☐

Remark 3.1

It follows from (3.10), (3.11) that

$$A(y+\rho) = u, \quad \frac{\partial(y+\rho)}{\partial\nu} = \lambda,$$
$$A^*p = 0, \quad \frac{\partial p}{\partial\nu_*} = y + \rho - z_d, \tag{3.31}$$
$$p + Nu = 0 \text{ in } \Omega.$$

If we think of λ as being *known*-and given by the solution of the full optimality system (3.10)...(3.13)-then one can verify that (3.31) is the optimality system of the following problem. *The state equation* is given by

$$A\eta = v, \quad \frac{\partial\eta}{\partial\nu} = \lambda, \tag{3.32}$$

which defines the state

$$\eta(v) = y(v) + \rho. \tag{3.33}$$

The cost function is taken to be

$$\mathcal{J}(v) = \int_\Gamma |y(v) - z_d|^2 d\Gamma + 2\int_\Gamma y(v)\rho d\Gamma + N\int_\Omega v^2 dx. \tag{3.34}$$

When $\lambda \in \hat{G} \subset G^{-5/2}(\Gamma)$, $\rho|_\Gamma \in H^{-3/2}(\Gamma)$ and since $y(v)|_\Gamma$ belongs to $H^{3/2}(\Gamma)$, (3.34) makes sense.

The optimality system of the problem

$$\inf. \mathcal{J}(v), \; v \in L^2(\Omega), \tag{3.35}$$

is then given by (3.31). If we observe that

$$\mathcal{J}(v) = J(v,\lambda) - J(u_o,\lambda) + \text{constant},$$

we see that u *is characterized* by

$$\inf. [J(v,\lambda) - J(u_o,\lambda)], \; v \in L^2(\Omega) = \mathcal{U}. \tag{3.36}$$

We emphasize that λ is not known a priori, but that it is given by the solution of the full optimality system (3.10)...(3.13).

Therefore this Remark is not constructive, but it gives a further information on the structure of the Pareto control relative to u_o. Of course the data u_o, G appear implicitly in λ. □

4. A PARABOLIC EXAMPLE

Let Ω and A be given as in Section 3. The state $y(v,g)$ is given by the solution of the *parabolic equation*:

$$\begin{aligned} &\frac{\partial y}{\partial t}(v,g) + Ay(v,g) = 0 \text{ in } Q = \Omega \times]o,T[, \\ &\frac{\partial y}{\partial \nu}(v,g) = v \text{ on } \Sigma = \Gamma \times [o,T[, \qquad (4.1) \\ &y(x,o;v,g) = g(x) \text{ in } \Omega. \end{aligned}$$

In (4.1)

$$v \in \mathcal{U} = L^2(\Sigma), \qquad (4.2)$$

$$g \in G \subset F = L^2(\Omega). \qquad (4.3)$$

Equation (4.1) admits a unique solution $y(x,t;v,g) = y(v,g)$. *The cost function* is given by

$$J(v,g) = \int_\Omega |y(x,T;v,g) - z_d(x)|^2 dx + N \int_\Sigma v^2 d\Sigma. \qquad (4.4)$$

We are looking for u *Pareto control relative to* u_o, u_o *being given in* $L^2(\Sigma)$.

As before, u exists and is unique.

We are going to give an optimality system which characterizes u. *The space* $\hat{G}$. □

Let us define ρ and σ by

$$\begin{aligned} &\rho' + A\rho = 0, \ -\sigma' + A^*\sigma = 0 \text{ in } Q \ (*), \\ &\rho(x,o) = g(x), \ \sigma(x,T) = \rho(x,T), \qquad (4.5) \\ &\frac{\partial \rho}{\partial \nu} = 0, \ \frac{\partial \sigma}{\partial \nu_*} = 0 \text{ on } \Sigma, \end{aligned}$$

and let us set

$$|||g||| = \|\sigma\|_{L^2(\Sigma)}. \qquad (4.6)$$

If $\sigma = 0$ on Σ, then–by the uniqueness of the solution of the *Cauchy problem*– $\sigma \equiv 0$ in Q, then $\rho(x,T) = 0$ and by virtue of the *backward uniqueness* property, $\rho \equiv 0$ so that $g = 0$ and (4.6) defines a *norm* on G.

We denote by $\hat{G}$ *the completion of* G for $|||g|||$. □

(*) We set $\frac{\partial \phi}{\partial t} = \phi'$.

Remark 4.1

If G is finite-dimensional, $\hat{G} = G$. If $G = L^2(\Omega)$, it does not seem straightforward to define $\hat{G}$ in "usual" terms. □

We can now state

Theorem 4.1. *The Pareto control relative to* u_o *is characterized by the solution of the optimality system:*

$$y' + Ay = 0,\ -\zeta' + A^*\zeta = 0,$$

$$-p' + A^*p = 0,\ \rho' + A\rho = 0 \text{ in } Q,$$

$$y(x,o) = 0,\ \zeta(x,T) = y(x,T), \tag{4.7}$$

$$p(x,T) = y(x,T) - z_d(x) + \rho(x,T),\ \rho(x,o) = \lambda(x) \text{ in } \Omega,$$

$$\frac{\partial y}{\partial \nu} = u,\ \frac{\partial \zeta}{\partial \nu_*} = 0,\ \frac{\partial p}{\partial \nu_*} = 0,\ \frac{\partial \rho}{\partial \nu} = 0 \text{ on } \Sigma,$$

$$p + Nu = 0 \text{ on } \Sigma, \tag{4.8}$$

$$\lambda \in \hat{G}, \tag{4.9}$$

$\pi(\zeta(x,o) - \zeta(x,o;y_o)) = 0$, where π denotes the

orthogonal projection $L^2(\Omega) \to G$. (4.10)

Example 4.1

If G is finite-dimensional, one can write the optimality system in a somewhat simpler way-analogous to the one given in Example 3.1. □

Example 4.2 $G = L^2(\Omega)$.

If $G = L^2(\Omega)$, then (4.10) gives

$$\zeta(x;o) = \zeta(x,o;u_o)$$

so that by the backward uniqueness (applied to $-\frac{\partial}{\partial t} + A^*$) one has $\zeta = \zeta(u_o)$ in Q, so that the optimality system "reduces" to

$$y' + Ay = 0,\ -p' + A^*p = 0,\ \rho' + A\rho = 0,$$

$$y(x,o) = 0,\ y(x,T) = \zeta(x,T;u_o),$$

$$p(x,T) = \zeta(x,T;u_o) - z_d(x) + \rho(x,T),\ \rho(x,o) = \lambda(x) \text{ in } \Omega, \tag{4.11}$$

$$\frac{\partial y}{\partial \nu} + \frac{1}{N}p = 0 \text{ on } \Sigma,\ \frac{\partial p}{\partial \nu_*} = 0,\ \frac{\partial \rho}{\partial \nu} = 0 \text{ on } \Sigma. \ \square$$

Sketch of proof of Theorem 4.1

The principle is analogous to the one used in Theorem 3.1. Only technical details change. One introduces the augmented state $y(v)$, $\zeta(v)$ by

$$y'(v) + Ay(v) = 0, \; -\zeta'(v) + A^*\zeta(v) = 0,$$

$$y(o;v) = 0, \; \zeta(T;v) = y(T;v), \tag{4.12}$$

$$\frac{\partial y}{\partial \nu} = v, \; \frac{\partial \zeta}{\partial \nu_*} = 0 \text{ on } \Sigma.$$

Then $v \in \mathcal{M}$ iff

$$\int_\Omega \zeta(x,o;v)g(x)dx = 0 \; \forall \; g \in G,$$

i.e.,

$$\pi(\zeta(x,o;v)) = 0.$$

Then the problem amounts to minimizing

$$J_o(v) = \int_\Omega (y(x,T:v) - z_d(x))^2 dx + N \int_\Sigma v^2 d\Sigma \tag{4.13}$$

subject to

$$\pi(\zeta(o;v) - \zeta(o;u_o)) = 0. \tag{4.14}$$

One then uses a penalty by introducing

$$K_\varepsilon(v) = J_o(v) + \frac{1}{\varepsilon}\|\pi(\zeta(o;v) - \zeta(o;u_o))\|^2_{L^2(\Omega)}, \tag{4.15}$$

and one minimizes $K_\varepsilon(v)$ over $L^2(\Sigma)$.

One can then pass to the limit on the optimality system for K_ε by using the space $\hat{G}$ as introduced in (4.5), (4.6).

Remark 4.2

If one considers the set of equations

$$
\begin{aligned}
&(y + \rho)' + A(y + \rho) = 0,\\
&(y + \rho)\,(x,o) = \lambda(x) \text{ in } \Omega,\\
&\frac{\partial}{\partial \nu}(y + \rho) = u \text{ on } \Sigma,\\
&-p' + A^*p = 0, \qquad\qquad (4.16)\\
&p(x,T) = (y + \rho)(x,T) - z_d(x),\\
&\frac{\partial p}{\partial \nu_*} = 0 \text{ on } \Sigma, \; p + Nu = 0 \text{ on } \Sigma,
\end{aligned}
$$

one can see that it is *the optimality system* of the following problem. We assume from now on *that λ is known*, as given through the solution of the full optimality system (4.7)...(4.10).

Let ρ be the solution of

$$\rho' + A\rho = 0, \; \rho(x,o) = \lambda(x), \; \frac{\partial \rho}{\partial \nu} = 0.$$

We have

$$\int_\Omega y(x,T;v)\rho(x,T)dx = \int_\Sigma v \, \sigma \, d\Sigma \tag{4.17}$$

and the last integral in (4.17) *makes sense* for $\lambda \in \hat{G}$ (which corresponds to $\sigma \in L^2(\Sigma)$).

We consider next *the state equation*

$$\begin{aligned} &\eta' + A\eta = 0, \\ &\eta(x,o) = \lambda(x) \text{ in } \Omega, \ \frac{\partial\zeta}{\partial\nu} = v \text{ on } \Sigma, \end{aligned} \tag{4.18}$$

which admits *a generalized solution* $\eta(v) = y(v) + \rho$.

The *cost function* is given by

$$\mathcal{J}(v) = \int_\Omega (y(x,T;v)-z_d)^2 dx + 2\int_\Omega y(x,T;v)\rho(x,T)dx + N\int_\Sigma v^2 d\Sigma \tag{4.19}$$

This functional makes sense by virtue of (4.17). Then *optimality system* for the problem

$$\inf \mathcal{J}(v), \ v \in L^2(\Sigma) \tag{4.20}$$

is indeed given by (4.16). But

$$\mathcal{J}(v) = J(v,\lambda) - J(u_o,\lambda) + \text{constant},$$

and we see *that* u *is characterized by*

$$\inf_{v \in \mathcal{U}} \ [J(v,\lambda) - J(u_o,\lambda)], \tag{4.21}$$

where λ is given by the solution of the full optimality system. Therefore this remark is *not* constructive. Compare to Remark 3.1. □

BIBLIOGRAPHY

J.L. Lions [1] "Contrôle de Pareto des Systèmes Distribués. (Pareto Control of Distributed Systems) - "Le cas stationaire," (Stationary case) C.R.A.S. Paris 1986. "Le cas d'évolution," (Evolution case) C.R.A.S. Paris 1986.

[2] "Lectures in the College de France". Fall 1985.

[3] "Control of Distributed Systems with incomplete Data." AMS Colloquium, Berkeley, 1983.

J.L. Lions, et E. Magenes

[1] " *Problèms aux limites nonhomogenes et applications* ." (Nonhomogenous Boundary Value Problems) Paris, Dunod - Vol. 1, 1968.

PART II

INVITED PAPERS

ABSTRACT STOCHASTIC EVOLUTION EQUATIONS AND RELATED CONTROL AND STABILITY PROBLEMS

N.U. Ahmed

University of Ottawa, Canada

INTRODUCTION

In this paper we consider the questions of optimal feedback control and stability of abstract stochastic evolution equations on Banach spaces.

In section 1, the questions of existence and regularity of solutions of certain semilinear stochastic evolution equations are studied. In section 2, associated optimal feedback control problems giving rise to Bellman equations on Banach spaces are discussed. In section 3 a general linear stochastic initial boundary value problem is formulated and certain stability questions are discussed.

1. Existence and Regularity of Solutions of Abstract Evolution Equations

(a) Semilinear Stochastic Evolution Equations (Non-coercive)

Let $(\Omega, F, F_{t\geq 0}$ rt. cont. $\uparrow, P)$ be a complete probability space, and X an arbitrary real Banach space. Consider the system

$$\begin{aligned} dx &= A(t)x\,dt + f(x)dW \quad , \quad t \in I = [0,T], \\ x(0) &= x_0 \end{aligned} \tag{1}$$

in Banach space X, where W is an F_t-Wiener process with values in a separable Banach space E. Let G(X) denote the space of generators of c_0-semigroup in X. We prove the existence, uniqueness and regularity of solutions under the following broad hypotheses.

(A1) $A:I \to G(X)$ is quasi stable with stability index $(M, \beta(\cdot))$, β upper integrable in the Lebesgue sense,

(A2) there exists a Banach space $Y \hookrightarrow X$, the embedding being continuous and dense, and a family $C(t) \in \mathrm{Iso}(Y,X)$, $t \in I$, such that
$(C(t)A(t) - A(t)C(t))C^{-1}(t) = \hat{C}(t) \subset L(X)$ a.e.,
$t \to \hat{C}(t)$ is strongly measurable and $t \to \|\hat{C}(t)\|_{L(X)}$ upper integrable.

(A3) $Y \subset D(A(t))$, $t \in I$; $A \in L_1(I, L(Y,X))$.

Under the above assumptions and certain additional hypotheses the following result holds.

THEOREM 1.

Suppose

(a1) (A1)-(A3) hold,

(a2) $f:X \to L(E,X)$ and there exists a $K \in R$ such that, for $x,y \in X$, $\|f(x)\|^2_{L(E,X)} \leq K^2(1+|x|^2_X)$,

$$\|f(x) - f(y)\|^2_{L(E,X)} \leq K^2|x-y|^2_X ,$$

(a3) W is an F_t-Wiener process with values in E and there exists a $Q \in L_n(E^*,E)$, the space of nuclear operators from E^* to E, such that, for $t > s$, $E(\langle W(t)-W(s),e^*\rangle)^2 = (t-s)\langle Qe^*,e^*\rangle_{E,E^*}$ for all $e^* \in E^*$.

Then, for each $x_0 \in L_2(F_0,X) \equiv \{F_0$-measurable, X valued random variables ξ with $E|\xi|^2_X < \infty\}$, the following conclusions hold:

(c1) the system (1) has a unique mild solution $x \in B(X)$, where $B(X) \equiv \{$The Banach space of all F_t-progressively measurable processes on I with values in X furnished with the norm topology $\|x\|_{B(X)} \equiv (\sup\{E|x(t)|^2_X, t \in I\})^{\frac{1}{2}}$,

(c2) x is weakly right continuous on I P-a.s and if $U^*(t,\tau)$, the dual of the evolution operator $U(t,\tau)$ corresponding to A, is strongly continuous on $\Delta \equiv \{t,\tau):0 \leq \tau \leq t \leq T\}$ to $L(X^*)$ then $x \in C(I,X_w)$ P-a.s where X_w denotes the space X carrying the weak topology. □

Proof.

For mild solutions one writes the integral equations

$$x(t) = U(t,0)x_0 + \int_0^t U(t,\tau)f(x(\tau))d\tau, \quad t \in I = [0,T],$$

where $U(t,\tau)$, $0 \leq \tau \leq t \leq T$, is the evolution operator corresponding to the generator $A(t)$, $t \in I$. Under the assumptions (A1)-(A3), $U \in C_s(\Delta, L(X)) \cap C_s(\Delta, L(Y))$. Hence under the hypotheses (a1)-(a3), the result follows from Banach fixed point theorem in $B(X)$. For details see Ahmed [1].

(b) Semilinear Stochastic Evolution Equations (Coercive)

Under coercive hypothesis on A, the assumptions on the non-linear operator f can be considerably relaxed. Consider the system

$$\begin{aligned} dx &= -A(t)x\,dt + f(x)dW, \quad t \in I = [0,T] \\ x(0) &= x_0. \end{aligned} \tag{2}$$

Let $V \hookrightarrow H = H^* \hookrightarrow V^*$ with continuous and dense embeddings and V having the structure of a reflexive Banach space. Let $A \in L_\infty(I, L(V,V^*))$ and suppose there exist constants $c > 0$, $\alpha > 0$, $\lambda \in R$ such that

(b1) $\quad \left|\langle A(t)u,v\rangle_{V^*,V}\right| \leq c\|u\|_V\|v\|_V \quad , \quad t \in I,$

(b2) $\quad \langle A(t)u,u\rangle_{V^*,V} + \lambda|u|_H^2 \geq \alpha\|u\|_V^2 \quad , \quad t \in I,$

for all $u,v \in V$. Then it is well known that $A = \{-A(t), t \in I\}$ is the generator of a strongly continuous evolution operator $U(t,\tau)$, $0 \leq \tau \leq t \leq T$, both in H and V^*. Further, for suitable constants $c_i (i=1-4)$, $U(t,\tau)$ satisfies the following inequalities,

(P1) $\quad |U(t,\tau)\phi|_H \leq C_1|\phi|_H \quad ,$ for all $\phi \in H$

(P2) $\quad \|U(t,\tau)\phi\|_{V^*} \leq C_2\|\phi\|_{V^*} \quad ,$ for all $\phi \in V^*$

(P3) $\quad |U(t,\tau)\phi|_H \leq (C_3/\sqrt{t-\tau})|\phi|_{V^*},$ for all $\phi \in V^*$

(P4) $\quad \|U(t,\tau)\phi\|_V \leq (C_4/\sqrt{t-\tau})|\phi|_H,$ for all $\phi \in H$.

Considering the abstract stochastic integral equation,

$$x(t) = U(t,0)x_0 + \int_0^t U(t,\tau)f(x(\tau))dW(\tau) .$$

One may then suspect that, for any E-valued Wiener process, if $f:H \to L(E,V^*)$ and satisfies

$$\text{(i)}\quad \|f(x)\|^2_{L(E,V^*)} \leq K^2(1 + |x|^2_H)$$

and

$$\text{(ii)}\quad \|f(x)-f(y)\|^2_{L(E,V^*)} \leq K^2|x-y|^2_H$$

then one may have solutions in B(H). This is generally not true unless further assumptions are imposed on A. In the linear case, however, if f is replaced by $\sigma \in L_\infty(I, L(E,V^*))$, then one can prove existence of solutions in B(H) which follows from a general result of the author [3]. For the nonlinear case the following result was proved in [1], where more general results including abstract initial boundary value problems are available [see also 4].

THEOREM 2.

Suppose

(b1) A satisfy (B1) and (B2)

(b2) there exists a reflexive Banach space W_ρ for $0<\rho<1$, such that $V \hookrightarrow W_\rho \hookrightarrow H$ with continuous and dense embeddings satisfying $W_0 = V$, $W_1 = H$,

(b3) $f:H \to L(E,W^*_\rho)$ for $0 < \rho < 1$ satisfying $\|f(x)\|^2_{L(E,W^*_\rho)} \leq K^2(1+|x|^2_H)$ for all $x \in H$, and

$\|f(x)-f(y)\|^2_{L(E,W^*_\rho)} \leq K^2_r|x-y|^2_H$ for all $x,y \in B_r \equiv \{\xi \in H: |\xi|_H < r\}$, where K, K_r, r are finite positive numbers,

(b4) W is an F_t-Wiener process with values in E having covariance operator $Q \in L_\infty(I,L^+_n(E^*,E))$.

Then the following conclusions hold:

(C1) For each $x_0 \in L_2(F_0,H)$, the system (2) has a unique mild solution $x \in B(H)$,

(C2) x is a weakly right continuous H-valued process P-a.s.

(C3) if $U^*(\cdot,\cdot) \in C_s(\Delta, L(H))$, then $x \in C(I,H_w)$ P-a.s. □

From the above result it follows that there exists a family of strongly measurable nonlinear evolution operators (2-parameter semigroups) $\{T_{t,s},\ 0 \leq s \leq t < \infty\}$ in $L_2(F,H)$ such that $T_{t,s}(\xi) \equiv x(t)$, $t \geq s$, is the solution of the problem

$$\begin{aligned} dx &= -A(t)xdt + f(x)dW \quad , \quad t > s \\ x(s) &= \xi \quad , \quad \xi \in H, \end{aligned}$$

and $\{T_{t,s}(\xi)\ 0 \leq s \leq t < \infty\}$ satisfy the following properties:

(p1) $T_{t,s}(\xi) \in L_2(F_t,H)$ whenever $\xi \in L_2(F_s,H)$, $s < t$,

(p2) $\text{s-lim}\ T_{t,s} = I$, i.e., $\lim_{t \downarrow s} E|T_{t,s}(\xi) - \xi|_H^2 = 0$ for all $\xi \in L_2(F_s,H)$,

(p3) $T_{t,\theta}T_{\theta,s} = T_{t,s}$ for $0 \leq s \leq \theta \leq t < \infty$.

Remark 1:

For the Banach space W_ρ, one can choose the domain of an appropriate fractional power of the operator $A(t)$ with the corresponding graph norm. For example, $W_\rho \equiv D(A(t)^{(1-\rho)/2})$, where $A(t)$ is considered as an unbounded closed operator in H with $D(A(t)) = D \subset H$ dense in H.

Remark 2:

The preceeding results hold also for time dependent f. As a corollary to Theorem 2 we have the following result. Consider the controlled system

$$\begin{aligned} dx &= -A(t)x\ dt + \eta(t)dt + f(x)dW \\ x(s) &= \xi \quad , \quad s \in (0,T),\ \xi \in L_2(F_s,H). \end{aligned} \tag{3}$$

COROLLARY 1:

Under the assumptions of Theorem 2, for every $\eta \in L^{\omega}_{2,p}(I,D) \equiv \{F_t$-progr. meas. D-valued processes f satisfying $E(\int_I |f(t)|^p_D dt)^{2/p} < \infty\}$, where $D = V^*$ for $p > 2$, or $D = W^*_\rho$ for $p > 2/(1+\rho)$, the system (3) has a unique mild solution in B(H) satisfying the regularity properties (c2) and (c3).

2. Optimal Feedback Control and Bellman Equations

Consider the control problem in X,

$$\begin{aligned} dy &= A(t)y\,dt + B(t)u\,dt + f(y)\,dW \\ y(0) &= y_0 \\ J(u) &= E\{\int_0^T \ell(y,u)dt + \phi_0(y(T))\} \equiv \min , \end{aligned} \tag{4}$$

and suppose $A(t) = A \in L(Y,X)$, $B(t) = B \in L(U,X)$, where U is a real separable uniformly convex Banach space and Γ is a closed convex subset of U and $\mathcal{U}$ is the class of admissible controls which are F_t-progressively measurable assuming values from Γ.

For simplicity we choose $\ell(x,v) = g(x) + \frac{1}{2}|v|^2_U$ and $\Gamma = U$, and assume that ϕ_0 and g are twice Frechêt differentiable on X. Further we assume that $\{X,U,E\}$ are real separable Banach spaces and $\{P_n,Q_n,R_n\}$ are an increasing family of projectors with the property that $P_n \to I_X$, $Q_n \to I_U$ and $R_n \to I_E$ strongly. Using these projectors the problem (4) is reduced to a sequence of finite dimensional problems $(4)_n$. Then, using Ito's formula and Bellman's principle of optimality for the finite dimensional system and letting $n \to \infty$, one obtains the Bellman's equation for the problem (4):

$$\begin{aligned} \frac{\partial\phi}{\partial t} &= \tfrac{1}{2}\,tr(fQf^*\cdot\phi_{xx}) + \langle Ax, \phi_x\rangle_{X,X^*} - \tfrac{1}{2}|B^*\phi_x|^2_{U^*} + g(x) \\ \phi(0,x) &= \phi_0(x) \quad , \quad (t,x) \in I \times Y \quad , \end{aligned} \tag{5}$$

with optimal control $u_0 = \nu(-B^*\phi_x)$, where ν is the duality map $\nu: U^*\setminus\{0\} \to U$ satisfying $\langle \nu(\xi),\xi\rangle_{U,U^*} = |\xi|^2_{U^*} = |\nu(\xi)|^2_U$.

Since U is uniformly convex, ν is continuous and hence, whenever the Frechêt derivative ϕ_x is continuous on $I \times Y$, u_0 is continuous. Further note that, for the linear quadratic problem with $f(x) \equiv \sigma \in L(E,X)$, $g(x) = \frac{1}{2} <Rx,x>$ with $R \in L^+(X,X^*)$, and $\phi_0(x) = \frac{1}{2} < P_0 x,x >$ with $P_0 \in L^+(X,X^*)$, equation (5) yields the operator Riccati equation,

$$\frac{d}{dt} < P(t)x,x >_{X^*,X} = <A^*P(t)x,x>_{Y^*,Y} + <P(t)Ax,x>_{X^*,X}$$
$$+ <Rx,x>_{X^*,X} - <B^*P(t)x, \nu(B^*P(t)x)>_{U^*,U} \quad (6)$$
$$P(0) = P_0 \ , \quad (t,x) \in I \times Y \ ,$$

with $$\phi(t,x) = \tfrac{1}{2} <P(t)\,x,x>_{X^*,X} + \int_0^t tr(\sigma Q \sigma^* P) d\theta \ . \quad (7)$$

In the general case, the question of existence of a solution for (5) poses a major problem. In case X = H (Hilbert), $f \equiv I$(identity), U = H and g and ϕ_0 convex, Barbu and DaPrato [6] solved the problem. Later DaPrato [7] extended this result to the nonconvex case, again with $X = U = H$, $f = I$ and $\Gamma = B_r$, where $Br = \{\xi \in H: |\xi|_H \leq r\}$. DaPrato used the theory of m-dissipative operators and Crandall-Ligget generation theorem.

Consider the system (4) with $u \equiv 0$ and $A(t) \equiv A$ (constant) and let T_t denote the strongly (F_t) measurable (nonlinear) semigroup giving $y(t,\xi) = T_t(\xi)$, $y_0 = \xi$, $\xi \in X$. Define the Banach space $Z \equiv C_{bu}(X)$, the space of bounded uniformly continuous real valued functions on X, furnished with the norm topology, $\|\psi\| = \sup\{|\psi(x)|, x \in X\}$. Then, for $\psi \in Z$ and $\lambda > 0$, the expression,

$$(G_\lambda \psi)(x) \equiv \int_0^\infty e^{-\lambda t} E\{\psi(T_t(x))\} dt, \quad (8)$$

is well defined and there exists a linear operator $\mathcal{A} : \mathcal{D}(\mathcal{A}) \subset Z \to Z$ such that $G_\lambda = (\lambda I - \mathcal{A})^{-1}$. Clearly $\|(\lambda I - \mathcal{A})^{-1}\| \leq \frac{1}{\lambda}$ for $\lambda > 0$, and $\mathcal{A}$ is m-dissipative in Z having the realization

$$\mathcal{A}\psi = \tfrac{1}{2} tr(fQf^*\psi_{xx}) - <Ax,\psi_x>_{X,X^*} \ , \quad \psi \in \mathcal{D}(\mathcal{A}) \quad (9)$$

and it is the generator of a semigroup in Z which is not c_0. In the general case, defining

$$F(x,q) = \inf_{u \in \Gamma} \{<B^*q,u>_{U^*,U} + \ell(x,u)\}, \ q \in X^* \ ,$$

and introducing the nonlinear operator,

$$(\mathcal{B}\psi)(x) = F(x,\psi_x) , \tag{10}$$

we can write equation (5) as an abstract evolution equation in $\mathcal{Z}$,

$$\begin{aligned} \frac{d\phi}{dt} &= \mathcal{A}\psi + \mathcal{B}\phi \quad , \quad (t,x) \in I \times Y \\ \phi(0,x) &= \phi_0(x) \quad , \quad x \in Y. \end{aligned} \tag{11}$$

In case $X = U = E = H$, $f = I$, $\ell(x,u) = g(x) + \frac{1}{2}|u|_H^2$, DaPrato [7] proved that $\mathcal{A} + \mathcal{B}$ is m-dissipative and, hence, by use of Crandall-Ligget generation theorem, concluded the existence of a (noninear) semigroup S_t, $t > 0$, on $\mathcal{Z}$ such that $\phi(t,x) = (S_t\phi_0)(x)$, $t > 0$. For the proof, DaPrato made a crucial use of the fact that the Martingale term is Gaussian. In the nonlinear case this is obviously not possible.

Classical results guaranteeing m-dissipativity of $\mathcal{A} + \mathcal{B}$ require, in the case of a general Banach space $\mathcal{Z}$, that $\mathcal{A}$ be m-dissipative and $\mathcal{B}$ everywhere defined, continuous and m-dissipative. In case $\mathcal{Z}^*$ is uniformly convex, then $\mathcal{A} + \mathcal{B}$ is m-dissipative if $\mathcal{A}$, $\mathcal{B}$ are m-dissipative and $\mathcal{B}$ is locally-relatively $\mathcal{A}$ bounded, that is,

$\|\mathcal{B}(\phi)\|_{\mathcal{Z}} < K_{\phi_0}\|\mathcal{A}\phi\|_{\mathcal{Z}} + \beta\|\phi\|_{\mathcal{Z}}$, $0 < K_{\phi_0} < 1$, $\beta > 0$, $\phi \in N(\phi_0) \cap \mathcal{D}(\mathcal{A})$,
where $N(\phi_0)$ is a neighbourhood of ϕ_0.

Unfortunately, in our situation neither $\mathcal{B}$ is everywhere defined nor $\mathcal{Z}^*$ is uniformly convex and hence neither of the above results is applicable. Thus, in the general case, the question of existence of a solution of equation (11) remains an outstanding problem for both the noncoercive and coercive systems.

Remark 3:

Under the general hypotheses of Theorem 2, it appears that there is substantial difficulty interpreting the Bellman's equation. However this difficulty disappears if $f: H \to \mathcal{L}(E,H)$ and satisfies (b3) with W_ρ^* replaced by H. Obviously this limits the class of nonlinear operators f admitted under the setup of Theorem 2.

3. Stochastic Initial-Boundary Value Problems

(a) A Second Order Parabolic Problem

Consider the system,

$$\frac{\partial y}{\partial t} + A(t)y = f, \qquad \text{on } I \times \Sigma \quad , \quad I = (0,T),$$

$$\frac{\partial y}{\partial \nu_A} = g + N \qquad \text{on } I \times \partial\Sigma \tag{12}$$

$$y(0) = y_0 \qquad \text{on } \Sigma ,$$

where A(t) is a second order elliptic operator on an open bounded connected set $\Sigma \subset R^n$, $\frac{\partial}{\partial \nu_A}$ is the outward normal derivative compatible with the operator A. The processes f, g, N are generalized random processes and y_0 a generalized random element. Let $Y \equiv H^1(\Sigma)$ or $\Xi^1(\Sigma) \equiv \{\phi \in L_2(\Sigma) : \rho\nabla\phi \in L_2(\Sigma)$, where $\rho \in \mathcal{D}(\bar{\Sigma})$ is any non-negative function of the same order as the distance function $d(x,\partial\Sigma)\}$, with Y* its dual, and $X = H^{\frac{1}{2}}(\partial\Sigma)$ with dual $X^* = H^{-\frac{1}{2}}(\partial\Sigma)$. We use $<\cdot,\cdot>$ and $<<\cdot,\cdot>>$ to denote the duality pairings in $\{Y^*,Y\}$ and $\{X^*,X\}$ respectively. Let W_t, $t \in \bar{I}$, be a generalized Wiener process with values in X* such that for each $\xi \in X$, $W_t(\xi) \equiv <<W_t,\xi>>$ is a scaler valued Wiener process satisfying all the standard properties and that there exists a nuclear operator $Q \in L_n^+(X,X^*)$ such that $E\{W_t(\xi)W_s(\eta)\} = \min(t,s)(Q\xi,\eta)$ for all $\xi,\eta \in X$. The generalized derivative of W is called the white noise N. That is, for each $\xi \in X$ and $\beta \in C_0^\infty(0,\infty)$, we have

$$\int_0^\infty N_t(\xi)\beta(t)dt = (-1)\int_0^\infty W_t(\xi)\dot{\beta}(t)dt.$$

Suppose A satisfy the following properties:

(A1) Associated to the operator A, there exists a sesquilinear form $a : I \times Y \times Y \to R$ such that, $t \to a(t,\phi,\psi)$ is measurable for $\phi,\psi \in Y$ and there exists a constant $0 < c < \infty$ such that

$$|a(t,\phi,\psi)| \leq c\|\phi\|_Y \|\psi\|_Y \text{ for all } \phi,\psi \in Y,$$

(A2) There exists a $\lambda \in R$, and $\alpha > 0$ such that $a(t,\phi,\phi) + \lambda|\phi|_H^2 \geq \alpha\|\phi\|_Y^2$ for all $\phi \in Y$.

Without further notice, we shall assume throughout the rest of the paper that all the random processes involved are based on the probability space $(\Omega, F, F_{t \geq 0} \text{ rt.cont.} \uparrow, P)$

DEFINITION 1.

The problem (12) is said to have a weak solution if there exists a measurable process y defined on $I \times \Omega$ with values in Y such that for each $t \in I$, y_t is F_t-measurable and

$$(y_t, \upsilon)_H + \int_0^t a(\theta, y_\theta, \upsilon) d\theta = (y_0, \upsilon)_H + \int_0^t \{ \langle f_\theta, \upsilon \rangle + \langle\langle g_\theta, \tau\upsilon \rangle\rangle \} d\theta + W_t(\tau\upsilon) \quad \text{P-a.s} \tag{13}$$

for all $\upsilon \in Y$, $t \in \bar{I}$, where τ denotes the trace operator $\upsilon \to \upsilon|_{\partial\Sigma}$ from Y to X.

In symbolic form we can write this as a stochastic differential equation in the weak form given by

$$d(y_t, \upsilon) + a(t, y_t, \upsilon) dt = (\langle f_t, \upsilon \rangle + \langle\langle g_t, \tau\upsilon \rangle\rangle) dt + dW_t(\tau\upsilon), \tag{14}$$

for $t \in I$, $\upsilon \in Y$.

THEOREM 3:

Suppose A satisfy (A1) and (A2), $y_0 \in L_2(F_0, H)$, $f \in L_2^\omega(I, Y^*)$, $g \in L_2^\omega(I, X^*)$, W_t is an X*-valued Wiener process with covariance operator $Q \in L_n^+(X, X^*)$, then the system (12) has a unique weak solution $y \in L_2^\omega(I, Y) \cap L_\infty^\omega(I, H)$ (in the sense of definition 1) and further y is weakly continuous with values in Y* P-a.s and $y \in C(\bar{I}, L_2(\Omega, H))$. □

For the proof and more general results involving nonlinear monotone hemicontinuous operators see [2,4].

Under the assumptions: $A(t) = A$ constant, $f \equiv 0$, $g \equiv 0$, $\{y_t, t \geq 0\}$ is an H-valued Feller process generating a Feller semigroup S_t, $t \geq 0$, on $Z \equiv C_{bu}(H)$ with the infinitesimal generator G given by

$$(G\phi)(x) \equiv -a(x, \phi_x) + \tfrac{1}{2} tr(\tau^* Q \tau \phi_{xx}), \quad x \in D(A). \tag{15}$$

Hence we obtain the evolution equation on Z

$$\frac{d\phi}{dt} = G\phi, \quad (t,x) \in I \times D(A)$$
$$\phi(0) = \phi_0, \quad x \in D(A) \tag{16}$$

which has a strong solution $\phi(t) = S_t \phi_0$ for $\phi_0 \in D(G)$.

DEFINITION 2:

The system (12) with $A(t) \equiv A$ (constant), $f \equiv 0$, $g \equiv 0$, is said to be globally almost surely asymptotically stable with respect to a closed set $K \subset H$ if, for all $\xi \in H$,

$$P\{\lim_{t\to\infty} y_t(\xi) \in K\} = 1,$$

where $y_t(\xi)$ denotes the solution of (12) corresponding to the initial state ξ.

The following stability result, based on supermartingale arguments, was proved in [2].

THEOREM 4:

The system (12), with $A(t) \equiv A$, $f \equiv 0$, $g \equiv 0$ and $\lambda = 0$, is globally almost surely asymptotically stable with respect to the closed ball $B_r \subset H$ for $r = \sqrt{trQ/2\alpha}$. □

(b) A General Problem

In general a linear stochastic initial boundary value problem can be formulated as follows:

Consider the system

$$\begin{aligned} \frac{\partial y}{\partial t} &= Ly + g_1 && \text{on } I \times \Sigma \\ By &= g_2 && \text{on } I \times \partial\Sigma \\ y(0) &= y_0 && \text{on } \Sigma \end{aligned} \qquad (17)$$

where L is a spatial differential operator and B is a boundary operator with $D(L) \subset D(B)$. Let Y denote a separable Banach space of functions or generalized functions on Σ and X a real separable Banach space of functions or generalized functions defined on $\partial\Sigma$. Define the operator A by

$$D(A) \equiv \{\upsilon \in Y\colon L\upsilon \in Y \text{ and } B\upsilon = 0\}$$

and set

$$A\phi = L\phi \text{ for } \phi \in D(A).$$

Assume that A is the generator of C_0-semigroup $T(t)$, $t > 0$, in Y and there exists an operator $\Gamma \in L(X,Y)$ such that for all $\phi \in X$, $B\Gamma\phi = \phi$. Let $\rho(A)$ denote the resolvent set of A and define $\Pi = (A-\lambda I)$ for

$\lambda \in \rho(A)$. Under the above assumptions it follows from a standard procedure introduced by Balakrishnan [5], that the system (17) is equivalent to the abstract Cauchy problem

$$\frac{dz}{dt} = Az + \Lambda_1 g_1 + \Lambda_2 g_2$$
$$z(0) = z_0 = \Pi^{-1} y_0 \ , \ y(t) = \Pi z(t), \ t > 0, \tag{18}$$

where $\Lambda_1 \equiv \Pi^{-1}$ and $\Lambda_2 \equiv \Pi^{-1}(L\Gamma - \lambda\Gamma) - \Gamma$ with $\Lambda_1 \in L(Y)$, and $\Lambda_2 \in L(X,Y)$. If the distributed data g_1 and the boundary data g_2 are given by sums of two generalized random processes, $g_1 = f_1 + \xi_1$ and $g_2 = f_2 + \xi_2$, where ξ_1 and ξ_2 are the generalized derivatives of two independent Wiener processes W_1 and W_2 with values in Y and X respectively, then we can write (18) in the form

$$dz = (Az + \Lambda_1 f_1 + \Lambda_2 f_2)dt + \Lambda_1 dW_1 + \Lambda_2 dW_2$$

with $\qquad z_0 = \Pi^{-1} y_0$ and $y(t) = \Pi z(t)$, $t > 0$.

Defining $E = Y \times X$ with the natural topology, and letting Λ denote the mapping $E \to Y$ given by $\Lambda f = \Lambda\{f_1, f_2\} = \Lambda_1 f_1 + \Lambda_2 f_2$, $f = \{f_1, f_2\} \in E$, we can write the system equation in the canonical form

$$dz = (Az + \Lambda f)dt + \Lambda \, dW$$
$$z(0) = \Pi^{-1} y_0 \ , \ y(t) = \Pi z(t), \ t > 0. \tag{19}$$

Thus a general linear stochastic initial boundary value problem can be formulated as in (19) where $\Lambda \in L(E,Y)$ and W is an E-valued Wiener process with covariance operator $Q = \begin{pmatrix} Q_{11} & 0 \\ 0 & Q_{22} \end{pmatrix}$, $Q_{11} \in L_n^+(Y^*,Y)$ and $Q_{22} \in L_n^+(X^*,X)$.

Let $F \equiv (D(A^*), \|\cdot\|_{D(A^*)})$ denote the linear space with the topology induced by the graph norm $\|\phi\|_{D(A^*)} = \|\phi\|_{Y^*} + \|A^*\phi\|_{Y^*}$. Since A is a closed densely defined linear operator from $D(A) \subset Y$ to Y, F is a Banach space contained in Y^*. Letting F^* denote the dual of F one can prove the following result [see 1].

THEOREM 5:

The system (19) has a unique mild solution $z \in B(Y)$ [see (C1), Theorem 1] and hence $y \in B(F^*)$. Further, if Y is reflexive, then y is weakly continuous with values in F^* P-a.s and $y \in L_2^{\omega}(I,F^*) \cap C(\bar{I}, L_2(\Omega,F^*))$. □

Consider the system (19) with the observation $y = \Pi z$ which is obviously unbounded in Y. However, according to theorem 5, the control problem

$$J(f) \equiv E \int_0^T \{\|y(t)\|_{F^*}^2 + \langle Mf, f\rangle_{E^*,E}\}dt \equiv \min \tag{20}$$

is well posed for $M \in L^+(E,E^*)$ and $f \in L_2^{\omega}(I,E)$. The question of existence of optimal controls can be resolved by standard techniques. However, the construction of a feedback control through the operator Riccati differential equations is rather cumbersome in the present situation. Instead, one may wish to design an approximate stationary feedback control law $f = Ky$ such that the error covariance of the process z is minimum.

This, in effect, improves the "signal-to-noise" ratio which is often what is desired in engineering applications. The problem becomes somewhat simplified if Y is assumed to be reflexive. In that case it is easy to verify that $F = (D(A^*), \|\cdot\|_{D(A^*)})$ is also a reflexive Banach space since its unit ball is weakly compact. Further $\Pi \in L(Y,F^*)$ and hence, for $K \in L(F^*,E)$, $\Lambda K\Pi \in L(Y)$. Thus, $\Lambda K\Pi$ is a bounded perturbation of the generator A of a C_0-s.g. $T(t)$, $t \geq 0$, in Y. Hence, for each $K \in L(F^*,E)$, $\tilde{A} \equiv A + \Lambda K\Pi$ is also a generator of a C_0-s.g. $\tilde{T}(t)$, $t \geq 0$, in Y. Hence, maximum noise rejection is achieved by solving the following optimization problem: find $K \in L(F^*,E)$ such that

$$J(K) \equiv \int_0^T e^{-\beta t} tr(C(t))dt, \ \beta > 0, \ 0 < T \leq \infty, \tag{21}$$

is minimum subject to the constraint,

$$\frac{d}{dt}(C(t),\xi,\eta)_{Y,Y^*} = (C(t)(A + \Lambda K\Pi)^*\xi,\eta)_{Y,Y^*} + (C(t)\xi,(A + \Lambda K\Pi)^*\eta)_{Y,Y^*}$$

$$+ (Q\Lambda^*\xi, \Lambda^*\eta)_{E,E^*}, \ t \in [0,T], \tag{22}$$

$$C(0) = C_0, \quad \text{for } \xi,\eta \in F,$$

where $C_0 \in L_n^+(Y^*,Y)$ is the covariance of the initial state z_0 which is given. The problem (21) - (22) also makes sense for variable K. For time invariant case it is equally satisfactory to consider the problem: find $K \in L(F^*,E)$ such that trC is minimum subject to $C(A + \Lambda K\Pi)^* + (A + \Lambda K\Pi)C = -\Lambda Q\Lambda^*$.

REFERENCES

Ahmed, N.U. "Abstract Stochastic Evolution Equations on Banach Spaces." *Stochastic Analysis and Applications* 3, 4 (1985): 397-432.

Ahmed, N.U., and S.K. Biswas. "Existence of Solutions and Stability of a Class of Parabolic Systems Perturbed by Generalized White Noise on the Boundary." *Stochastic Analysis and Applications* 2, 4 (1984): 347-368.

Ahmed, N.U. "Stochastic Control on Hilbert Space for Linear Evolution Equations with Random Operator Valued Coefficients." *SIAM J. Control and Optimization* 19, 3 (1981): 401-30.

———— "Nonlinear Integral Equations on Reflexive Banach Spaces with Applications to Stochastic Integral Equations and Abstract Evolution Equations." *Journal of Integral Equations* 1 (1979): 1-15.

Balakrishnan, A.V. *Applied Functional Analysis.* New York Berlin Heidelberg: Springer-Verlag, 1976.

Barbu, V. and G. DaPrato. "Solution of the Bellman Equation Associated with Infinite Dimensional Stochastic Control Problem and Synthesis of Optimal Control." *SIAM J. Control and Optimization* 21, 4 (1983): 531-50.

DaPrato, G. "Some Results on Bellman Equations in Hilbert Spaces." *SIAM J. Control and Optimization* 23, 1 (1985): 61-71.

VIABILITY THEOREMS FOR CONTROL SYSTEMS WITH FEEDBACKS

Jean-Pierre Aubin*

Résumé

On propose dans cet article des théorèmes de viabilité et d'invariance pour des inclusions aux dérivées partielles paraboliques et des inclusions différentielles opérationnelles. Ces théorèmes sont appliqués à la régulation par viabilité des systèmes contrôlés avec rétroactions et aux problèmes d'obstacles.

INTRODUCTION

Let Ω be an open subset of R^n with boundary Γ, Y be a Hilbert space, f be a map from $R \times Y$ to R, U be a set-valued map from $\Omega \times R$ to Y, the feedback map, and $x_0 \in L^2(\Omega)$ be a given initial data.

We consider the following model of a control problem with feedbacks for distributed systems:

(1) $\partial x/\partial t\ (t,\omega) = \Delta x(t,\omega) + f(x(t,\omega),u(t,\omega))$ (state equation).

(2) $u(t,\omega) \in U(\omega,x(t,\omega))$ (feedbacks).

(3) for all $t \in [0,T]$, $x(t,\omega)|_{\partial\Omega} = 0$ (Dirichlet boundary condition).

(4) $x(0,\omega) = x_0(\omega)$ for almost all $\omega \in \Omega$ (initial condition).

We introduce now a viability domain K, which is a closed subset of the state space $L^2(\Omega)$, describing constraints that the state $x(t,\cdot)$ must obey for each $t \in [0,T]$.

---

* CEREMADE, Université de Paris-Dauphine and Centre de recherches mathématiques, Université de Montréal.

Ce rapport a été publié en partie grâce à une subventio du Fonds FCAR pour l'aide et le soutien à la recherche.

We define the:

Viability Property.
$\forall x_0 \in K$, closed subset of $L^2(\Omega)$, there exists a solution to (1), (2), (3), (4) which is viable in the sense that

(5) $$\forall t \in [0,T],\ x(t,\cdot) \in K.$$

Δ

Viability theorems provide conditions linking the dynamics of the problem (described by Δ, f, and U) and the viability constraints (described by K) which imply the viability property.

We refer to Aubin-Cellina [1984] for an exposition of viability theory for ordinary differential inclusions and control problems and a list of references.

The main idea is to "differentiate" in some sense the viability condition (5): If $t \to x(t,\cdot)$ is absolutely continuous from $[\cdot,T]$ to $L^2(\Omega)$, then (5) implies that

(6) for almost all $t \in [0,T]$, $\partial x/\partial t\ (t,\cdot) \in T_K(x(t,\cdot))$
where $T_K(x)$ is the "contingent cone to K at x", characterized by:

(7) $v \in T_K(x)$ if and only if

$$\liminf_{h \to 0+} d(x + hv, K)/h = 0$$

As far as the model problem is concerned, we shall prove the following theorem.

Theorem 1.

We assume that

(8) Ω is open, regular and bounded

(9) K is a closed subset of $L^2(\Omega)$

(10) f is continuous from $R \times Y$ to Y, $U^{(1)}$ is upper hemi-continuous$^{(2)}$ from K to Y and the images

$$F(x) = \{f(x,u)\}_{u \in U(x)}$$

are closed and convex in $L^2(\Omega)$ and remain in a bounded set of $L^2(\Omega)$.

If the "tangenitial condition"

(11) $\forall x \in K \cap H_0^1(\Omega,\Delta),\ \Delta x \in T_K(x) - F(x)$

holds true, then:

a) the set $S(x_0)$ of viable solutions to (1), (2), (3), (4) is not empty and the solution map $S : x_0 \to S(x_0)$ is upper hemi-continuous from K to $L^\infty(0,T;L^2(\Omega))$,

b) the "viable controls" (regulating viable trajectories) evolve according the regulation law:

(12) for almost all $t \in [0,T]$, $u(t) \in U(x(t)) \cap R(x(t))$

where

(13) $R(x) := \{u \in L^2(0,T;Y) \mid f(x,u) \in T_K(x) - \Delta x\}$.

Furthermore, if we assume that

(14) the viability domain is convex, there exists a viable equilibrium (x,u) of the system, a solution to

(15)
i) $-\Delta x = f(x,u),\ x|_{\partial\Omega} = 0$
ii) $x \in K$ and $u \in U(x)$.

Δ

We mention here an application to obstacle problems due to Shi Shuzhong [1986]. We consider a function $\phi : \Omega \times R$ to R to which we associate

(16) $\underline{\phi}(\omega,x) := \liminf_{y \to x} \phi(\omega,y),\ \overline{\phi}(\omega,x) := \limsup_{y \to x} \phi(\omega,y)$.

For example, dealing with obstacle problems, we can take

$$(17)\qquad \phi(\omega,x) := \begin{cases} \min(f(\omega), -\Delta\psi(\omega)) & \text{if } x > \psi(\omega) \\ f(\omega) & \text{if } x < \psi(\omega) \end{cases}$$

We assume that

(18) ϕ, $\underline{\phi}$ and ϕ map measurable functions to measurable functions

and

(19) $\exists\, s \in [1, n + 2/n - 2[$ such that $|\phi(\omega,x)| < a(\omega) + e|x|^s$ where $a \in L^2(\Omega)$.

Theorem 2. (Shi Shuzhong, [1986])

Let assumptions (8), (18) and (19) hold true. Then

a) For all $x_0 \in L^2(\Omega)$, there exists a solution $x(t,\omega)$ to the parabolic differential inclusion

$$(20)\qquad \begin{aligned} &\text{i)}\quad \partial x/\partial t\,(t,\omega) \in \Delta x(t,\omega) + [\underline{\phi}(\omega,x(\omega)),\overline{\phi}(\omega,x(\omega))] \\ &\text{ii)}\quad x(t,\omega)\big|_{\partial\Omega} = 0 \text{ and } x(0,\omega) = x_0(\omega). \end{aligned}$$

b) There exists a solution $x(\omega)$ to the elliptic differential inclusion

$$(21)\qquad \begin{aligned} &-\Delta x(\omega) \in [\underline{\phi}(\omega,x(\omega)),\overline{\phi}(\omega,x(\omega))] \\ &x(\omega)\big|_{\partial\Omega} = 0. \end{aligned}$$

Δ

This result follows from the theorem by taking $K = L^2(\Omega)$ (so that (11) is satisfied), $f(x,u) = u$ and

$$U(x) := \{u \,|\, u(\omega) \in [\underline{\phi}(\omega,x(\omega)),\overline{\phi}(\omega,x(\omega))]\}.$$

These results follow from more abstract theorems we are about to describe.

1. Operational-differential inclusions with convex-valued right-hand side.

The Laplace operator Δ, together with the Dirichlet boundary conditions, is an example of an "unbounded operator" A defined on the Hilbert space $H = L^2(\Omega)$. Therefore, problem (1), (2), (3), (4) is a particular case to a control problem with feedbacks

(1.1) $x'(t) = Ax(t) + f(x(t)), u(t))$

(1.2) $u(t) \in U(x(t))$

(1.3) $x(0) = x_0$.

We can bring also a further notational simplification by introducing the subsets

(1.4) $F(x) := \{f(x,u)\}_{u \in U(x)}$

of velocities. Hence problem (1.1), (1.2), (1.3) is a particular case of an "Operational Differential Inclusion"

(1.5)
i) $x'(t) \in Ax(t) + F(x(t))$
ii) $x(0) = x_0$ is given.

We shall provide viability theorems for the class of Operational Differential Inclusion when F is upper hemi-continuous and convex-valued. For that purpose, we introduce two Hilbert spaces X and H satisfying

(1.6) $X \subset H = H^* \subset X^*$, the injections being compact and dense,

(1.7) a continuous linear oeprator $A \in L(X,X^*)$, satisfying $A = A^*$ and $\langle -Ax,x\rangle \geq c\|x\|_X^2$ ($-A$ is X-elliptic),

(1.8) a closed subset K of H,

(1.9) a bounded upper hemi-continuous set-valued map F from K to H with closed convex images.

Theorem 1.1

Assume the assumptions (1.6), (1.7), (1.8) and (1.9). If the tangential condition

$$(1.10) \qquad \forall x \in K,\ Ax \in T_K(x) - F(x)$$

holds true, then, for any $x_0 \in K$, the subset $S(x_0)$ of viable solutions to (1.5) is not empty and the solution map $S : x_0 \to S(x_0)$ has a compact graph in $K \times L^{\infty}(0,T;H)$.

Furthermore, if K is convex, there exists a viable equilibrium x, a solution to

$$(1.11) \qquad \begin{array}{ll} \text{i)} & x \in K \\ \text{ii)} & 0 \in Ax + F(x) \end{array}$$

Δ

This theorem is an extension to the infinite dimensional case of the Haddad viability theorem (see Haddad [1981]) for ordinary differential inclusions. There are other theorems providing different sufficient conditions implying the viability property for this class of Operational Differential Inclusions (see Shi Shuzhong, [1986]), for instance. In this paper, the author uses Galerkin approximations and Haddad's theorem for solving the approximate problems).

2. Operational-differential inclusions with Lipschitz right-hand side.

We consider now another class of Operational Differential Inclusions, when we do not assume anymore that F has convex values. The price to pay is to replace the weak continuity requirement (F is upper hemi-continuous) by the much stronger assumption that F is Lipschitz, in the sense that there exists a constant $c > 0$ such that

$$(2.1) \qquad \forall x,y \in \text{Dom } F,\ F(x) \subset F(y) + c\|x - y\|B$$

where B denotes the unit ball of H.

The celebrated Filippov theorem (see Filippov [1967], Aubin-Cellina [1984], p. 120) solves ordinary differential inclusions with Lipschitz right-hand sides and implies that the solution map S is Lipschitz. This problem has been extensively applied for studying the Pontryagin maximum principle and local controllability of differential inclusions (see Frankowska, [1986] a), [1986] b).)

We quote here the adaptation of Filippov's theorem to the case of Operational Differential Inclusions.

Theorem 3.1 (Frankowska [to appear])

Let H be a Hilbert space, A be the infinitesimal generator of a semi-group of contractions and F be a Lipschitz set-valued map from H to H with weakly compact images (not necessarily convex).

Let $z(\cdot)$ be a (mild) solution to the Operational Differential Equation

$$(2.2) \qquad z'(t) = Az(t) + g(t),\ z(0) = z_0$$

and let us set

$$(2.3) \qquad p(t) := d(g(t),F(z(t))),\ \delta := \|z_0 - x_0\|.$$

If $p(\cdot)$ belongs to $L^2(0,T,H)$, then there exists a solution $x(\cdot)$ to the Operational Differential Inclusion

$$(2.4) \qquad x'(t) \in Ax(t) + F(x(t)),\ x(0) = x_0$$

satisfying

$$(2.5) \qquad \|x(t) - z(t)\| \leq e^{ct}(\delta + \int_0^t p(s)ds).$$

Furthermore, the solution map S is Lipschitz from H to C(0,T;H).

Δ

Let us consider now a closed subset K of H. We define the

Invariance Property.

For any $x_0 \in K$, all solutions to Operational Differential Inclusion (2.4) (defined on H) are actually viable in K.

Let G(t) the contraction semi-group whose infinitesimal generator is A. We say that K is <u>invariant by A + F</u> if

$$(2.6)\qquad \begin{cases} \forall x \in K,\ \forall f \in F(x), \\ \liminf_{h \to 0^+} d(G(h)x + h,K)\,|\,h = 0 \end{cases}$$

When x belongs to $K \cap D(A)$, condition (2.6) can be written in the form

$$(2.7)\qquad Ax + F(x) \subset T_K(x)$$

(all velocities are contingent to K at x).

We complete Frankowska's theorem by the following invariance theorem.

<u>Theorem 2.2.</u>

We assume the assumptions of Theorem 2.1. Let K be a closed subset of H, invariant by A + F. Then K enjoys the invariance property.

Δ

<u>Notes</u>

(1) Actually, we extend U to the set-valued map from $L^2(\Omega)$ to $L^2(\Omega,Y)$ defined by $U(x) := \{u \in L^2(\Omega,Y) \mid \text{for almost all } \omega \in \Omega,\ u(\omega) \in U(\omega,x(\omega))\}$.

(2) We say that U is upper hemi-continuous from K to Z if and only if for all $p \in Z^*$, the function

$$x \to \sup_{v \in U(x)} \langle p,v \rangle$$

is upper semi-continuous. (see Aubin-Ekeland, [1984], chapter 2).

References

[1] Aubin, J.-P. and Cellina, A. [1984] Differential Inclusions, Springer-Verlag.

[2] Aubin, J.-P., and Ekeland, I. [1984] Applied Nonlinear Analysis, Springer-Verlag.

[3] Chang, K. C. [1980] The obstacle problem and parital differential equations with discontinuous nonlinearities. Comm. Pure Appl. Math. 33, 117-146.

[4] Filippov, A. F. [1967] Classical solutions of differential equations with multi-valued right-hand side. Westnik Moscow Un. 22, 16-26.

[5] Frankowska, H. [1986] a) Local controllability and infinitesimal generators of semi-groups of set-valued maps. SIAM J. Control.

[6] Frankowska, H. [1986] b) The maximum principle for an optimal solution to a differential inclusion with end points constraints. SIAM J. Control.

[7] Frankowska, H. [to appear] Filippov's theorem for partial differential inclusions.

[8] Haddad, G. [1981] Monotone trajectories of differential inclusions with memory, Israel J. Math. 39, 83-100.

[9] Lions, J.-L. [1968] Contrôle optimal de systèmes gouvernés par des équations aux dérivées partielles. Dunod.

[10] Lions, P.-L. [1982] On the existence of positive solutions of semilinear elliptic equations. SIAM Review 24, 441-467.

[11] Shi, Shuzhong [186] Théorèmes de viabilité pour les inclusions aux dérivées partielles. C.R.A.S.

[12] Shi, Shuzhong [to appear] Viability theory for partial differential inclusions.

[13] Zolezzi. To appear.

ON COMPACTNESS OF ADMISSIBLE PARAMETER SETS:

H.T. Banks and D.W. Iles

Brown University, Providence, Rhode Island, USA

ABSTRACT

We report on a series of numerical examples and compare several algorithms for estimation of coefficients in differential equation models. Unconstrained, constrained and Tikhonov regularization methods are tested for their behavior with regard to both convergence (of approximation methods for the states and parameters) and stability (continuity of the estimates with respect to perturbations in the data or observed states).

In this brief note we summarize some of our findings [3] from numerical studies on certain aspects of ill-posedness in inverse or parameter estimation problems involving differential equation constraints. There is a vast literature (which we shall not attempt to discuss here) on a number of questions (e.g., lack of existence and/or uniqueness of solutions, lack of continuous dependence of solutions on data) related to the estimation of parameters even when the constraining systems are algebraic equations or ordinary differential equations. Additional difficulties arise when one is attempting to estimate functional (i.e. time and/or spatially dependent) coefficients in partial differential equation or distributed parameter systems. Here we focus on the role that compactness of the admissible parameter or coefficient set plays in such problems. Due to the limitations of space, our presentation will be sketchy, with all but the expert reader most likely wishing to consult some of our references for further elaboration.

Consider a general least-squares inverse or parameter estimation problem: Minimize $J(q) = |\mathcal{C}u(q) - z|_Z$ over $q \in Q_{AD}$, $Q_{AD} \subset Q$, subject to the constraints $\mathcal{A}(q,u(q)) = \mathcal{F}$. Here $\mathcal{C}$ is an observation operator from the state space X to an observation or data space Z, Q_{AD} is the admissible subset of parameters in the space Q, and the parameter dependent operator $\mathcal{A}$ defines the dynamics that constrain the problem. In the problems of interest to us, $\mathcal{A} = \mathcal{F}$ represents a partial differential equation (elliptic, parabolic, hyperbolic) with parameter functions q which depend on time and/or spatial coordinates (or even the state u itself in some nonlinear system problems). It is now well-understood (e.g. see [1] for a discussion) that a compactness hypothesis (Q_{AD} compact in some Q topology) for Q_{AD} plays an important theoretical role in both convergence (of approximating solutions) and stability (continuity of the estimated parameters with respect to the data or observations). Here we wish to demonstrate that this compactness also plays an important computational role in such problems. To do this, we illustrate the basic ideas with one dimensional elliptic systems (so we actually have an ordinary differential equation with spatially dependent coefficient). We wish to emphasize, however, that our findings are most certainly relevant to problems with more complex system dynamics (second order parabolic or hyperbolic equations, or higher order equations of elasticity). Indeed we have observed the difficulties and phenomena we discuss here in a number of these technically more challenging problems.

Turning to a class of concrete examples, we consider minimization of

$$J(q) = \int_0^1 |u(q) - \hat{u}|^2 dx \tag{1}$$

over $q \epsilon Q_{AD} \subset Q = C[0,1]$ subject to

$$D(qDu) = f, \quad u(0) = u(1) = 0. \tag{2}$$

Here f is assumed known, $D = \frac{\partial}{\partial x}$, and $\hat{u}$ are given observations for $u = u(q)$ the solution of (2). For computational purposes, we replace this original problem by a sequence of approximating problems where the states u are replaced by Galerkin approximations u^N (for example, here we take u^N in the N-1 dimensional space of linear splines with grid size 1/N and satisfying the boundary conditions $u^N(0) = u^N(1) = 0$) and the approximate parameters q^M are chosen from an approximating set Q^M for Q_{AD}. That is, our algorithms are used to seek $q^M \epsilon Q^M \subset Q = C[0,1]$ that minimizes

$$J^N(q^M) = \int_0^1 |u^N(q^M) - \hat{u}|^2 dx. \tag{3}$$

A convergence theory (as the dimensions of the approximating spline spaces increase i.e., $N \to \infty$, $M \to \infty$) can be given where one may use either linear or cubic splines for the state approximations and for the parameter approximations (e.g. see [2], [4] for the ideas). An essential feature of these particular convergence proofs is that the admissible parameter set Q_{AD} and its approximations Q^M lie in some compact subset of $C[0,1]$. This same compactness assumption plays a fundamental role in proving stability (e.g., continuity of the inverse of the mapping from the parameter estimates to the observations or data) as is discussed in [1], for example.

Perhaps the most direct way to interpret the compactness requirements is in terms of constraints on the parameters. For example, in the computations reported on herein, we imposed compactness in $C[0,1]$ by putting pointwise upper and lower bounds on the parameter function values as well as an upper bound on the absolute values of the slope of the functions. In practice it is common to ignore functional constraints, imposing the pointwise upper and lower bounds to insure that the optimization algorithms perform satisfactorily. The results summarized in this note illustrate the apparent necessity in many examples of including the full compactness constraint in computational algorithms. Examples are given here and in [3] where both stability and convergence properties are as expected whenever a constrained estimation procedure is employed whereas instability and divergence are in evidence when unconstrained techniques are used. It is safe to speculate that similar behavior occurs in problems with parabolic and hyperbolic as well as elliptic systems. In our own work and in that reported in the literature - e.g. see Yoon and Yeh [6], one sometimes encounters severe problems with oscillations in the estimates for q as one pushes the algorithms for increased accuracy in the parameter estimates (i.e. as one lets $M \to \infty$). As the examples in this note and [3] demonstrate, these difficulties can to some extent be alleviated by imposition of compactness constraints.

An alternative but essentially theoretically equivalent approach involves the use of Tikhonov regularization as formulated by Kravaris and Seinfeld in [5]. One restricts the parameter set to $Q_R \subset Q$ with Q_R compactly imbedded in Q and then modifies the original least squares criterion J to minimize $J_\beta = J + \beta|q|_R^2$ where $|\cdot|_R$ is the norm in Q_R and β is a regularization parameter. Thus minimizing sequences for J_β are bounded in Q_R and hence compact in Q; this is, in some sense, roughly equivalent to minimizing J over a restriction of Q which is compact even though the minimization of J_β only produces (hopefully) an approximation to the minimizer for the original criterion J. In the cases considered below, we use $Q_R = H^1$ while $Q = C$ (which corresponds to $\Lambda = C^1$ and $\mathfrak{X} = H^2$ in the notation of [5]).

As we shall see below, each approach has inherent difficulties in choosing related imbedding parameters: in the first, the estimates produced are sensitive to the constraints (the bound L on the derivatives of the parameters in the computations summarized here) while the estimates produced using regularization are quite sensitive to the regularization parameter β.

We carried out a series of numerical tests to compare spline based algorithms (linear spline approximations for both the states and parameters) on a number of examples for three cases: the unconstrained minimization of J; the constrained minimization of J; and unconstrained minimization of a regularized criterion J_β. Details of our packages and the algorithms are given in [3]. Here we only note that for the constrained minimization we used a reduced gradient algorithm with a corresponding gradient algorithm for the unconstrained minimization. For the compactness constraints we used $|Dq^M(x)| \leq L$ and $.5 \leq q^M(x) \leq 10.0$ in all our examples.

Our algorithms were compared on examples for which we knew the true solutions, i.e. we used "true" parameter values q^* to generate data $\hat{u}$ (in some cases with noise) as described in [3]. We summarize our findings and present representative results.

CONVERGENCE

In Figure 1 we compare estimates for several values of approximation indices N, M produced for an example (Example 2 of [3]) with

$$q^*(x) = \begin{cases} 2 + x & 0 \leq x \leq 1/3 \\ 8/3 - x & 1/3 \leq x \leq 2/3 \\ 4/3 + x & 2/3 \leq x \leq 1, \end{cases} \tag{4}$$

$$u(x) = \sqrt{x}(1 - \sqrt{x}), \tag{5}$$

in the system (2). We chose L = 1 (the exact maximum of $|Dq^*(x)|$) for the constrained algorithms. For low values of N (often a desirable situation in practice) compared to M, the unconstrained estimates are totally useless. For larger values of N, the unconstrained estimates are improved with only small oscillations appearing at each end. In all cases, convergence took

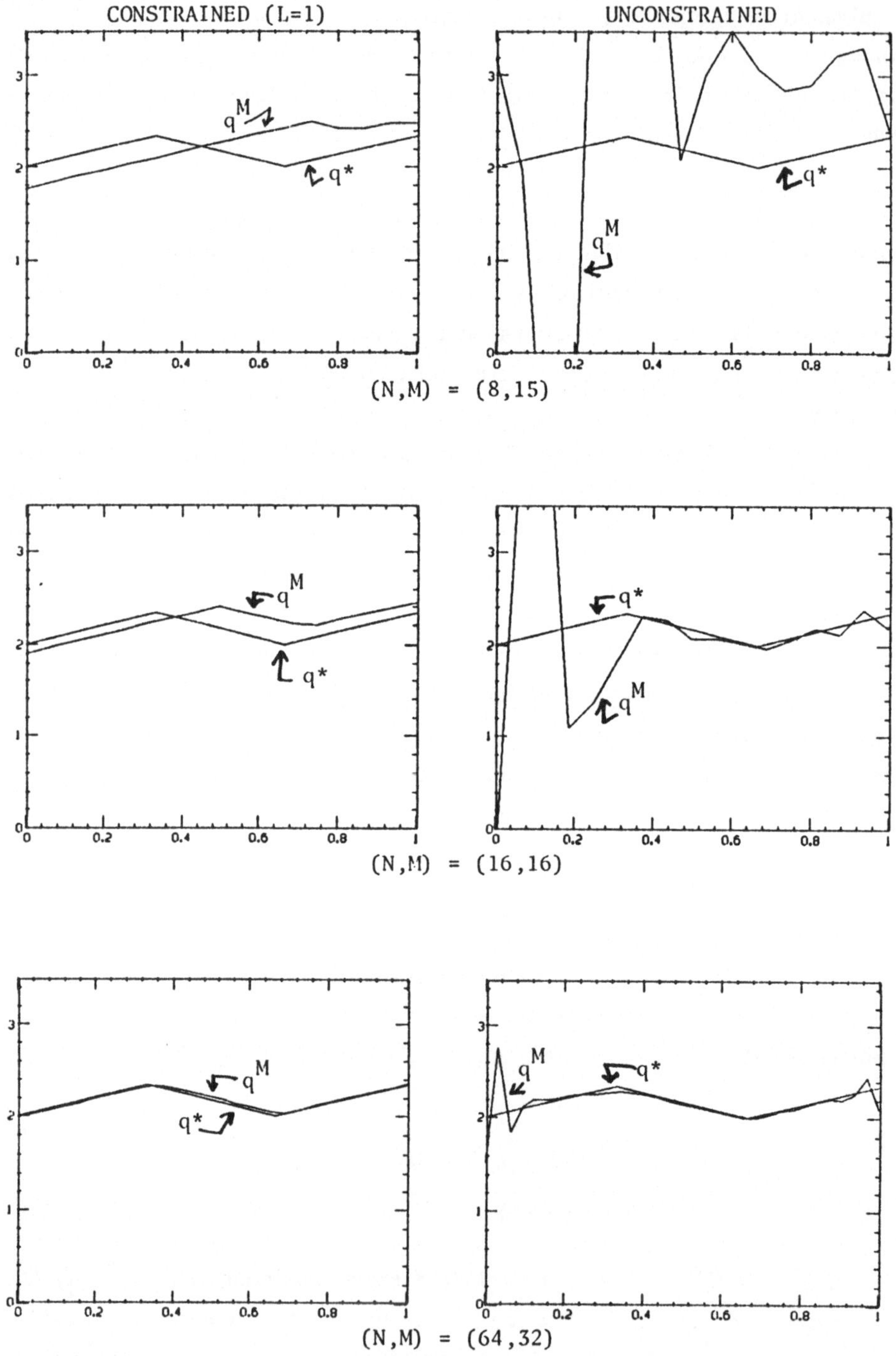

FIGURE 1

much longer for the unconstrained package. For this same example we depict in Figure 2 the estimates obtained using Tikhonov regularization with N = 64, M = 16 and several values of the regularization parameter β. Results for a slightly different example (Example 3 of [3]) with the same u but $q^*(x)$ piecewise linear as in (4) except with slopes ∓ 2 are depicted in Figure 3. Here we illustrate, for N = 64, M = 16, the typical performance of Tikhonov regularization as β changes and that of the constrained minimization as L varies. As one might expect, the estimates begin to resemble unconstrained estimates as β → 0 and L → ∞.

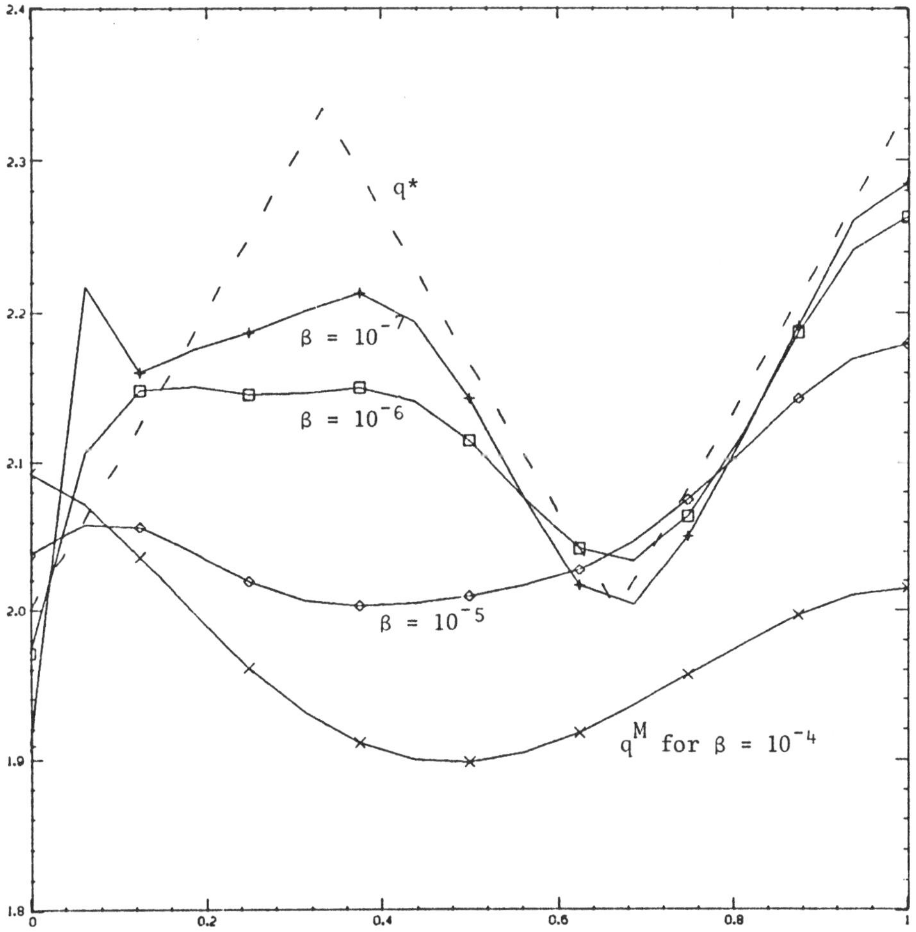

TIKHONOV ESTIMATES
AS β VARIES

FIGURE 2

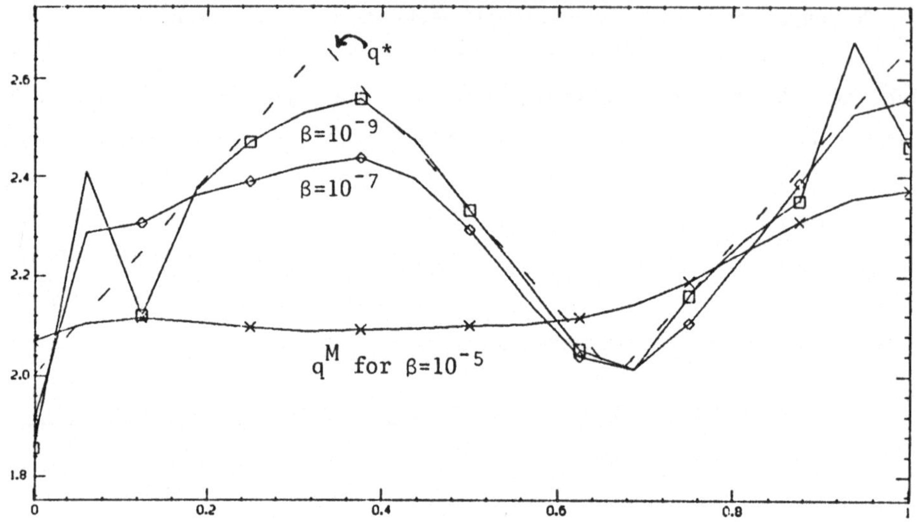

TIKHONOV ESTIMATES AS β VARIES

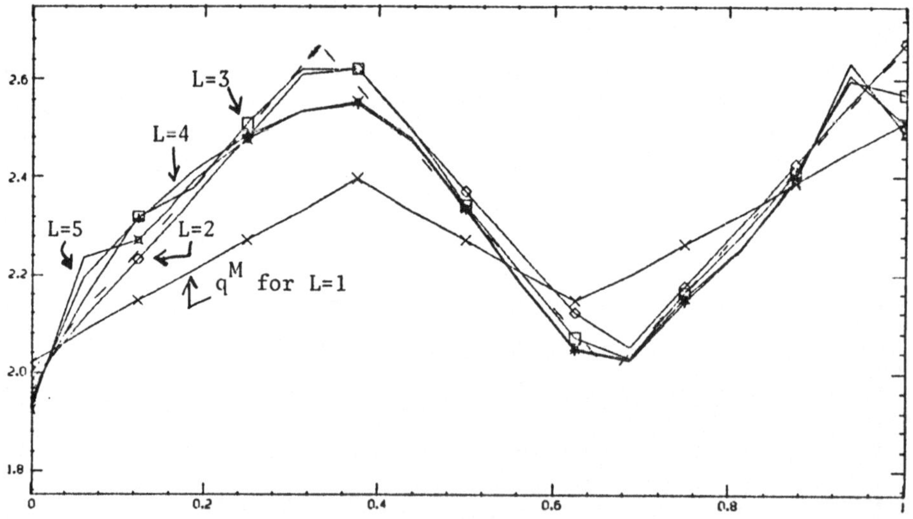

CONSTRAINED ESTIMATES AS L VARIES

(N,M) = (64,16)

FIGURE 3

STABILITY

To investigate stability with respect to noise in the data, we took the true u, q^* and f associated with (4), (5) above but perturbed u to produce data $\hat{u} = u_p$ for the least squares criterion J^N of (3). We used perturbations of the form $u_p(x) = u(x) + p(x)/K$ where p(x) is a perturbing function and K can be varied to control the size of the perturbation. As $K \to \infty$, we have $u_p(x) \to u(x)$ for bounded perturbations p. In our numerical experiments we used two different perturbation functions: $p_1(x) = x(1 - x)$, $p_2(x) = 1$. Note that p_1 (like u) satisfies the homogeneous boundary conditions while p_2 does not. The unconstrained, constrained and Tikhonov estimates for several values of K with perturbation function p_1 in the data and N = 8, M = 15 are given in Figure 4. The depicted behavior is typical: For all values of N the behavior of the constrained and Tikhonov methods are similar, with the estimates improving steadily as $u_p(x) \to u(x)$, i.e. as the noise in the observations tends to zero. In Figure 5, we present results obtained using the perturbation p_2 with the unconstrained, constrained and Tikhonov estimation procedures for several values of (N,M) : (8,15), (16,16) and (64,16). The L^∞ norm of the error (from the true values q^*) in the final estimates is graphed versus the values of K.

SUMMARY REMARKS

The results here and in [3] demonstrate severe problems in some instances with using an unconstrained algorithm to estimate the parameter q in examples such as (1), (2). When modified, either by regularizing the problem using Tikhonov regularization or by constraining the estimate set as in this note, the algorithm does give good estimates.

Unlike the unconstrained algorithm, both the Tikhonov and constrained algorithms are stable with respect to increasing M while holding N fixed. However as N is increased the estimates from the Tikhonov algorithm do not improve as much as do those of the constrained algorithm. The Tikhonov estimates are biased by the regularization of the cost functional, and never show all the detail of q when q has significant variation.

Both the constrained and Tikhonov estimation algorithms are stable with respect to systematic errors in the observation data, while, except when N is large, the unconstrained algorithm fails to give good results on even the exact data.

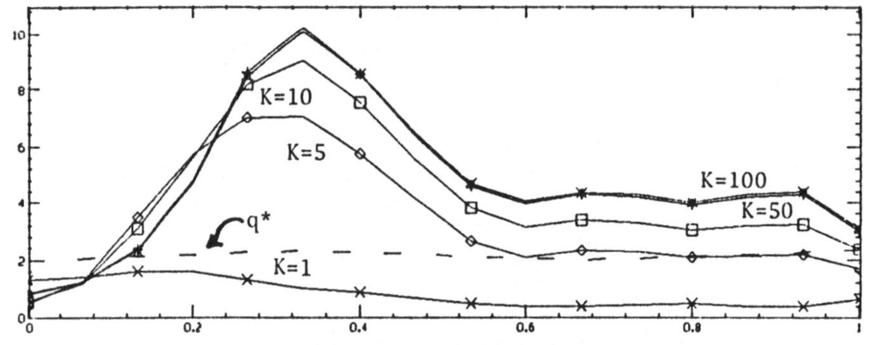

UNCONSTRAINED ESTIMATES

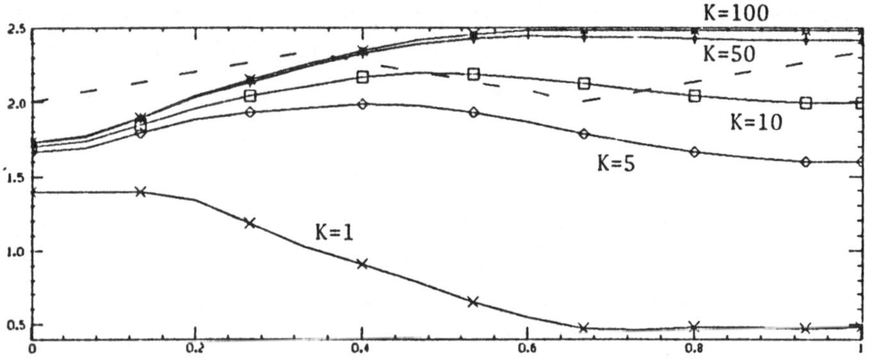

TIKHONOV ESTIMATES

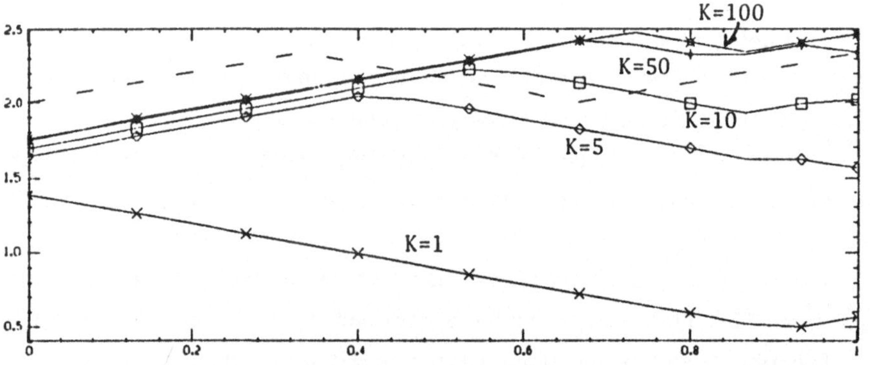

CONSTRAINED ESTIMATES

(N,M) = (8,15), PERTURBATION p_1

FIGURE 4

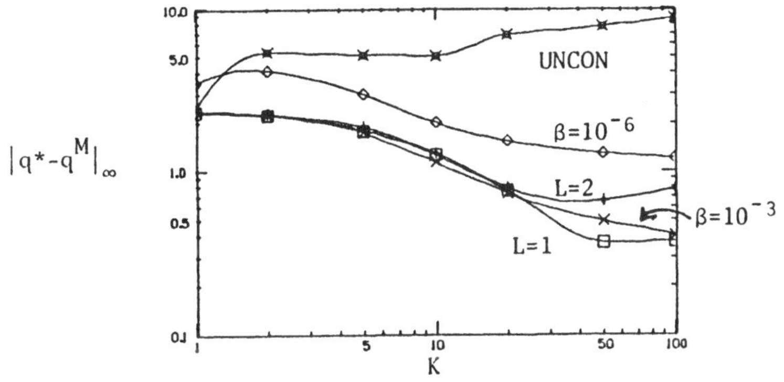

(N,M) = (8,15)

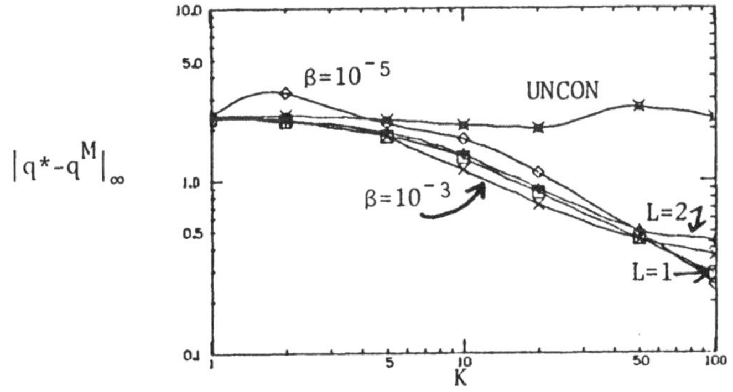

(N,M) = (16,16)

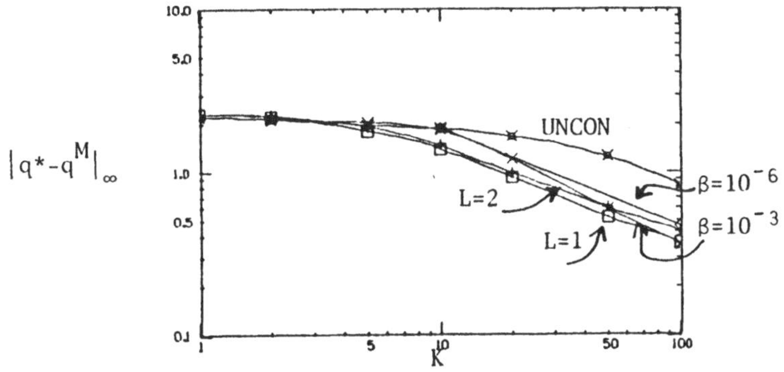

(N,M) = (64,16)

UNCONSTRAINED, CONSTRAINED, TIKHONOV, PERTURBATION p_2

FIGURE 5

For both the Tikhonov and constrained algorithms there are parameters which affect the algorithm's performance. For the constrained algorithm suitable constraints must be found while for the Tikhonov algorithm suitable values of β must be found. The constrained algorithm has the advantage that the constraints used here, i.e. limits on the slope of q, have an obvious meaning, and so may well be (at least approximately) known in advance. In the Tikhonov algorithm β has no obvious meaning. It must be chosen by looking at the change in the estimate behavior as β changes and perhaps using some a priori knowledge about the shape of q to choose values of β that give an estimate that is neither too flat, nor too oscillatory. For the constrained algorithms, the estimates are sensitive to the slope constraint parameter L. We have begun investigations into how one might use this sensitivity in some type of adaptive manner in algorithms to choose a "best" value of L (and hence a good parameter estimate). In Figure 6 we depict some of our initial findings. In this figure we graph the square of the H^1 norm of the final estimate versus the constraint L for several different examples and various values of N (with M fixed at 16). All but the second of the examples involve true parameters q^* of the form (4), differing only in the slope of the piecewise linear functions. In the first example $Dq^*(x) = \mp 1$, while in the last two $Dq^*(x) = \mp 2$ and $Dq^*(x) = \mp 5$ respectively. The second is made up of piecewise linear and parabolic segments satisfying $|Dq^*(x)| \leq 1$. Note that it is not necessary to know the true values q^* in order to obtain the graphs in this figure. Furthermore, we observe a striking separation in the values of the H^1 norm of the estimate for q^M at the value of L corresponding to the desired value of L to be used with each example. We are continuing our investigations into how these and other features of some of our results might be used to develop "adaptive" constrained parameter estimation algorithms.

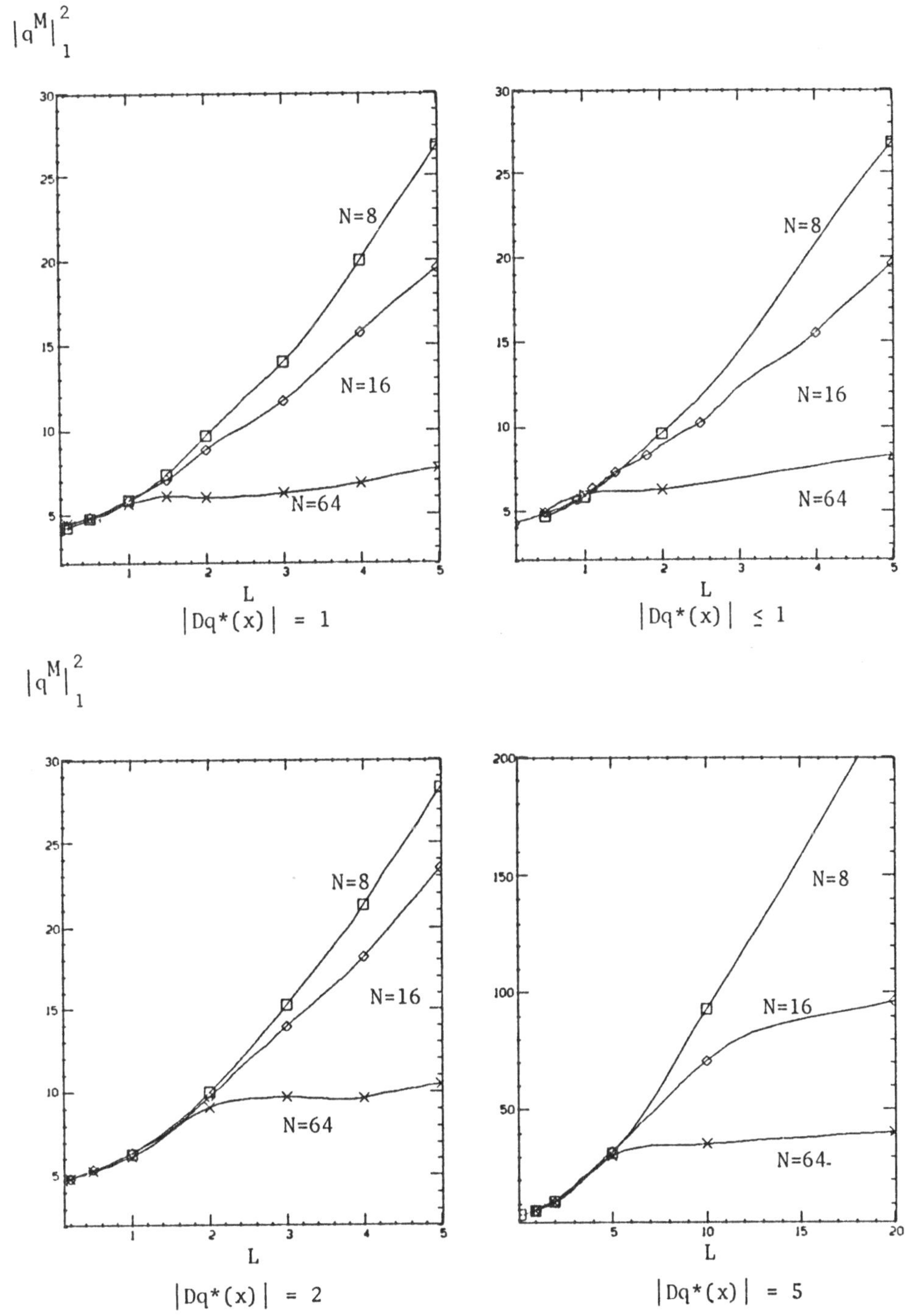

CONSTRAINED ESTIMATES, M = 16

FIGURE 6

ACKNOWLEDGMENTS

This research was supported in part by the National Science Foundation under NSF Grant MCS-8504316, the Air Force Office of Scientific Research under Contract AFOSR-84-0398, and the National Aeronautics and Space Administration under NASA Grant NAG-1-517. Part of this research was carried out while the first author was a visiting scientist at the Institute for Computer Applications in Science and Engineering (ICASE), NASA Langley Research Center, Hampton, VA, which is operated under NASA Contracts NAS1-17070 and NAS1-18107.

REFERENCES

[1] H.T. Banks, On a Variational Approach to Some Parameter Estimation Problems, ICASE Rep. No. 85-32, NASA Langley Res. Ctr, Hampton, VA June 1985.

[2] H.T. Banks, P.L. Daniel, and E.S. Armstrong, A Spline Based Parameter and State Estimation Technique for Static Models of Elastic Surfaces, ICASE Rep. No. 83-25, NASA Langley Res. Ctr, Hampton VA, June 1983.

[3] H.T. Banks and D.W. Iles, A Comparison of Stability and Convergence Properties of Techniques for Inverse Problems, LCDS Tech Rep. No. 86-3, Brown University, Providence, January, 1986.

[4] H.T. Banks, P.K. Lamm, and E.S. Armstrong, Spline-Based Distributed System Identification with Application to Large Space Antennas, AIAA J. Guidance, Control and Dynamics, May-June, 1986, to appear.

[5] C. Kravaris and J.H. Seinfeld, Identification of Parameters in Distributed Parameter Systems by Regularization, SIAM J. Control and Optimization 22, (1985), 217-241.

[6] Y.S. Yoon and W.W-G. Yeh, Parameter Identification in an Inhomogeneous Medium with the Finite-Element Method, Soc. Pet. Engr. J, 16, (1976), 217-226.

OPTIMALITY CONDITIONS FOR OPTIMAL CONTROL PROBLEMS OF VARIATIONAL INEQUALITIES

A. Bermúdez

University of Santiago de Compostela, Spain

C. Saguez

SIMULOG S.A. Av. du Centre, St. Quentin en Yvelines, France

1. INTRODUCTION

The optimal control of a variational inequality is a nonconvex non-differentiable optimization problem in a functional space. Some examples are given in Lions |8| where existence results are proved. A recent book on this subject is Barbu |2|.

The nondifferentiablity comes from the fact that the mapping giving the state of the system from the control is not differentiable. As a consequence of this, to get optimality conditions and to build numerical algorithms are difficult tasks.

These subjects have been studied in several articles by using different techniques. In Yvon |13|, Barbu |1| and Saguez |12| regularization-penalty methods are used to approximate the original problem by a differentiable optimal control problem. Optimality conditions are the obtained by passing to the limit.

On the other hand, Mignot in |9| consider a weak derivative, the so-called conical derivative. Using this concept Mignot and Puel |10| proved optimality conditions in the case of the obstacle variaional inequality.

In Bermúdez-Saguez |5| a new method is introduced to deal with this type of problems. It consists on transforming the original optimal control problem into another one which is linear-quadratic but state constrained, one of the constraints being nonconvex. Then optimality conditions can be written by analyzing the ones corresponding to this transformed problem.

Using this method Mignot and Puel |11| obtained optimality conditions for the optimal control of a parabolic variational inequality of the obstacle type. The proof uses a regularization technique to get optimality conditions for the transformed problem. Some complicated a priori estimates are necessary to pass to the limit.

In this paper we use another technique which consists on writing optimality conditions only for two special directions of the control space along which the functional to be minimized is convex.

The interest of this method is that it only needs tools of classical convex analysis. While it is quite general here it is applied to a particular case quoted as an open problem in Mignot-Puel |10|: the state variational inequality is similar to that modelling the elastic-plastic torsion of a cylindrical bar. This problem presents some technical difficulties because the Lagrange multiplier associated to the state variational inequality does not belong to a classical functional space.

2. The optimal control problem

Let Ω be an open bounded subset of $\mathbb{R}^N$ with smooth boundary $\Gamma = \partial\Omega$ and let f be a given function in $L^2(\Omega)$.

For $v \in L^2(\Omega)$ we define the state of the system y(v) as the unique solution of the first kind variational inequality

$$a(y, z-y) \geq (f+v, z-y), \qquad \forall z \in K \quad (1)$$
$$y \in K$$

where a is the bilinear form given by

$$a(y,z) = \sum_{i=1}^{N} \int_\Omega \frac{\partial y}{\partial x_i} \frac{\partial z}{\partial x_i} dx, \quad (2)$$

K is the closed convex subset of $H_0^1(\Omega)$ defined by

$$K = \{v \in H_0^1(\Omega) : |\nabla v(x)|_2 \leq 1 \quad \text{a.e. in } \Omega\} \quad (3)$$

and (,) represents the usual scalar product in $L^2(\Omega)$.

Let J be the cost function given by

$$J(v) = \frac{1}{2} \| y(v) - z_d \|^2_{L^2(\Omega)} + \frac{\nu}{2} \| v \|^2_{L^2(\Omega)} \quad (4)$$

The optimal control problem is to find $u \in L^2(\Omega)$ such that

$$J(u) \leq J(v), \qquad \forall v \in L^2(\Omega) \quad (5)$$

PROPOSITION 1. The optimal control problem has a solution.

Proof. Let $\{u_n\}$ be a minimizing sequence. From (4) we deduce

$\{u_n\}$ is bounded in $L^2(\Omega)$.

By taking $v=u_n$ and $z=0$ in (1) we get

$$a(y_n,y_n) \leq (f+u_n,y_n) \tag{6}$$

where $y_n=y(u_n)$.

Inequality (6) implies

$\{y_n\}$ is bounded in $H_0^1(\Omega)$

because a is coercive.

Therefore there exist subsequences $\{u_{n_k}\}$ and $\{y_{n_k}\}$ such that

$\{u_{n_k}\} \rightarrow u$ in $L^2(\Omega)$ weakly

$\{y_{n_k}\} \rightarrow y$ in $H_0^1(\Omega)$ strongly

Passing to the limit in the state variational inequality gives $y=y(u)$.

Finally we have

$$J(u) \leq \liminf_{k\to\infty} J(u_{n_k})$$

and then u is an optimal control.

3. Optimality conditions

We first introduce some functional spaces.

Let X denote the subspace of $H^2(\Omega)$ given by

$$X=H^2(\Omega)\cap H_0^1(\Omega) \tag{7}$$

Let Y be the vector space defined by

$$Y=\{\varphi=\nabla z \ : \ z\in X\} \tag{8}$$

and

$$\|\varphi\|_Y=\|z\|_X \qquad \text{if} \quad \varphi=\nabla z. \tag{9}$$

We denote by W the vector space

$$W = \{\eta \in Y' : \nabla.\eta \in L^2(\Omega)\} \tag{10}$$

Where $-\nabla.$ represents the adjoint of the gradient operator considered as an operator from X into Y.

When endowed with the norm

$$\|\eta\|_W = (\|\eta\|^2_{Y'} + \|\nabla.\eta\|^2_{L^2(\Omega)})^{\frac{1}{2}} \tag{11}$$

W is a Hilbert space.

Finally let C be the closed convex subset of Y defined by

$$C = \{\varphi \in Y : |\varphi(x)|_2 \leq 1 \quad \text{a.e.} \quad \text{in } \Omega\} \tag{12}$$

We denote by χ_C the indicator function of C in Y, i.e.

$$\chi_C(\varphi) = \begin{cases} 0 & \text{if} \quad \varphi \in C \\ \infty & \text{if} \quad \varphi \in Y \setminus C \end{cases} \tag{13}$$

and by supp_C the support function of C which is defined by

$$\text{supp}_C(\eta) = \sup_{\delta \in C} \{(\eta, \delta)_{Y'Y}\} \tag{14}$$

We have the following:

PROPOSITION 2. Let u be an optimal control. Then there exist $p \in L^2(\Omega)$, $\xi \in W$ and $\theta \in Y'$ such that

$$\begin{aligned} & -\Delta y - \nabla.\xi = f+u \\ & y|_\Gamma = 0 \end{aligned} \tag{15}$$

$$\xi \in \partial\chi_C(\nabla y) \tag{16}$$

$$(p, \Delta z) + (y - z_d, z) + (\theta, \nabla z)_{Y'Y} = 0, \quad \forall z \in X \tag{17}$$

$$(\theta, -\nabla y)_{Y'Y} \leq 0, \quad \forall \in \tilde{D}_\xi \tag{18}$$

$$(p, \nabla.\eta - \nabla.\xi) \geq, \quad \forall \eta \in D_Y \tag{19}$$

$$p + \nu u = 0 \tag{20}$$

where

$$\tilde{D}_\xi = C \cap \{\delta \in Y : \text{supp}_C(\xi) \leq (\xi, \delta)_{Y'Y}\} \tag{21}$$

and

$$D_y = \{\eta \in W: \operatorname{supp}_C(\eta) \leq (\eta, \nabla y)_{Y'Y}\} \tag{22}$$

Proof. We first prove the following.

LEMMA 1. A function y is the solution of the variational inequality (1) if and only if there exists $\xi \in W$ such that

$$\begin{aligned} &- \Delta y - \nabla.\xi = f+u \\ &y|_\Gamma = 0 \end{aligned} \tag{23}$$

$$\nabla y \in C \tag{24}$$

$$\operatorname{supp}_C(\nabla y) \leq (\xi, \nabla y)_{Y'Y} \tag{25}$$

Proof of the lemma 1. Following Brezis-Stampacchia |6| the solution of (1) belongs to $H^2(\Omega)$ and then it also satisfies

$$\begin{aligned} &a(y,z-y) + \chi_C(\nabla z) - \chi_C(\nabla y) \geq (f+u, z-y), \ \forall z \in X \\ &y \in X \end{aligned} \tag{26}$$

Therefore there exists $g \in X'$ such that

$$-\Delta y + g = f+u$$
$$g \in \partial(\chi_C . \nabla)(y)$$

Moreover, we have (see Bermúdez |4|)

$$\partial(\chi_C.\nabla)(y) = -\nabla.(\partial\chi_C(\nabla y))$$

which implies the existence of $\xi \in Y'$ such that

$$\xi \in \partial\chi_C(\nabla y) \tag{27}$$

and $g = -\nabla.\xi$.

The equivalence between (24) (25) and (27) follows from the fact that $\operatorname{supp}_C = \chi_C^*$ and well known results of convex analysis (see for instance Ekeland-Temann |7| or Barbu-Precupanu |3|).

We return to the proof of the Proposition 2. From the previous lemma it is easily deduced that $u \in L^2(\Omega)$ is an optimal control if and only if there exists $\xi \in W$ such that (u,ξ) is a solution of the following optimal control problem:

. Control space: $U\times W$ $(U= L^2(\Omega))$ (28)

. State equation:

$$-\Delta y= f+v+\nabla.\eta \qquad (y= y(v,\eta)) \tag{29}$$

$$y|_{\Gamma}= 0$$

. Cost function:

$$G(v,\eta) = \frac{1}{2}\ \| y-z_d\|^2_{L^2(\Omega)} + \frac{\nu}{2}\ \| v\|^2_{L^2(\Omega)} \tag{30}$$

. Constraints on the state:

$$\nabla y\in C \tag{31}$$

$$\text{supp}_C(\eta) - (\eta,\ \nabla y)_{Y'Y} \leq 0 \tag{32}$$

Let F be the functional from $U\times W$ into $(-\infty,\infty]$ given by

$$F(v,\eta) = G(v,\eta) + \Psi(\nabla y,\eta) \tag{33}$$

where Ψ: $Y\times W \rightarrow (-\infty,\infty]$ is defined by

$$\Psi(\varphi,\eta) = \chi_C(\varphi) + \chi_{\mathbb{R}^-}\{\text{supp}_C(\eta) - (\eta,\varphi)_{Y'Y}\} \tag{34}$$

Then (u,ξ) is a solution of the optimal control problem (28)-(32) if and only if

$$F(u,\xi) \leq F(v,\eta) \quad , \quad \forall (v,\eta)\in L^2(\Omega)\times W \tag{35}$$

In order to write optimality conditions for this optimization problem we prove the following.

LEMMA 2. Assume that (35) holds. Then we have:

i) The functional

$$v\in L^2(\Omega) \rightarrow F(v,\xi)\in(-\infty,\infty]$$

is convex lower semicontinuous (l.s.c.) and has a minimum at $v=u$.

ii) The functional

$$\eta\in W \rightarrow F(u-\nabla.\eta,\ \xi+\eta)\in(-\infty,\infty]$$

is convex l.s.c. and has a minimum at $\eta= 0$.

Proof of the lemma 2.

i) The result follows from the fact that

$$F(v,\xi) = \frac{1}{2}\ \| y(v,\xi)-z_d\|^2_{L^2(\Omega)} + \frac{\nu}{2}\ \| v\|^2_{L^2(\Omega)} + \chi_{\tilde{D}_\xi}(\nabla y(v,\xi)) \tag{36}$$

and $\tilde{D}_\xi$ as defined in (21) is a closed convex subset of Y.

ii) In this case we have

$$F(u-\nabla.\eta,\ \xi+\eta)=\frac{1}{2}\ \|y(u,\xi)-z_d\|^2_{L^2(\Omega)}+\|u-\nabla.\eta\|^2_{L^2(\Omega)} \tag{37}$$

$$+\ \chi_{D_y}(\xi+\eta)$$

and D_y is a closed convex subset of W.

We return again to the proof of the Proposition 2. From the lemma 2 we deduce

$$0\in\partial_v F(v,s)\big|_{v=u} \tag{38}$$

and

$$0\in\partial_\eta F(u-\nabla.\eta,\xi+\eta)\big|_{\eta=0} \tag{39}$$

Relation (38) implies the existence of $\theta\in Y'$ such that

$$\nu(u,-\Delta z)+(y-z_d,z)+(\theta,\nabla z)_{Y'Y}=0 \quad , \quad \forall z\in X \tag{40}$$

$$\theta\in\partial\chi_{\tilde{D}_\xi}(\nabla y) \tag{41}$$

because the mapping given by

$$v\in X \longrightarrow \nabla z\in Y \ ,$$

z being the unique solution of the problem

$$-\Delta z=v$$
$$z\big|_\Gamma=0$$

is surjective.

Similarly from (39) we deduce the existence of $q\in W'$ such that

$$-\nu(u,\nabla.\eta)\ +\ (q,\eta)_{W'W}=0 \qquad \forall\eta\in W \tag{42}$$

$$q\in\partial\chi_{D_y}(\xi) \tag{43}$$

Let $p=-\nu u$. Then from (40) we obtain (17).

Finally (18) can be easily deduced from (41) and (19) from (42) and (43).

Remark 1. If one can prove the following regularity result:

$$u\in H_0^1(\Omega) \tag{44}$$

(which is true if one regularizes the problem) we get:

$$p\in H_0^1(\Omega) \tag{45}$$
$$\nabla.\theta\in H^{-1}(\Omega) \tag{46}$$

and (17) is equivalent to the boundary value problem

$$-\Delta p = y - z_d - \nabla . \theta$$
$$p|_\Gamma = 0 \tag{47}$$

Remark 2. Assuming that all functions in (15)-(20) are smooth enough we can give an interpretation of the optimality conditions obtained in the Proposition 2.

Let Ω_i (i= 1,2,3) be the following subsets of Ω:

$$\Omega_1 = \{x \in \Omega : |\nabla y(x)|_2 = 1 , \quad \xi(x) = 0\} \tag{48}$$

$$\Omega_2 = \{x \in \Omega : |\nabla y(x)|_2 < 1 , \quad \xi(x) = 0\} \tag{49}$$

$$\Omega_3 = \{x \in \Omega : |\nabla y(x)|_2 = 1, \quad \xi(x) = \lambda \nabla y(x) , \lambda > 0\} \tag{50}$$

Then from (16) we deduce

$$\Omega = \Omega_1 \cup \Omega_2 \cup \Omega_3 \tag{51}$$

and conditions (18), (19) can be interpreted as follows

$$(\theta, \nabla y) \geq 0 , \quad (\nabla p, \nabla y) \leq 0 \quad \text{in } \Omega_1 \tag{52}$$

$$\theta = 0 \quad \text{in } \Omega_2 \tag{53}$$

$$(\nabla p, \nabla y) = 0 \quad \text{in } \Omega_3 \tag{54}$$

Indeed, in Ω_1 we have

$$\tilde{D}_\xi = C, \quad D_y = \{\mu \nabla y : \mu \geq 0\}$$

and then (18) gives

$$\theta = \alpha \nabla y \quad \text{with} \quad \alpha \geq 0 \tag{55}$$

and from (19) we deduce

$$(\nabla p, \mu \nabla y) \leq 0 \quad , \quad \forall \mu \geq 0 \tag{56}$$

which implies (52).

In Ω_2 we have

$$D_y = \{0\}$$

and then (18) easily implies (53).

Finally, in Ω_3 D_y is defined by

$$D_y = \{\mu \nabla y : \mu \geq 0\}$$

and then from (19) we get

$$(\nabla p,\ (\mu-\lambda)\nabla y) \leq 0 \qquad \forall \mu \geq 0 \tag{57}$$

where

$$\xi = \lambda \nabla y \tag{58}$$

The inequality (57) gives (54) because λ is strictly positive in Ω_3.

Remark 3. The lemma 2 suggests to use the following relaxation algorithm in order to solve numerically the optimal control problem:

Step 1. Let ξ^1 be arbitrarily given in W.

Take u^1 to be a minimum of the convex function:

$$v \in L^2(\Omega) \to \frac{1}{2} \| y(v,\xi^1) - z_d \|^2_{L^2(\Omega)} + \frac{\nu}{2} \| v \|^2_{L^2(\Omega)} + \chi_{\tilde{D}_{\xi^1}}(\nabla y(v,\xi^1))$$

Actually this minimization problem is equivalent to solve a linear-quadratic optimal control problem with the convex constraint on the state :

$$\nabla y \in \tilde{D}_{\xi^1}$$

Step n+1. ξ^n is given from the previous step.

i) Take u^{n+1} to be a solution of the optimal control problem:

Control space: $L^2(\Omega)$

State equation:

$$-\Delta y = f + v + \nabla . \xi^n$$

$$y|_\Gamma = 0$$

Cost function:

$$v \to G(v,\xi^n) = \frac{1}{2} \| y - z_d \|^2_{L^2(\Omega)} + \frac{\nu}{2} \| v \|^2_{L^2(\Omega)}$$

Constraint on the state:

$$\nabla y \in \tilde{D}_{\xi^n}$$

Observe that it is a linear-quadratic optimal control problem and the constraint on the state is convex.

ii) Let η^n be a minimum of the convex function:

$$\eta \in W \rightarrow \frac{\nu}{2} \| u^{n+1} - \nabla . \eta \|^2_{L^2(\Omega)}$$

under the constraint

$$\eta + \xi^n \in D_{y(u^{n+1}, \xi^n)}$$

then take

$$\xi^{n+1} = \xi^n + \eta^n \quad .$$

References

|1| Barbu, V. "Necesaary conditions for distributed control problems governed by parabolic variational inequalities". SIAM J. on Cont. and Opt. 19 (1981): 64-86.

|2| Barbu, V. Optimal control of variational inequalities. London: Pitman, 1984.

|3| Barbu, V., and Precupanu, Th. Convexity and optimization in Banach spaces. Bucarest: Sijthoff & Noordhoff, 1978.

|4| Bermúdez, A. 1981. Mixed finite elements for nonlinear problems (in Spanish). In Proceedings of the 4th CEDYA. University of Sevilla, Department of Mathematics.

|5| Bermúdez, A., and Saguez, C., 1985. Optimal control of variational inequalities: optimality conditions and numerical methods. In Free boundary problems: Theory and Applications, ed. A. Bossavit, A. Damlamian and M. Frémond, 478-487. London: Pitman.

|6| Brézis, H. and Stampacchia, G. "Sur la regularité de la solution d'inéquations elliptiques". Bull. Soc. Math. France 96 (1968): 153-180.

|7| Ekeland, I. and Temam, R. Analyse convexe et problèmes variationnels. Paris: Dunod, 1974.

|8| Lions, J.L. Contrôle optimal de systèmes gouvernés par des équations aux dérivées partielles. Paris: Dunod, 1968.

|9| Mignot, F. "Contrôle dans les inéquations variationnelles elliptiques". J. of Funct. Anal. 22 (1976): 130-185.

|10| Mignot, F. and Puel, J.P. "Optimal control in some variational inequalities". SIAM J. on Cont. and Opt. 22 (1984): 466-476.

|11| Mignot, F. and Puel, J.P. "Contrôle optimal d'un système gouverné par une inéquation variationnelle parabolique". C.R. Acad. Sc. Paris, t. 298, Série I, 12 (1984): 277-280.

|12| Saguez, C. "Contrôle optimal de problèmes à frontière libre". Thèse d'Etat. University of Technology of Compiegne (France). 1981.

|13| Yvon, J.P. "Contrôle optimal de systèmes gouvernés par des inéquations variationnelles". LABORIA Report (INRIA, France) 53 (1974).

EQUIVALENT CONTROL PROBLEMS AND APPLICATIONS

J.F. Bonnans

INRIA, Rocquencourt, France

D. Tiba

INCREST, Bucuresti, Romania

1. INTRODUCTION

It is known that optimal control problems governed by variational inequalities may be equivalently transformed into control problems with linear state equation and mixed type (state-control) constraints. This method was used in C.Saguez and A.Bermudez [9], F.Mignot and J.P.Puel [8] with applications to the optimality conditions for control problems governed by variational inequalities.

In [11], [12] it is shown that, conversely, state constrained control problems may be efficiently approximated by control problems governed by variational inequalities. The basic idea originates from the unstable systems control theory developed by J.L.Lions [7] and J.F.Bonnans [3].

It is the aim of this paper to give some equivalence results by a different approach than in [8], [9] and extending [12], [11] in a certain sense. As an application we obtain bang-bang controls for state constrained problems governed by variational inequalities. This may be compared with some recent work of A.Friedman [5].

2. EQUIVALENCE

Consider the following control problem:

(P) $$\text{Min}\,\{ F(y) + G(u)\}, \tag{2.1}$$

$$y' + Ay = Bu + f \qquad \text{in } [0,T], \tag{2.2}$$

$$y(0) = y_0, \tag{2.3}$$

$$y(t) \in C. \tag{2.4}$$

Here V, H, U are Hilbert spaces with $V \subset H \subset V^*$ densely, continuously, $F : L^2(0,T;H) \to]-\infty, +\infty]$ is a convex, continuous functional with $F(y) \geq c$, for $y \in L^2(0,T;H)$ and $G : L^2(0,T;U) \to]-\infty, +\infty]$ is convex, lower semicontinuous, satisfying

$$\lim_{|u|_{L^2} \to \infty} G(u) = +\infty. \tag{2.5}$$

We assume that $B : U \to V^*$ is linear, continuous, $f \in L^2(0,T;V^*)$, $A : V \to V^*$ is linear, continuous, coercive and symmetric and $C \subset V$ is a bounded, closed, convex subset.

Therefore problem (2.1) - (2.4) is an abstract model for distributed or boundary control problems governed by parabolic equations. Control constraints may be also considered, by $u \in \mathrm{dom}(G)$.

If $y_0 \in H$, then (2.2), (2.3) have a unique solution $y \in L^2(0,T;V) \cap C(0,T;H)$, $y' \in L^2(0,T;V^*)$. Under the usual admissibility assumption, there is at least one optimal pair $[y^*, u^*]$ for (2.1) - (2.4).

Let $B^* : V \to U^*$ be the adjoint of B. As $C \subset V$ is bounded, then $B^*(C) \subset U^*$ is closed, convex. We define

$$\tilde{C} = \{ v \in V, B^* v \in B^*(C) \} \tag{2.6}$$

and ϕ, $\tilde{\phi}$, ψ to be the indicator functions of C, $\tilde{C}$, $B^*(C)$, respectively.

We associate with (P) the problem ($\tilde{P}$), where (2.4) is replaced by

$$y(t) \in \tilde{C}, \tag{2.7}$$

and the problem (P_1):

$$(P_1) \qquad \mathrm{Min}\, \{ F(y) + G(u - w) + \tfrac{1}{2} |w|^2_{L^2(0,T;U)} \}, \tag{2.8}$$

$$y' + Ay + Bw = Bu + f, \quad w \in \partial \psi(B^* y), \tag{2.9}$$

$$y(0) = y_0.$$

Remark. Let $S : V \to V^*$ be the (multivalued) operator $Sy = B \partial \psi(B^* y) = Bw$. Generally $Sy \subset \partial \tilde{\phi}(y)$ with equality for certain conditions on $\mathrm{dom}(\psi) \cap \mathrm{range}(B^*)$ (see [10], [2]). By monotonicity, (2.9) has at most one solution and if y is a solution of (2.9) it also satisfies

$$y' + Ay + \partial \tilde{\phi}(y) \ni Bu + f \tag{2.10}$$

and (2.3).

Therefore, the problem (P_1) may be viewed as governed by variational inequalities (without state constraints) and has to be interpreted as a singular control problem since it is possible that (2.9) has no solution or that $w \notin L^2(0,T;U)$. However any admissible control for (P) is also admissible for (P_1). Moreover, if ψ is regularized by ψ_λ, then the corresponding problem (P_1^λ) is well posed.

THEOREM 2.1. *The problems* $(\tilde{P})$ *and* (P_1) *are equivalent.*

Proof. Let $\tilde{J}$, J_1 denote the cost functionals of $(\tilde{P})$, (P_1). If $[y^*, u^*]$ is a solution of $(\tilde{P})$, then $0 \in \partial\psi(B^*y^*)$ and $[y^*, u^*]$ is admissible for (P_1) with $J_1(y^*, u^*) = \tilde{J}(y^*, u^*)$.

If $[\hat{y}, \hat{u}]$ is optimal pair for (P_1), then $B^*\hat{y} \in B^*(C)$ in $[0,T]$, that is $\hat{y}(t) \in \tilde{C}$ in $[0,T]$ and the pair $[\hat{y}, \hat{u} - \hat{w}]$, where $\hat{w} \in \partial\psi(B^*\hat{y})$, is admissible for $(\tilde{P})$. Moreover,

$$\tilde{J}(\hat{y}, \hat{u} - \hat{w}) \le J_1(\hat{y}, \hat{u}) \le J_1(y^*, u^*) = \tilde{J}(y^*, u^*).$$

This gives $\hat{w} = 0$ and $[\hat{y}, \hat{u}]$ is also a solution of $(\tilde{P})$.

We conclude that $(\tilde{P})$ and (P_1) have the same optimal values and optimal pairs.

COROLLARY 2.2. *If* $C = \tilde{C}$, *then* (P) *is equivalent with* (P_1). *Moreover, if* B^{-1} *is bounded, then* (P) *is equivalent with*

$$(P_2) \qquad \mathrm{Min}\,\{ F(y) + G(u - w) + \tfrac{1}{2}\,|w|^2_{L^2(0,T;U)} \},$$

$$y' + Ay + \partial\phi(y) \ni Bu + f, \quad w \in \partial\psi(B^*y),$$

$$y(0) = y_0.$$

Proof. $C = \tilde{C}$ gives $\phi = \psi \circ B^*$. If B^{-1} is bounded, by a result from [10], we get $\partial\phi = B\,\partial\psi B^*$. Therefore $(\tilde{P})$ may be rewritten as (P) and (P_1) may be rewritten as (P_2), and we may apply *Theorem 2.1* .

In order to make clear the above abstract setting, we briefly discuss the following example of boundary control:

$$(P_3) \qquad \mathrm{Min}\,\{ \tfrac{1}{2}\int_0^T\int_\Omega (y - y_d)^2 dxdt + \tfrac{1}{2}\int_0^T\int_\Gamma u^2 d\sigma dt \}, \tag{2.11}$$

$$\partial y/\partial t - \Delta y = 0 \qquad \text{in } \Omega \times]0,T[, \tag{2.12}$$

$$y(0,x) = y_0(x) \qquad \text{in } \Omega, \tag{2.13}$$

$$\partial y/\partial n = u \qquad \text{in } \Sigma = \Gamma \times [0,T], \tag{2.14}$$

with state constraints:

$$-\alpha \leq y|_\Gamma \leq \alpha \qquad t \in [0,T]. \tag{2.15}$$

The Ω is a bounded domain in $\mathbf{R}^n$. Let $V = H^1(\Omega)$, $H = L^2(\Omega)$, $A : H^1(\Omega) \to H^1(\Omega)^*$ be the Laplace operator. We define $B : L^2(\Gamma) \to H^1(\Omega)^*$ by

$$(Bu, v)_{V \times V^*} = \int_\Gamma uvd\sigma, \quad v \in V. \tag{2.16}$$

We remark that $B^* : V \to L^2(\Gamma)$ is given by $B^*v = v|_\Gamma$, $v \in V$. Therefore, we have

$$C = \{ v \in H^1(\Omega);\ -\alpha \leq v|_\Gamma \leq \alpha\}$$

$$= \{ v \in H^1(\Omega);\ -\alpha \leq B^*v \leq \alpha\} = \tilde{C}$$

by the definition of C. Then (P_3) is equivalent to the problem (P_1). On this example (P_1) reads as follows (see [1], p.145):

(P_4) $$\text{Min}\ \{ \tfrac{1}{2} \int_0^T\int_\Omega (y - y_d)^2 dxdt + \tfrac{1}{2} \int_\Sigma w^2 d\sigma dt + \tfrac{1}{2} \int_\Sigma (u - w)^2 d\sigma dt \},$$

$$\partial y/\partial t - \Delta y = 0 \qquad \text{in } \Omega \times]0,T[,$$

$$y(0,x) = y_o(x) \qquad \text{in } \Omega,$$

$$\partial y/\partial n = u - w, \qquad w \in \partial\gamma(y|_\Gamma),$$

where $\gamma : L^2(\Gamma) \to]-\infty, +\infty]$ is given by

$$\gamma(y) = \begin{cases} 0 & -\alpha \leq y \leq \alpha \text{ a.e. } \Gamma, \\ +\infty & \text{otherwise.} \end{cases}$$

If control constraints $u \in U_{ad}$ are imposed in (P_3), they become $u - w \in U_{ad}$ in (P_4).

3. BANG-BANG CONTROL

In this section we discuss state constrained control problems governed by variational inequalities. First we give a new equivalence result, next we obtain the optimality conditions, and finally we deduce the existence of bang-bang controls.

For the sake of simplicity we deal with a more specific case:

(Q) $\quad$ Min { F(y) + G(u) } , (3.1)

$$\partial y/\partial t - \Delta y + \beta(y) \ni u \quad \text{in } \Omega \times]0,T[, \tag{3.2}$$

$$y(0,x) = y_o(x) \quad \text{in } \Omega , \tag{3.3}$$

$$y(t,x) = 0, \quad \text{in } \Gamma \times [0,T] \tag{3.4}$$

and the constraints:

$$y(t,x) \leq b \quad \text{a.e. } \Omega \times]0,T[, \tag{3.5}$$

$$c \leq u(t,x) \leq d \quad \text{a.e. } \Omega \times]0,T[. \tag{3.6}$$

The G, F, y_o are as in the preceding section and $\beta \subset \mathbf{R} \times \mathbf{R}$ is the maximal monotone graph

$$\beta(y) = \begin{cases}]-\infty, 0] & y = a, \\ 0 & y > a, \\ \emptyset & y < a. \end{cases} \tag{3.7}$$

We assume that a, b, c, d are real constants such that $0 \in [c,d]$ (the null control is allowed) and $0 \in [a,b]$ (compatibility with boundary data).

The usual admissibility assumption implies $y_o(x) \in [a,b]$ a.e. Ω. If $y_o \in W^{2,\infty}(\Omega)$, by the general theory of parabolic variational inequalities [6], (3.2)-(3.4) has a unique solution $y \in W^{2,1,\infty}(\Omega \times]0,T[)$. It is easy to infer the existence of at least one optimal pair for the problem (Q).

Let $\alpha \subset \mathbf{R} \times \mathbf{R}$ be the maximal monotone graph given by

$$\alpha(y) = \begin{cases} 0 & y < b, \\ [0, +\infty[& y = b, \\ \emptyset & y > b. \end{cases} \tag{3.8}$$

Obviously $\alpha + \beta \subset \mathbf{R} \times \mathbf{R}$ is maximal monotone. We associate with the problem (Q) the problem

(Qa) $\quad$ Min { F(y) + G(u) } ,

$$\partial y/\partial t - \Delta y + \alpha(y) + \beta(y) \ni u \quad \text{in } \Omega \times]0,T[, \tag{3.9}$$

$$y(0,x) = y_o(x) \quad \text{in } \Omega, \qquad y(t,x) = 0 \quad \text{in } \Sigma .$$

We start with the following fundamental lemma

LEMMA 3.1. *Assume that* $G(u) = h(|u|)$ ($|\cdot|$ *is the modulus) and* $h : L^2(\Omega \times]0,T[) \to \mathbf{R}$ *is nondecreasing with respect to the a.e. order on the positive cone. Then there is an optimal pair for the problem* (Qa), $[y^*, u^*]$ *being such that* $\alpha(y^*) = 0$ a.e. $\Omega \times]0,T[$.

Proof. If $y^*(t,x) < b$, then $\alpha(y^*(t,x)) = 0$ by (3.8). If $y^*(t,x) = b$, then $\beta(y^*(t,x)) = 0$ by (3.7) and due to the regularity of y^* it yields $\partial y^*/\partial t - \Delta y^* = 0$. Therefore, on the set where $y^* = b$ we have $\alpha(y^*) = u^* \geq 0$. But replacing both $\alpha(y^*)$, u^* by 0 on this set, by hypothesis, the cost would decrease since y^* remains unchanged.

Remark. If h is strictly nondecreasing, all the optimal pairs have this extremal property.

Remark. Similarly, in the problem (Q) one may show that $\beta(\hat{y}) = 0$, where $[\hat{y}, \hat{u}]$ is some optimal pair of (Q).

THEOREM 3.2. *If* h *is strictly nondecreasing the problems* (Q) *and* (Qa) *are equivalent. If* h *is nondecreasing, any solution of* (Q) *is also a solution of* (Qa).

Proof. Let $[y^*, u^*]$ be an optimal pair of (Qa). Obviously $y^* \leq b$ a.e. $\Omega \times]0,T[$ and *Lemma 3.1* shows that $[y^*, u^*]$ is admissible for (Q), too.

Let J, J_a be the cost functionals of (Q), (Qa). Then $J(y^*, u^*) = J_a(y^*, u^*) \geq J(\hat{y}, \hat{u})$, where $[\hat{y}, \hat{u}]$ is an optimal pair for (Q).

As $\hat{y} \leq b$ a.e. $\Omega \times]0,T[$, we may take $\alpha(\hat{y}) = 0$, and hence $[\hat{y}, \hat{u}]$ is admissible for (Qa) with the same cost

$$J_a(\hat{y}, \hat{u}) = J(\hat{y}, \hat{u}) \geq J_a(y^*, u^*).$$

It yields that (Q) and (Qa) have the same optimal values and the same optimal pairs.

Remark. Under appropriate assumptions on G we may see that the problem (Q) is, in fact, equivalent (since $\beta(\hat{y}) = 0$ a.e. $\Omega \times]0,T[$) with

$$\text{Min} \{ F(y) + G(u) \},$$

$$\partial y/\partial t - \Delta y = u \qquad \text{in } \Omega \times]0,T[,$$

$$y(0,x) = y_0(x) \quad \text{in } \Omega, \qquad y(t,x) = 0 \quad \text{in } \Sigma,$$

$$a \leq y(t,x) \leq b \qquad \text{a.e. } \Omega \times]0,T[,$$

$$c \leq u(t,x) \leq d \qquad \text{a.e. } \Omega \times]0,T[,$$

See [4] for a general discussion of such problems.

If F or G are strict convex functions, then (Q) itself is strictly convex and we get the *uniqueness* of the optimal pair. The convexity of control problems governed by variational inequalities is a consequence of the results of Saguez and Bermudez [9] and of Mignot and Puel [8].

We return to the problem (Qa). In the same way as in Friedman [5], Barbu [1], we get the optimality conditions for (Qa) and, consequently, for (Q). For simplicity, we take G = 0.

THEOREM 3.3. *There is* $p^* \in L^\infty(0,T; L^2(\Omega)) \cap L^2(0,T; H_0^1(\Omega))$ *and* $\rho \in D'(\Omega \times]0,T[)$, $\text{supp}\, \rho \subset M = \{(t,x) \in]0,T[\times \Omega;\ y^*(t,x) \in \{a,b\}\}$ *(the coincidence set) such that:*

$$\partial p^*/\partial t + \Delta p^* - \rho = \partial F(y^*), \tag{3.10}$$

$$\int_0^T \int_\Omega p^*(u^* - v) \geq 0, \quad \forall v \in L^2(\Omega \times]0,T[), \quad c \leq v \leq d. \tag{3.11}$$

Moreover, $p^* \in W^{2,1,\infty}_{loc}(E)$, *where* E *is the complement of* M *in* $\Omega \times]0,T[$.

Remark. Optimality conditions for state constrained control problems governed by variational inequalities have been recently obtained by Zheng-Xu He [13] in a different form.

Now we are able to derive the structure of an optimal control u^* for the problem (Q). As shown by *Lemma 3.1*, we may take $u^* = 0$ a.e. M. By an appropriate choice of v in (3.11) we get

$$\iint_E p^*(u^* - v)dxdt \geq 0 \quad \forall v \in L^2(E),\ c \leq v \leq d.$$

Then obviously

$$u^* = d \text{ a.e. on } \{(t,x) \in E;\ p^*(t,x) > 0\},$$

$$u^* = c \text{ a.e. on } \{(t,x) \in E;\ p^*(t,x) < 0\}.$$

Assume that $F(y) = \frac{1}{2} |y|^2_{L^2(0,T;H)}$ and that the set $\{(t,x) \in E; p^*(t,x) = 0\}$ has positive measure. By (3.10) we get $y^* = 0$ on this set, therefore $u^* = 0$, too, due to *Lemma 3.1* and (3.9).

COROLLARY 3.4. *If* $G = 0$ *and* $F(y) = \frac{1}{2} |y|^2_{L^2(0,T;H)}$, *then* (Q) *has at least one bang-bang optimal control.*

Remark. It is possible to give examples with $G(u) = h(|u|)$ and h strictly nondecreasing. Then all the optimal controls are bang-bang.

REFERENCES

1) Barbu, V.. *Optimal Control of Variational Inequalities.* Research Notes in Mathematics 100. Boston-London-Melbourne: Pitman, 1984.
2) Barbu, V., and Precupanu, Th.. *Convexity and Optimization in Banach Spaces.* Dordrecht - Boston - Lancaster: D. Reidel Publishing Company, 1986.
3) Bonnans, J.F.. "Analysis and Control of Unstable Semilinear Distributed Systems". Ph. D. diss. (in French), Univ. de Technologie de Compiegne, 1982.
4) Bonnans, J.F., and Casas, E.. "On the Choice of the Function Spaces for some State Constrained Control Problems". *Numer. Funct. Anal. and Optimiz.* 7(1984-85): 333-48.
5) Friedman, A.. "Optimal Control for Parabolic Variational Inequalities". *SIAM J. Control and Optimization.* Forthcoming.
6) Friedman, A.. *Variational Principles and Free Boundary Problems.* New York: John Wiley and Sons, 1982.
7) Lions, J.L.. *Controle des Systemes Distribues Singuliers (Control of Singular Distributed Systems).* Paris: Dunod, 1983.
8) Mignot, F., and Puel, J.P.. "Controle Optimal d'un Systeme Gouverne par une Inequation Variationelle Parabolique" (Optimal Control of a System Governed by a Parabolic Variational Inequality). *C. R. A. S.* Paris 298, no.12 (1984): 277-80.
9) Saguez, C., and Bermudez, A.. "Optimal Control of Variational Inequalities". In *Proceedings of 23rd CDC* 249-51. Las Vegas, NV, 1984.
10) Tiba, D.. "Subdifferentials of Composed Functions and Applications in Optimal Control". *An. St. Univ. Iasi* XXIII, S.I, f.2, (1977): 381-86.
11) Tiba, D.. "Une Approche par Inequations Variationnelles pour les Problemes de Controle avec Contraintes" (A Variational Inequality Approach in Constrained Control Problems). *C. R. A. S.* 302, no.1 (1986): 29-31.
12) Tiba, D., and Neittaanmaki, P.. "A Variational Inequality Approach in Constrained Control Problems". *Appl. Math. and Optimiz.* Submitted.
13) Zheng-Xu He.. "State Constrained Control Problems Governed by Variational Inequalities". *SIAM J. Control and Optimization.* Forthcoming.

MATHEMATICAL METHODS FOR THE CONTROL OF INFECTIOUS DISEASES*

V. Capasso

Universita di Bari, Italy

1. INTRODUCTION

The control of epidemic systems is not usually an easy task since in real situations it is rather difficult to implement the control policies suggested by the mathematical analysis. Anyhow we can give suggestions to the public health authorities about the effects of a particular control policy with respect to others, and in this context the analysis and simulation of mathematical models may become a powerful tool in the hands of the above authorities [3 , 11 , 12].

An important class of epidemics is the one related to the man-environment-man (MEM) mechanism of transmission. For such a class, the epidemic is sustained by the environment via a positive feedback mechanism due to the infectious agent which directly or indirectly is produced by the infectious human population. The mathematical theory of MEM epidemic diseases is not yet well developed as such [1] , even if many attempts have been made to describe particular examples of such diseases (see e.g. the brilliant work that has been done on schistosomiasis and helminthic infections [14, and its references]). The interest of the author and his coworkers has been mainly devoted to the analysis of fecal-oral transmitted diseases,

---

* Work performed in the context of the Special Program "Control of Infectious Diseases" of the National Research Council (C.N.R.) of Italy.

such as cholera, typhoid fever, infectious hepatitis, etc. which may be regarded as a particular class of MEM diseases; anyhow we like to point out the general structure of these systems. In fact if we compare the basic model (see Section 2) proposed to describe the mechanism of transmission of fecal-oral transmitted epidemics in the European Mediterranean regions [7 , 9], with the basic models used to describe helminthic infections and schistosomiasis [ ] , it turns out that all these models can be considered as particular cases of positive feedback systems which can be defined, in the space homogeneous case, by the following ODE system

$$\frac{dz_1}{dt} = f_1(z_1,z_2)$$
$$\frac{dz_2}{dt} = f_2(z_1,z_2) \tag{1.1}$$

where $f = (f_1,f_2)'$ is a quasi-monotone nondecreasing vector field. This means that $f_1(z_1,z_2)$ is monotone non decreasing with respect to z_2, while $f_2(z_1,z_2)$ is monotone non decreasing with respect to z_1. Models of this kind may be extended to include more than only two interacting population with few technical modifications.

For this class of systems it is usually possible to introduce a threshold parameter which is of great use for the prevention of future epidemics: its value discriminates situations in which an emerging epidemic always tends to extinction, from situations in which it will tend to a nontrivial endemic level.

Relevant modifications of the threshold parameter are due to the time and space heterogeneity of an epidemic system; the first case is due for example to the seasonal variation of the parameters [8], while the second one is due to the spatial structure of the habitat with respect to the mechanisms of transmission of the infectious agent from the human population to the environment, and vice versa [7 , 10].

In Section 2 the basic model and the corresponding threshold parameter are discussed. In Section 3 a reaction-diffusion system which extends the above model to include spatial heterogeneities is presented. Finally in Section 4 an optimal control problem is considered for the above system, corresponding to the implementation of a program of sanitation of the environment; the cost function opposes the cost of the epidemic itself to the cost of the sanitation program [2].

The author wishes to emphasize at this point that the present paper contains a review of the results obtained in the different papers to which he has referred, and which are due to various authors collaborating with himself.

2. THE BASIC ODE MODELS

The basic mathematical model suggested in [7 , 9] is given by the following ordinary differential equation system

$$\begin{aligned} \frac{dz_1}{dt} &= - a_{11}z_1(t) + a_{12}z_2(t) \\ \frac{dz_2}{dt} &= - a_{22}z_2(t) + g(z_1(t)) \end{aligned} \qquad (2.1)$$

for $t > 0$, with the following interpretation; $z_1(t)$ denotes the (average) concentration of the infectious agent in the environment, at time t; $z_2(t)$ denotes the infective human population; $1/a_{11}$ is the mean lifetime of the agent in the environment; $1/a_{22}$ is the mean infectious period of the human infectives; a_{12} is the rate of transfer of the infectious agent from the human infective population to the environment; $g(z_1)$ is the "force of infection" of the human population due to the agent.

A more detailed account of the epidemiological meaning of the various terms in (2.1) can be found in [9].

It is assumed that a_{11}, a_{22}, a_{12} are all positive quantities, and that $g: \mathbb{R}_+ \to \mathbb{R}_+$ is a twice continuously differentiable function satisfying the following assumptions:

(i) $0 < g(z') < g(z'')$, if $0 < z' < z''$; (ii) $g(0) = 0$; (iii) $g''(z) < 0$, for any $z > 0$; (iv) $0 < g'_+(0) < +\infty$; (v) $\lim_{z \to +\infty} \frac{g(z)}{z} < \frac{a_{11}a_{12}}{a_{12}}$.

Due to the assumptions (i)-(v) on $g(z)$, system (2.1), which always has the trivial equilibrium solution, may have or not a nontrivial equilibrium solution depending upon the value of the following "threshold parameter"

$$\theta := \frac{g'_+(0)\, a_{12}}{a_{11}a_{22}} \tag{2.2}$$

It can in fact be shown that

(a) if $0 < \theta < 1$, then system (2.1) admits only the trivial equilibrium solution in the positive quadrant $\mathbb{R}_+ \times \mathbb{R}_+$, which is globally asymptotically stable; (b) if $\theta > 1$ the origin is unstable, while a nontrivial steady state appears which is globally asymptotically stable in $\mathbb{R}_+ \times \mathbb{R}_+ - \{0\}$.

These facts have an obvious epidemiological interpretation; when $\theta < 1$ any epidemic eventually tends to extinction.On the other hand, if $\theta > 1$ then a nontrivial endemic state appears to which any epidemic eventually tends. Expression (2.2) may suggest preventive measure to maintain θ below one. In particular we shall refer to it when considering the optimal control of the epidemic.

It may be worth mentioning here that similar considerations arise when considering other infectious diseases of the MEM type. If we refer for example to schistosomiasis [14], and denote by $z_1(t)$ the human infective population, and by $z_2(t)$ the snail infective population, the ODE system modelling the epidemic is the following

$$\frac{dz_1}{dt} = -\psi z_1(t) + \alpha z_2(t)(1-z_1(t))$$

$$\frac{dz_2}{dt} = -\delta z_2(t) + \beta z_1(t)(1-z_2(t)) \qquad (2.3)$$

for $t > 0$.

For this model the threshold parameter is $\theta = \frac{\alpha\beta}{\psi\delta}$. It can be easily observed that both model (2.2) and model (2.3) belong to the class (1.1). We shall refer from now on to the case (2.1), for the sake of simplicity.

3. A REACTION-DIFFUSION MODEL WITH BOUNDARY FEEDBACK

In a real situation such as the one related to fecal-oral diseases in a town on the Mediterranean shore, spatial heterogenities must be taken into account.

In these areas, the sewage produced by the human infective population usually goes untreated into the sea which is then contaminated; it is likely to think that somehow the infectious agent is sent back to the habitat via some diffusion mechanism, infecting then other human population due to its particular eating habits.

A mathematical model which takes into account random dispersal of the infectious agent in the habitat is the following [5,10] modification of system (2.1),

$$\frac{\partial u_1}{\partial t}(x;t) = \Delta u_1(x;t) - a_{11}u_1(x;t)$$

$$\frac{\partial u_2}{\partial t}(x;t) = -a_{22}u_2(x;t) + g(u_1(x;t)) \qquad (3.1)$$

with $x \in \Omega$, $t > 0$. The habitat is mathematically represented by Ω , an open bounded subset of $\mathbb{R}^2$, whose boundary $\partial\Omega$ is assumed to be sufficiently smooth. $u_1(x;t)$ is the concentration of the infectious agent at point $x \in \Omega$ and time $t \geq 0$; $u_2(x;t)$ is the spatial density at time $t \geq 0$ of the human infective population.

As far as the boundary conditions are concerned we shall assume first that the boundary $\partial\Omega$ is made of two disjoint parts Γ_1 and Γ_2 ($\partial\Omega = \Gamma_1 \cup \Gamma_2$), such that Γ_1 represents the sea shore and Γ_2 the boundary on the land side. Hence the feedback mechanism of the infectious agent due to the human infective population in the habitat is assumed to occur at the boundary Γ_1 as follows

$$\frac{\partial}{\partial\nu} u_1(x;t) + \alpha(x)u_1(x;t) = \int_\Omega K(x,x')u_2(x';t)dx' \qquad (3.2a)$$

for $x \in \Gamma_1$, $t > 0$; moreover we assume complete isolation along Γ_2:

$$\frac{\partial}{\partial\nu} u_1(x;t) = 0 \qquad (3.2b)$$

for $x \in \Gamma_2$, $t > 0$.

Here $\partial/\partial\nu$ denotes the outward normal derivative

The function $\alpha : \Gamma_1 \to \mathbb{R}_+$ and the kernel $K : \Gamma_1 \times \Omega \to \mathbb{R}_+$ are sufficiently smooth functions.

As far as the existence of a solution of system (3.1), (3.2a), (3.2b), subject to suitable initial conditions

$$\begin{aligned} u_1(0;x) &= u_1^o(x) \\ u_2(0;x) &= u_2^o(x) \end{aligned} \qquad (3.3)$$

for $x \in \Omega$, is concerned, it can be shown [6], for the time homogeneous case, that under suitable regularity assumptions on the data the above initial value problem admits a unique classical solution $\{u(t),\ t \in \mathbb{R}_+\}$, $u(t) = (u_1(t), u_2(t))'$ in the sense that

$$u \in (C^{1,2}((0,+\infty) \times \Omega, \mathbb{R}) \cap C^{0,1}((0,+\infty) \times \bar{\Omega}, \mathbb{R})) \times$$

$$\times C^{1,0}((0,+\infty) \times \bar{\Omega}, \mathbb{R}).$$

The asymptotic behaviour of system (3.1), (3.2a), (3.2b) can be analyzed in the ordered Bonach space $X:=C(\bar{\Omega}) \times C(\bar{\Omega})$, with supnorm and partial order induced by the positive cone $X_+:=\{u \in X,\ u=(u_1,u_2)' \mid u_1(x) \geq 0,\ u_2(x) \geq 0 \text{ in } \bar{\Omega}\}$.

To do this we consider the linear operator

$$Au:= (A_1u,\ A_2u)', \qquad u = (u_1,u_2)' \tag{3.4a}$$

with

$$\begin{aligned} A_1u &:= \Delta u_1 - a_{11}u_1 &&, \text{ in } \Omega \\ Bu_1 &= Hu_2 &&, \text{ in } \partial\Omega \\ A_2u &:= a_{21}u_1 - a_{22}u_2 &&, \text{ in } \bar{\Omega} \end{aligned} \tag{3.4b}$$

It can be shown that [10]

PROPOSITION 3.1. - The operator A admits a dominant simple real eigenvalue $\lambda_1 > -a_{22}$ $(\forall \lambda \in \sigma(A): \mathrm{Re}\lambda \leq \lambda_1)$ in X; the associated eigenvector $\Phi \in X$ can be chosen to be strongly positive, i.e. $\Phi >> 0$ in $\bar{\Omega}$, and with norm $\|\Phi\|_X = 1$.

For system (3.1), (3.2a), (3.2b) a threshold theorem can be proved which suggests a new threshold parameter.

THEOREM 3.1. [10] . Under the "basic" assumptions on the parameters, if the dominant eigenvalue λ_1 of A is negative, then the trivial solution is globally asymptotically stable in X_+ for the evolution system. If $\lambda_1 > 0$ then it is unstable.

An explicit estimate of the sign of λ_1 can be given in the particular case in which $\alpha(x) \equiv 0$.

<u>COROLLARY 3.1.</u> [10]. Under the above assumptions if furthermore in (3.2a), $\alpha(x) \equiv 0$ then

(a) $\lambda_1 < 0$ if $g'_+(0) \sup_{y\in\Omega} \int_{\partial\Omega} K(x,y)d\sigma(x) < a_{11}a_{22}$ (3.5a)

while

(b) $\lambda_1 > 0$ if $g'_+(0) \inf_{y\in\Omega} \int_{\partial\Omega} K(x,y)d\sigma(x) > a_{11}a_{22}$ (3.5b)

Moreover if

$$g'_+(0) \inf_{y\in\Omega} \int_{\partial\Omega} K(x,y)d\sigma(x) > a_{11}a_{22} \qquad (3.6a)$$

and

$$g'(+\infty) \sup_{y\in\Omega} \int_{\partial\Omega} K(x,y)d\sigma(x) < a_{11}a_{22} \qquad (3.6a)$$

then a strictly positive steady state an endemic state exists which is globally asymptotically stable in $X_+ -\{0\}$ for the evolution system (here $g'(+\infty) = \lim_{z\to+\infty} \frac{g(z)}{z}$).

We may compare the new "threshold parameter" suggested by (3.6a)

$$\tilde{\theta} := \frac{g'_+(0) \inf_{y\in\Omega} \int_{\partial\Omega} K(x,y)d\sigma(x)}{a_{11}a_{22}}$$

with the one defined in (2.2) for the space homogeneous case.

4. THE OPTIMAL CONTROL PROBLEM *(in collaboration with V. Arnautu and V. Barbu).*

An optimal control problem arises if one wishes to reduce the epidemic phenomenon described by the above model by reducing the boundary feedback along the sea shore, i.e. by reducing the "strength" of the kernel $K(x,y)$, $x \in \Omega$, $y \in \partial\Omega$.

This corresponds to the implementation of a sanitation program by means of a treatment of the sewage before sending it to the sea

[11]. The sanitation program implies a cost that has to be compared with the cost of the epidemic itself.

In the control problem we shall assume that the kernel K has a time varying strength, and in fact that it has the following structure

$$K(t,x,y) = \sum_{i=1}^{N} w_i(t) K_i(x,y) \tag{4.1}$$

for $t \in [0,T] \subset \mathbb{R}_+$, $x \in \Omega$, $y \in \Gamma_1$.

If we denote then by $w(t) := (w_1(t),\dots,w_N(t))$, the optimal control problem we consider will be the following

<u>PROBLEM (P)</u> : For any fixed $T > 0$, minimize

$$\int_0^T \int_\Omega f(u_2(t,x))\,dx\,dt + \int_0^T h(w(t))\,dt + \int_\Omega \ell(u_2(T,x))\,dx \tag{4.2}$$

for all (u_1, u_2, w) subject to the state system

$$\begin{aligned} &\frac{\partial u_1}{\partial t} - \Delta u_1 + a_{11} u_1 = 0 && \text{in } Q := (0,T) \times \Omega \\ &\frac{\partial u_2}{\partial t} - a_{22} u_2 - g(u_1) = 0 && \text{in } Q \end{aligned} \tag{4.3}$$

$$\frac{\partial u_1}{\partial \nu} + \alpha u_1 = K * u_2 := \int_\Omega K(t,x,y) u_2(t;x)\,dx, \text{ on } \Sigma_1 = (0,T) \times \Gamma_1 \tag{4.4a}$$

$$\frac{\partial u_1}{\partial \nu} = 0 \quad , \text{ on } \Sigma_2 = (0,T) \times \Gamma_2 \tag{4.4b}$$

$$\begin{aligned} &u_1(0;x) = u_1^0(x) && , \text{ in } \Omega \\ &u_2(0,x) = u_2^0(x) && , \text{ in } \Omega \end{aligned} \tag{4.5}$$

The following assumptions will be in effect throughout in the sequel:

$$K_i \in L^\infty(\Omega \times \Gamma_1), \quad w_i \in L^\infty(0,T), \quad i=1,\dots,n \qquad (4.6)$$

(H1) $f,g,\ell \in C^1(\mathbb{R})$, $f,g \geq 0$ in $\mathbb{R}$; $|g(r)| \leq c_1|r|+c_2$, $r \in \mathbb{R}_+$

(H2) $h: \mathbb{R}^N \to]-\infty,+\infty] \equiv \bar{\mathbb{R}}$ lower semicontinuous; $\exists M \in \mathbb{R}^N$, M bounded and closed s.t. $h(w)=+\infty$ for $w \in M$.

EXAMPLE 4.1. $h(w) = \sum_{i=1}^{N} h_i(w_i)$ with $h_i(r) = \lambda/r^2$ if $0 < r \leq a$, $h_i(r) = +\infty$ if $r > a$.

By classical results [13] it can be shown that

PROPOSITION 4.1. - Under the above assumptions, if $\overset{\circ}{u}_1, \overset{\circ}{u}_2 \in L^\infty(\Omega)$, the control system (4.3), (4.4), (4.5) admits a unique solution (u_1, u_2) satisfying $u_1 \in L^2(0,T;H^1(\Omega)) \cap C([0,T];L^2(\Omega)) \cap L^\infty(Q)$; $\frac{\partial u_1}{\partial t} \in L^2(0,T;(H^1(\Omega))')$; $u_2 \in C^1([0,T]; L^\infty(\Omega))$. If in addition $\overset{\circ}{u}_1 \in H^2(\Omega)$ and w is Lipschitzian then $\frac{\partial u_1}{\partial t} \in L^2(Q)$.

It has been shown [2 , 4] that

PROPOSITION 4.2. - Problem (P) admits at least one solution (u_1^*, u_2^*, w^*), with u_1^*, u_2^* as in Proposition 4.1 and $w^* \in L^\infty(0,T; \mathbb{R}^N)$.

For the optimally conditions we have the following theorem [2].

THEOREM 4.1. - Let (u_1^*, u_2^*, w^*) be optimal in Problem (P). Then there exist

$p_1 \in L^2(0,T;H^1(\Omega)) \cap C([0,T];L^2(\Omega))$ with $\frac{\partial p_1}{\partial t} \in L^2(0,T;(H^1(\Omega))')$

and $p_2 \in C([0,T]; L^2(\Omega))$ with $\frac{\partial p_2}{\partial t} \in L^2(0,T; L^2(\Omega))$

such that

$$\frac{\partial p_1}{\partial t} + \Delta p_1 - a_{11}p_1 + g_1'(u_1^*)p_2 = 0 \quad , \text{ in } Q$$

$$\frac{\partial p_2}{\partial t} - a_{22}p_2 + \int_{\Gamma_1} K^*(t,x,y)p_1(t,y)\,d\sigma(y) = f'(u_2^*), \quad \text{in } Q$$

$$p_1(T,x) = 0 \qquad , \quad x \in \Omega$$

$$p_2(T,x) = -\ell'(u_2^*(T,x)) \qquad , \quad x \in \Omega$$

$$\frac{\partial p_1}{\partial \nu} + \alpha p_1 = 0 \qquad , \quad \text{in } \Sigma_1$$

$$\frac{\partial p_1}{\partial \nu} = 0 \qquad , \quad \text{in } \Sigma_2$$

and

$$w^*(t) = (\partial h)^{-1}\left(\int_{\Omega\times\Gamma_1} K_i(x,y)p_1(t,y)u_2^*(t,x)\,dx\,d\sigma(y)\right)_{i\in\{1,\dots,N\}} \qquad (4.7).$$

Here ∂h is the subdifferential of h [4].

If we choose h as in Example 4.1, then (4.7) gives

$$w_i^*(t) = \begin{cases} -\left(\int_\Omega K_i(x,y)p_1(t,y)u_2^*(t,x)\,dx\,d\sigma(y)\right)^{-1/3}(2\lambda)^{1/3} & \text{if} \quad \int_\Omega K_i p_1 u_2^*\,dx\,d\sigma(y) < \dfrac{2\lambda}{a^3} \\ a & \text{if} \quad \int_\Omega K_i p_1 u_2^*\,dx\,d\sigma(y) > \dfrac{2\lambda}{a^3}, \end{cases}$$

for $i = 1,\dots,N$.

In paper [2] the authors apply the above procedure to particular cases and suggest algorithms for the numerical evaluation of the optimal control. Moreover the following state identification problem has been faced. Usually the initial state $u_1^0(x)$, $x \in \Omega$ which gives the initial distribution of the infectious agent in the habitat cannot be measured directly; anyhow continuous measures can be obtained of the above distribution along the sea shore. We may suppose then that $u_1(t,x) = \eta(t,x)$, $(t,x) \in \Sigma_1 = [0,T]\times\Gamma_1$ is given. A least square procedure to identify $u_1^0(x)$ based on the knowledge of $\eta(t,x)$ has been proposed (see[2]for details).

REFERENCES

[1] Anderson, R.M. and May, R.M., eds. *Population Dynamics of Infectious Diseases Agents.* Dahlem Konferenzen. Heidelberg: Springer-Verlag, 1982.

[2] Arnautu, V., Barbu, V., Capasso, V., Controlling the spread of a class of epidemics. Submitted.

[3] Bailey, N.T.J. *The Mathematical Theory of Infectious Diseases.* London: Griffin 1975.

[4] Barbu, V. *Optimal Control of Variational Inequalities,* Research Notes in Mathematics 100 London: Pitman, 1983.

[5] Capasso, V. and Kunisch, K. A reaction-diffusion system modelling man-environment epidemics. *Annals of Differential Equations* (R.P. China). 1 (1985): 1-12.

[6] Capasso, V. and Kunisch, K. A nonlinear semigroup in L^1 associated with a reaction-diffusion system with positive feedback through the boundary. To appear.

[7] Capasso, V. and Maddalena, L. Convergence to equilibrium states for a reaction-diffusion system modelling the spatial spread of a class of bacterial and viral diseases. *J. Math. Biology* 13 (1981): 173-184.

[8] Capasso, V. and Maddalena, L. Periodic solution for a reaction- diffusion system modelling the spread of a class of epidemics. *SIAM J. Appl. Math.* 43 (1983): 417-427.

[9] Capasso, V. and Paveri-Fontana, S.L. A mathematical model for the 1973 cholera epidemic in the European Mediterranean region. *Rev. Epidem. Santé Publ.* 27 (1979): 121-132.

[10] Capasso, V. and Thieme, H. A threshold theorem for an epidemic system with a boundary feedback. To appear.

[11] Cvjetanovic, B., Grab B. and Uemura, K. *Dynamics of Acute Bacterial Diseases. Epidemiological Models and their Application in Public Health.* Suppl. N°1 to Vol. 56 of the *Bulletin of the World Health Organization.* Geneve: WHO, 1978.

[12] Hethcote, H.W. and Yorke, J.A. *Gonorrhea Transmission Dynamics and Control*. Lecture Notes in Biomathematics, 56 Heidelberg: Springer-Verlag, 1984.

[13] Lions, J.L. and Magenes, E. *Problèmes aux Limites non Homogènes et Applications*. Paris: Dunod, 1968.

[14] Nåsell, I. *Hybrid Models of Tropical Infections*. Lect. Notes in Biomathematics, 59. Heidelberg: Springer-Verlag, 1985.

SOLUTION OF THE 3-D STATIONARY EULER EQUATION BY OPTIMAL CONTROL METHODS

T. Chacon

University of Sevilla, Spain

O. Pironneau

University of Paris; INRIA, Rocquencourt, France

Abstract

An hyperbolic nonlinear PDE arising in turbulence is solved by the techniques of optimal control theory because of its non standard boundary conditions. Because there are 4 unknown functions of 3 variables, this method of solution yieldsextremely large optimal control problems. Conjugate Gradient Algorithms and Finite Element Discretization where employed with satisfactory results. However, a Quasi-Newton method failed to improve computer time.

INTRODUCTION

Nonlinear PDE's may be solved by least-squares provided one works with the right norms. This is demonstrated by the simple example below :

Consider the problem on a domain Ω :

$$-\Delta\phi = f(\phi) \qquad \phi \in H^1_0(\Omega) \tag{1}$$

it is equivalent to the least-square problem

$$\min_{\phi\in H^1_0(\Omega)} \| \Delta\phi+f(\phi) \|^2_{H^{-1}} \tag{2}$$

It is also equivalent to

$$\min_{\phi\in H^2(\Omega)\cap H^1_0(\Omega)} \| \Delta\phi+f(\phi) \|^2_0 \tag{3}$$

when (1) has a solution in $H^2(\Omega)$. However (3) is a numerically dangerous form to use because its optimality conditions involve a biharmonic problem :

$$(\Delta + f'(\phi).(\Delta\phi+ f(\phi))= 0 \quad \text{in} \quad \Omega \tag{4}$$

$$\phi= 0, \quad \Delta\phi + f(\phi) = 0 \quad \text{on } \partial\Omega \quad ; \tag{5}$$

so numerical methods based on (3) must be so that $\Delta\phi+f(\phi) = 0$ on $\partial\Omega$ least (1) will not be recovered.

Successful implementation of this technique of "abstract" least-square can be found in [2], [4], [6], for the Navier-Stokes equations and the transonic equation.

Problems like (2) are easy to transform into Optimal Control Problems by introducing ε solution of

$$-\Delta\varepsilon = \Delta\phi + f(\phi), \qquad \varepsilon \in H_0^1(\Omega). \tag{6}$$

Then it is easy to show that (2) is also

$$\min_{\phi\in H_0^1(\Omega)} \{ \|\nabla\varepsilon\|_0^2 \ : \ \varepsilon \text{ solution of (6)} \} \tag{7}$$

Optimization algorithms can be used to solve (7) but again experience shows that considerable speed up is achieved when these algorithms are set up with the natural scalar product of the optimization space, here $H_0^1(\Omega)$. Thus, in this example an iterative process of the type

$$\phi^{n+1} = \phi^n + \rho(\Delta\varepsilon^n + f'(\phi^n)\varepsilon^n) \tag{8}$$

requires typically a few thousand iterations while

$$\phi^{n+1} = \phi^n + \rho(-\Delta)^{-1}(\Delta\varepsilon^n + f'(\phi^n)\varepsilon^n) \tag{9}$$

takes less than a hundred. This is another reason for not using (3) because it would require the use of Δ^{-2}, an expensive item.

In turbulence theory [7] the following problem arise :

Find $u \in L^4(R^3)^3$, $p \in L^2(R^3)$ periodic on $X =]-\frac{1}{2}, \frac{1}{2}[^3$ such that

$$u\nabla u + C\nabla p = 0 \qquad \nabla.u = 0 \quad \text{in} \quad X \tag{10}$$

$$\int_X u^2 = 1 \qquad\qquad \int_X u.\nabla\times u = r \tag{11}$$

where C is a symmetric constant 3×3 matrix.
The abstract least-square technique was never tried on hyperbolic problems, this is the purpose of this paper. Our plan will be

1 To study the transformation of (10) into an optimal control problem.
2 To study the discretization and the algorithm
3 To report on the numerical results.

1. THE OPTIMAL CONTROL PROBLEM

We shall only consider the case r = 0 in (11) and replace this constraint by "u odd in X". Thus the problem is now :

Find a vector valued function u and a scalar value function p such that

$$u\nabla u + C\nabla p = 0 \qquad \nabla.u = 0 \quad \text{in } X =]-\frac{1}{2}, \frac{1}{2}[^3 \tag{1.1}$$

u and p are periodic on X and (1.2)

$$u(-x) = -u(x) \ , \ p(-x) = p(x) \quad \forall x \in X \tag{1.3}$$

$$\| u \|_0^2 = \int_X |u|^2 dx = 1 \tag{1.4}$$

This problem will be solved by a general least-square method, so we shall consider

$$\min_{u \in W^{s,4}} \{ \| \nabla.(u \otimes u) + C\nabla p \|^2_{s-1} : -\nabla.(C\nabla p) = \nabla.(u\nabla u) \quad \text{in} \quad X \tag{1.5}$$
$$\| u \|_0^2 = 1, \ u \text{ odd}, \ u \text{ X-periodic} \}$$

Here $W^{s,4}$ is the usual Sobolev space of order s defined from $L^4(X)^3$ and $\|.\|_{s-1}$ is the $H^{s-1}(X)$-norm.

To set up an optimal control problem like (1.5), one can play with the following options :

- The choice of the space in which u will minimize the functional, here $W^{s,4}$.
- The choice of a norm for the least-square of (1.1), here $^{s-1}$.

The PDE in (1.5) is a convenient way to insure the solenoïdal condition on u ; indeed if

$$\nabla.(u \otimes u) + C\nabla p = 0 \tag{1.6}$$

$$-\nabla. \ C\nabla p = \nabla.(u\nabla u) \tag{1.7}$$

then

$$u\nabla(\nabla.u) + |\nabla.u|^2 = 0 \tag{1.8}$$

and this equation can be integrated on the stream lines to give : $\nabla.u = 0$.

Notice that $u \in W^{s,4}$ implies $\nabla.(u \otimes u) \in H^{s-1}(X)$ and $\nabla p \in H^{s-1}(X)$ if C is positive definite.

In the case s = 0, problem (1.5) can be transformed further.

Let $\varepsilon \in H^1(X)^3$ be the periodic solution of

$$- \Delta\varepsilon = \nabla.(u \otimes u) + C\nabla p \tag{1.9}$$

then (1.5) is equivalent to

$$\min_{u \in L^4(X)^3} \int_X |\nabla\varepsilon|^2 dx : -\nabla. \ C\nabla p = \nabla.(u\nabla u) \ ; \tag{1.10}$$
$$-\Delta\varepsilon = \nabla.(u \otimes u) + C\nabla p \ ;$$
$$u \text{ odd} \ ; \ p, \ \varepsilon, \ u \ : \text{X-periodic}, \ \|u\|_0 = 1$$

because

$$\| \Delta\varepsilon \|_{-1} = \sup_{\eta \in H^1} \int_X \eta\Delta\varepsilon / \| \nabla\eta \|_0 = \sup_{\eta \in H^1} \int_X \nabla\eta\nabla\varepsilon / \| \nabla\eta \|_0 = \| \nabla\varepsilon \|_0$$

Now problem (1.10) has a serious difficulty due to the fact that $L^4(X)$ is not a Hilbert space and so differentiable optimization methods will fail. Thus we make an illegal approximation and replace $L^4(X)^3$ by $H^1(X)^3$. However, problem (1.10) so modified will be well posed if (1.1)-(1.4) has a solution in $H^1(X)^3$.

The numerical solution of (1.10) will be found by a gradient algorithm. If $E(\varepsilon(u))$ denotes the cost function in (1.10) then its H^1-gradient is E_1' such that

$$\int_X \nabla E_1'.\nabla\delta u = E(\varepsilon(u+\delta u))-E(\varepsilon(u))+o(\|\delta u\|) \quad \forall\delta u \in H^1(X)^3 \qquad (1.11)$$

To compute it, we denote $\delta\varepsilon = \varepsilon(u+\delta u)-\varepsilon(u)$ and proceed as usual [6][9]:

$$E(\varepsilon+\delta\varepsilon)-E(\varepsilon) = 2\int_X \nabla\varepsilon\,\nabla\delta\varepsilon + o(\|\delta\varepsilon\|_1) \qquad (1.12)$$

$$\int_X \nabla\varepsilon\,\nabla\delta\varepsilon = \int_X \nabla.(\delta u\otimes u + u\otimes\delta u)\varepsilon + \int_X C\nabla\delta p.\varepsilon + o(\|\delta u\|_1) \qquad (1.13)$$

The last integral requires to set up an adjoint ; let q be the periodic solution of

$$-\nabla.C\nabla q = \nabla.\,C\varepsilon \qquad (1.14)$$

then

$$\int_X C\nabla\delta p\varepsilon = -\int_X C\nabla q.\nabla\delta p = \int_X \nabla q.(u\nabla\delta u + \delta u\nabla u) + o(\|\delta u\|_1) \quad (1.15)$$

So finally,

$$E(\varepsilon(u+\delta u))-E(\varepsilon(u)) = -2\int_X (\delta u\otimes u + u\otimes\delta u):\nabla\varepsilon + 2\int_X \nabla q.(u\nabla\delta u+\delta u\nabla u) + o(\|\delta u\|_1) \qquad (1.16)$$

To handle the nonlinear constraints $\|u\|_0 = 1$, we have two options :

a) Use the fact that problem (1.10) is homogeneous in u. If (u,p) is a solution of (1.6)-(1.7), then so is λu and $\lambda^2 p$. Thus one may replace (1.10) by

$$\min_{u\,H^1(X)^3}\left\{\int_X |\nabla\varepsilon|^2 dx/\|u\|_0^4 \;:\; (1.6)\text{-}(1.7),\ u \text{ odd},\ u,p,\varepsilon\ X\text{-periodic}\right\} \qquad (1.17)$$

b) Use penalty and replace $E(\varepsilon(u))$ by

$$\tilde{E}(\varepsilon(u),u) = E(\varepsilon(u)) + a\|u\|_0^2 + b\|u\|_0^{-2} \qquad (1.18)$$

where a,b, are any positive numbers. This special type of penalization is valid because E is homogeneous in $\bar{u}$.

Let u be a solution of (1.10), then it must satisfy (see (1.11)) for some μ

$$\int_X \nabla E_1'(\bar{u}).\nabla v + 2\mu \int_X \bar{u}.v = 0 \qquad \forall v \in H^1(X)^3 \tag{1.19}$$

on the other hand if (1.18) is used and yields a solution u^* , it must satisfy

$$\int_X \nabla E_1'(u^*)\nabla v + 2a \int_X u^* v - 2b \int_X u\ .v/ \|u^*\|_0^3 = 0 \quad \forall v \in H^1(X)^3$$

which is (1.19) with $\mu = 2a-2b/\|u^*\|_0^3$. So $u^*/\|u^*\|$ is a solution of (1.19).

2. DISCRETIZATION AND ITERATIVE SOLUTION

Let Q_h and V_h be finite dimensional approximations of the space of X-periodic functions of $H^1(X)$ and $H^1(X)^3$ respectively. Then (1.17) is approximated by

$$\min_{u_h \in V_h^3} \bar{E}(u_h) = \int_X |\nabla\varepsilon_h|^2 / (\int_X u_h^2)^2 \tag{2.1}$$

where ε_h and p_h are solutions of

$$\int_X \nabla w_h^T C \nabla p_h = -\int_X (u_h \nabla u_h) w_h \qquad \forall w_h \in Q_h \ ; \ p_h \in Q_h \tag{2.2}$$

$$\int_X \nabla\varepsilon_h . \nabla v_h = \int_X (u_{hi} u_{hj} + C_{ij}) v_{hi,j} \quad \forall v_h \in V_h \ ; \ \varepsilon_h \in V_h \tag{2.3}$$

As in (1.11)-(1.16) one shows that

$$\begin{aligned} \bar{E}(u_h + \delta u_h) - \bar{E}(u_h) &= [-2 \int (\delta u_h \otimes u_h + u_h \otimes \delta u_h) : \nabla\varepsilon_h \\ &+ 2 \int_X \nabla q_h . (u_h \nabla \delta u_h + \delta u_h \nabla u_h)] / (\int_X u_h^2)^2 - \\ &- 4 [\int_X |\nabla\varepsilon_h|^2 / (\int_X u_h^2)^3] \int_X u_h . \delta u_h + o(\|\delta u_h\|_1) \end{aligned} \tag{2.4}$$

where q_h is the solution of

$$\int_X C \nabla q_h \nabla w_h = -\int_X C \varepsilon_h \nabla w_h \qquad \forall w_h \in Q_h \ ; \ q_h \in Q_h \tag{2.5}$$

Since V_h is finite dimensional, it has a basis $\{v^1\}$ and by writing u_h on it, one gets

$$u_h(x) = \sum_{i=1}^{N} u_i \, v_i(x) \tag{2.6}$$

thus $\bar{E}$ is really a function of $\{u_i\}$ and one can compute $\partial\bar{E}/\partial u_i$ from (2.4) by replacing δu_h by vi and dropping the $o(\ )$ terms.

Thus, a conjugate gradient algorithm in H^1 has a main loop like

$$u_i^{n+1} = u_i^n + \rho^n d_i^n \tag{2.7}$$

where ρ^n is the stepsize, d^n the descent direction, g^n the gradient

$$\rho^n = \arg\min_{\rho} \bar{E}\ (u_i + \rho d_i^n) \tag{2.8}$$

$$d_i^n = -\, g_i^n + \gamma\, d_i^{n-1}\ ,\ \gamma = \|g^n\|^2 / \|g^{n-1}\|^2 \tag{2.9}$$

To compute g_i^n, one must solve a PDE : if $\|u_h\| = 1$ then

$$\int_X \nabla g_h \nabla v_h = -2 \int_X (v_h \delta u_h + u_h \delta v_h) : \nabla\varepsilon_h + 2 \int_X \nabla q_h \cdot (u_h \nabla v_h + v_h \nabla u_h) \tag{2.10}$$
$$- 4 \int_X u_h v_h \ ; \quad \forall v_h \in V_h \ ; \ g_h = \Sigma\, g_i^n v^i \in V_h$$

Finally, at each iteration, we rescale u_h so that $\|u_h\| = 1$; this is allowed because rescalling does not change E . When (1.18) is used (penalty) a similar algorithm can be derived ; only (2.10) changes ; however step (2.8) is more difficult to carry out because while $\bar{E}$ is a polynomial of degree 4 in ρ , $\tilde{E}$ is a rational fraction of polynomials.

3. IMPLEMENTATION AND RESULTS

The most natural choices for V_h and Q_h, spectral approximations, proved unfeasible. Indeed the standard trick to compute nonlinear terms like $u_h \nabla u_h$ is to do it in the physical space and use FFTs to convert the results ; however aliasing errors [8] must be controled here least the gradient algorithms would soon fail to find directions of descents. On the other hand, direct computation of $u_h \nabla u_h$ from the Fourier modes of u_h is possible in 2-d [1] but too costly in 3-d.

Thus, V_h and Q_h were constructed from the usual finite element spaces for the Navier-Stokes equations [11] :

$Q_h = \{q_h$ continuous X-periodic and piecewise linear on a triangulation of X : $\mathcal{T}_h\}$

$V_h = \{v_h$ continuous X-periodic and piecewise linear on the triangulation of X, $\mathcal{T}_{h/2}$ obtained by dividing each tetraedron into 6 subtetraedra whose vertices are either mid edges or vertices of $\mathcal{T}_h\}$.

To preserve oddness of odd functions care must be taken to choose symmetric triangulation with respect to the origin.

The Fletcher-Reeve conjugate gradient method [10] with preconditionning due to the choice of the H^1-scalar product as explained above (see also [4]) was used.

Both cost functions (1.17) and (1.18) where tested and a Buckley-Lenir [3] quasi-Newton method was also tested for comparison on (1.18).

Since an exact calculation of the step size was not implemented in connection with (1.18) better results were usually achieved with (1.17).

The numerical tests are explained in the captions of Figures 1 to 4 ; they were carried out on a 17^3 triangulation which gives a V_h of dimension $17^3 \times 3 = 9639$ and 729 for Q_h.

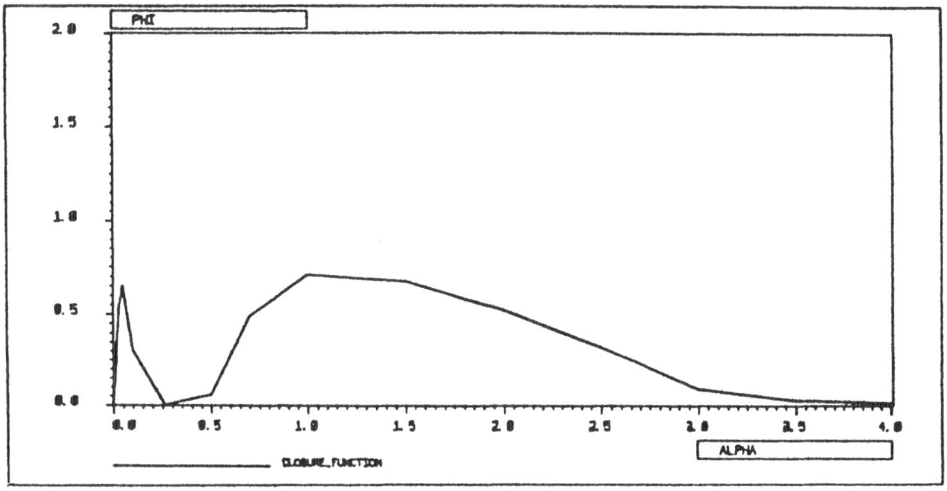

Figure 1

This graph shows $\int_X u_1 u_2 dx$ as a function of C_{12} when u is computed by (2.1). The smoothness of the curve (except near the origin) is an indication of the quality of the solution. Each run, i.e. each point on this curve, requires a few minutes of CRAY-1.

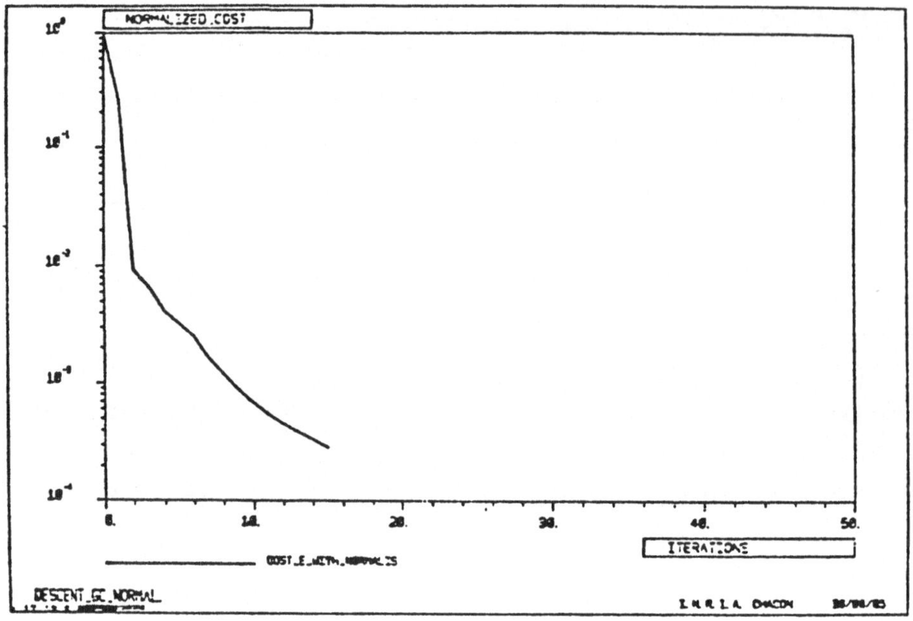

Figure 2

On this graph, we have plotted E given by (1.17) as a function of the number of iterations when a conjugate gradient method is used.

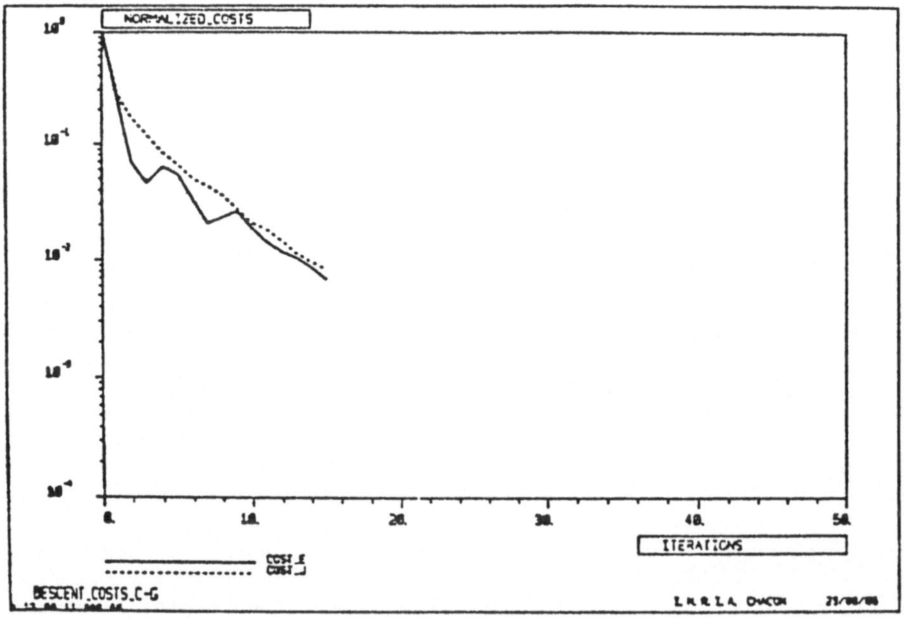

Figure 3

Same as above but with E given by (1.18).

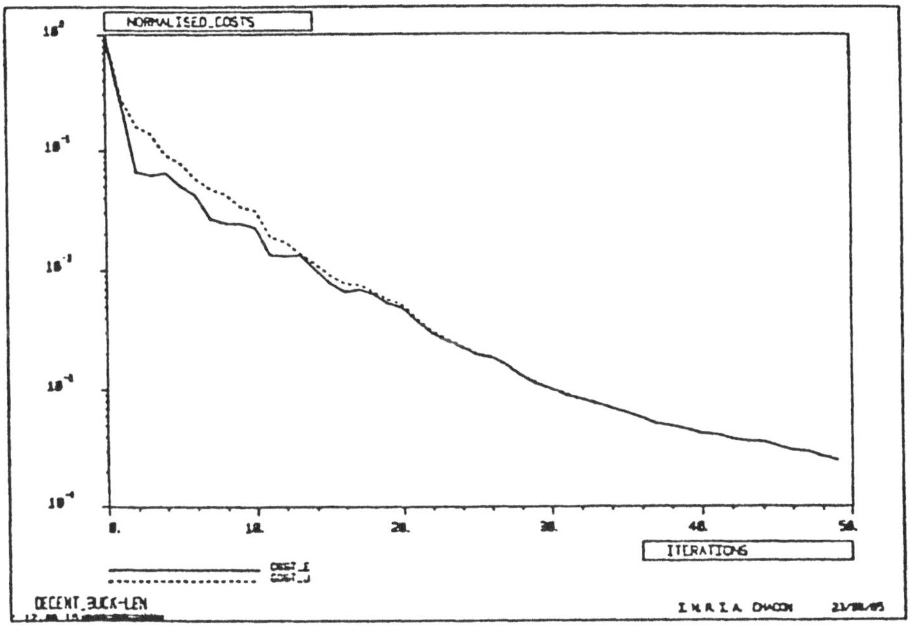

Figure 4

Same as above but with E given by (1.18) with a Buckley-Lenir Quasi-Newton method.

So in effect, we have solved a distributed elliptic optimal control problem, where the control space is of dimension 9639, the state space 729 ; the problem had also one nonlinear constraint on the control. It remains now to include the second constraints in (11).

References

[1] C. Bègue, O. Pironneau : Hyperbolic systems with periodic boundary conditions, Comp. & Maths with Appls., Vol. 11, Nos 1-3, pp.113-128, (1985).

[2] M.O. Bristeau, R. Glowinski, B. Mantel, J. Périaux, P. Perrier : Finite Element Methods for Solving the Navier-Stokes Equations for Compressible Unsteady Flows, Proc. of 5th International Conference on Finite Element and Flow Problems, University of Texas at Austin, U.S.A., 1984 (CAREY G.F. & ODEN J.T. Eds., pp. 449-462).

[3] A. Buckley, A. Lenir : ON-Like variable storage conjugate gradients. Mathematical Programming 27, 2, pp. 155-175 (1983).

[4] R. Glowinski : Numerical Methods for Nonlinear Variational Problems, Springer-Verlag, New-York, 1984.

[5] R. Glowinski, O. Pironneau : On a mixed finite element approximation of the Stokes problem (I). Numer. Math. 33, 397-424 (1979).

[6] J.L. Lions : Control Optimal des Systèmes gouvernés par des E.D.P. Dunod, Paris (1968).

[7] D.W. McLaughin, G. Papanicolaou and O. Pironneau : Convection of Microstructure and related problems. SIAM Appl. Math., Vol. 45, No. 5, Oct. 85.

[8] S.A. Orszag : Numerical simulation of the Taylor Green Vortex (Edited by R. Glowinski), Lecture Notes in Computer Sciences, Vol. 11, Part 2, p. 50, Springer Verlag, Berlin (1974).

[9] O. Pironneau : Optimal Shape design for elliptic systems. Springer Series in Comp. Physics, 1983.

[10] E. Polak : Computational Methods in Optimization. Academic Press (1971).

[11] F. Thomasset : Finite Element Solutions of the Navier-Stokes Equations. Springer Series in Comp. Physics (1980).

OUTPUT LEAST SQUARES STABILITY FOR ESTIMATION OF THE DIFFUSION COEFFICIENT IN AN ELLIPTIC EQUATION

F. Colonius*

Universität Frankfurt, The Federal Republic of Germany

K. Kunisch*

Technische Universität Graz, Austria

ABSTRACT

The estimation of unknown coefficients in partial differential equations is frequently studied as an output least squares problem involving an "observation" of the system for which the model is derived and the solution of the model equation as a function of the unknown parameter. We study the continuous dependence of the output least squares formulation on the observation of the system. There is no a-priori assumption on the uniqueness of the output least squares solutions.

OUTPUT LEAST SQUARES STABILITY

We study estimation of the diffusion coefficient $q = \mathrm{col}(q_1,\dots,q_n)$ in the elliptic equation

$$\begin{cases} -\sum_{i=1}^{n} (q_i u_{x_i})_{x_i} + c\,u = f \quad \text{in } \Omega \subset \mathbb{R}^n \\ u \mid \partial\Omega = 0 \end{cases} \tag{1}$$

where $f \in L^2$ and $c \in L^2$ with $c \geqq 0$. We assume that $n = 2$ or 3 but with the appropriate changes the case of arbitrary n can be treated with the same techniques. All function spaces are taken over the bounded domain Ω, which is assumed to have a smooth ($C^{2,-}$) boundary or to be a parallelepiped. Let $u = u(q)$ denote the solution of (1) corresponding to the diffusion coefficient q, and let $z^o \in L^2$ be

---

* Both authors acknowledge support from the Fonds zur Förderung der wissenschaftlichen Forschung, under grant S 3206.

an observation of the (e.g. physical) system for which (1) is a proposed model equation. Due to model and observation error there may or may not exist q^* in a set of admissible parameters Q_{ad} to be defined below, which satisfies $u(q^*) = z^o$. To estimate the unknown parameter q so that the corresponding solution $u(q)$ best fits the data we adopt the output least squares method

$$(OLS)_{z^o} \qquad \min |u(q) - z^o|^2$$

where q is chosen from the following set Q_{ad} of admissible parameters:

$$Q_{ad} = \{q \in Q : 0 < k_i(x) \leq q_i(x),\ x \in \Omega,\ i=1,\ldots,n,\ |q|_Q \leq \gamma\}$$

where $k_i \in H^2$ and $\gamma > |col(k_1,\ldots,k_n)|_Q$ are given. Here $Q = \bigotimes_{i=1}^{n} H^2$ is endowed with the Hilbert space product topology. Recall that $H^2 \subset C$ is a continuous embedding for $n \leq 3$.

Our objective in this note is to study the stability of the solution of $(OLS)_{z^o}$ on z^o. Solving $(OLS)_{z^o}$ involves the inversion of $\Phi : q \to u(q)$. Without strong assumptions on the problem data, Φ is not injective. Moreover, considering Φ^{-1} as a multivalued mapping, this inverse is not continuous unless the parameter space is endowed with a sufficiently weak topology. Along with $(OLS)_{z^o}$ we consider therefore the following regularized output least squares problem for $\beta \in \mathbb{R}$, $\beta > 0$:

$$(ROLS)^{\beta}_{z^o} \qquad \min |u(q) - z^o|^2 + \beta |q|^2_Q \quad \text{over } Q_{ad}.$$

Here and below we drop the index for the inner product and norm in L^2. The behaviour of the solutions of $(ROLS)^{\beta}_{z}$ as $\beta \to 0^+$ has been studied e.g.in [3-6].- We point out the fact that $(ROLS)^{\beta}_{z^o}$ has a solution q_{z^o} for every $\beta > 0$ without the necessity of a norm constraint in the definition of Q_{ad}, whereas $(OLS)_{z^o}$ may have no solution unless Q_{ad} is a bounded set. The results of this paper remain unchanged if the norm constraint in the definition of Q_{ad} is replaced by the assumption that $(OLS)_{z^o}$ has a solution.

We define the following stability concept for the solutions of $(ROLS)^{\beta}_{z^o}$.

DEFINITION. The parameter q is called Output Least Squares stable by Regularization (ROLS-stable) at $z^o \in L^2$ in Q_{ad} for β in the interval $I \subset (0,\infty)$ if for every $\beta \in I$ and every global solution $q^{\beta}_{z^o}$ of $(ROLS)^{\beta}_{z^o}$ there exists a constant $d > 0$ and neighborhoods $V(z^o)$ of z^o in L^2 and $V(q^{\beta}_{z^o})$ of $q^{\beta}_{z^o}$ in Q such that for all $z \in V(z^o)$ there exists a local solution $q^{\beta}_{z} \in V(q^{\beta}_{z^o})$ of $(ROLS)^{\beta}_{z}$ and for every solution $q^{\beta}_{z} \in V(q^{\beta}_{z^o})$ we have $|q^{\beta}_{z} - q^{\beta}_{z^o}|_Q \leq d\, |z-z^o|^{1/2}_{L^2}$.

Remark 1. The method which we shall propose to study ROLS-stability will also allow to analyse continuous dependence on other parameters in the problem, as for example on k_i and γ (see [5]).

Remark 2. ROLS-stability requires that for every $\beta \in I$ every solution of $(ROLS)^{\beta}_{z}$ depends continuously on z. It does not presuppose injectivity of the mapping Φ or uniqueness of the solutions of $(OLS)_{z^o}$ or $(ROLS)^{\beta}_{z^o}$. However, once ROLS-stability is established it follows that the solutions of $(ROLS)^{\beta}_{z^o}$ are isolated.

First we summarize some properties of equation (1). Let

$$U = \{q \in Q : q_i(x) \geq k_i(x)/2,\ x \in \Omega,\ i=1,\dots,n\}.$$

It is then wellknown [7] that for every $f \in L^2$ and $q \in U$ there exists a unique solution $u(q) \in H^2$ of (1). Moreover there exists a nondecreasing positive function $\tilde{c}_1(\cdot)$ depending only on c and k_i such that

$$|u(q)|_{H^2} \leq \tilde{c}_1(|q|_Q)\ |f|_{L^2} \tag{2}$$

for $q \in U$ and $f \in L^2$. For $q \in U$ we define operators $A(q)$ in L^2 by $D(A(q)) = H^2 \cap H^1_o$ and $A(q)u = -\sum_{i=1}^{n} (q_i u_{x_i})_{x_i} + cu$. These operators are densely defined, closed and selfadjoint. By the above considerations they also are homeomorphisms between $D(A(q))$ endowed with the H^2-norm and L^2.

We further put

$$A_1(q)u = -\sum_{i=1}^{n} (q_i u_{x_i})_{x_i} \quad \text{for} \quad u \in D(A(q)) \quad \text{and} \quad q \in U .$$

Observe that $A_1(q)$ is selfadjoint for every $q \in U$ and that for some constant K_1

$$|A_1(q)u| \leq K_1 |q|_Q \, |u|_{H^2} \quad \text{for all} \quad u \in H^2 \cap H^1_o \quad \text{and} \quad q \in Q . \qquad (3)$$

Finally one can show that

$$q \to A^{-1}(q)f = u(q)$$

from $U \subset Q$ to L^2 is continuous from the weak to the strong topology. In order to reformulate $(\text{ROLS})^{\beta}_{z_o}$ in an abstract setting we define $g : Q \to \bigotimes_{i=1}^{n} C \times \mathbb{R}$ by

$$g(q) = (\text{col}(k_1 - q_1, \ldots, k_n - q_n) \ , \ |q|_Q^2 - \gamma^2)$$

and a closed convex cone K in $\bigotimes_{i=1}^{n} C \times \mathbb{R}$ with vertex at the origin by

$$K = \bigotimes_{i=1}^{n} C^+ \times \mathbb{R}^+ ,$$

where C^+ is the natural nonnegative cone in C. Then $(\text{ROLS})^{\beta}_{z_o}$ is equivalent to

$$\min |A^{-1}(q)f - z|^2 + \beta|q|^2 \quad \text{over} \quad q \in Q_{ad} = \{q \in Q : g(q) \in -K\} .$$

In order to discuss ROLS-stability we use a stability result for optimization problems in infinite dimensional spaces. Let Q and W be Hilbert spaces, and Y a Banach space with an ordering induced by a closed convex cone K with vertex at the origin. Suppose that $f : D \times W \to \mathbb{R}$ and $g : X \times W \to Y$ and that $D \subset Q$ is an open set satisfying

$$Q_{ad} = \{q \in Q : g(q,w) \in -K\} \subset D .$$

Let w^o be a fixed reference parameter and for arbitrary $w \in W$ consider

$$(P)_w \qquad \min f(q,w) \text{ such that } g(q,w) \in -K .$$

A functional $\lambda^* \in Y^*$, the dual of Y , is called Lagrange multiplier for $(P)_{w^o}$ at q^o if

$$f_q(q^o,w^o) + \lambda^* g_q(q^o,w^o) = 0 , \quad \lambda^* \in K^+ \quad \text{and} \quad \lambda^* g(q^o,w^o) = 0 \quad (4)$$

and $F : Q \to \mathbb{R}$ given by $F(q,w^o) = f(q,w^o) + \lambda^* g(q,w^o)$ is called the associated Lagrange functional. Here K^+ is the dual cone of K. If q^o is a solution of $(P)_{w^o}$, then there exists a Lagrange multiplier, provided that q^o is a regular point with respect to the constraint, i.e.

$$0 \in \text{int } \{g(q^o,w^o) + g'(q^o,w^o)Q + K\} \quad .$$

PROPOSITION 1. [1]. Let q^o be a regular point and suppose that f and g are twice continuously Fréchet-differentiable with respect to q at (q^o,w^o) and that there exist constants $\nu > 0$ and $\gamma > 0$ such that for a Lagrangian functional F

$$F_{qq}(q^o,w^o)(h,h) \geq \gamma |h|^2$$

holds for all $h \in g_q^{-1}(-K + \mathbb{R}\, g(q^o)) \cap \{h : \lambda^* g_q(q^o,w^o)h \leq \nu |h|\}$.
Moreover assume that there exists a neighborhood $U = U_q \times U_w$ of (q^o,w^o) and constants L_q and L_g such that

$$|f(q,w) - f(q',w^o)| \leq L_f(|q-q'| + |w-w^o|)$$

$$|g(q,w) - g(q,w^o)| \leq L_g \, |w-w^o|$$

for all $(q,w) \in U$ and $q' \in U_q$.
Then there exist $r > 0$, $d > 0$ and a neighborhood V of w^o such that:

(i) The local extremal value function

$$\mu_r(w) = \inf \{f(q,w) : g(q,w) \in -K, \ |q-q^o| \leq r\}$$

is Lipschitz continuous at w^o .

For every $w \in V$ the following additional statements hold:

(ii) For every sequence $\{q_n\}$ with $g(q_n,w) \in -K$, $|q-q_n| \leq r$ and $\lim\limits_n f(q_n,w) = \mu_r(w)$ it follows that $|q_n-q^o| < r$ for all n sufficiently large.

(iii) If there exists q_w with $g(q_w,w) \in -K$, $|q_w-q^o| \leq r$ and $f(q_w,w) = \mu_r(w)$, then $|q_w-q^o| < r$ and $|q_w-q^o| \leq d\, |w-w^o|^{1/2}$.

Clearly, $(\text{ROLS})_z^\beta$ can be considered as a special case of $(P)_w$. Furthermore existence of solutions q_z^β as required in (iii) is guaranteed.

LEMMA 1. The first and second derivatives $\eta = u_q(q)h$ and $\xi = u_{qq}(q)(h,h)$ of the solution u at q in directions $h \in \mathcal{Q}$ are characterized by the property that η and $\xi \in H^2 \cap H_o^1$ and

$$A(q)\eta = -A_1(h)u(q) \ ,$$

$$A(q)\xi = -2A_1(h)\eta \ .$$

Here $q \to u(q)$ is taken as a mapping from Q to $H^2 \cap H_o^1$.

In order to formulate our first main result we introduce

$$Q^\beta = \{q^\beta \in Q : q^\beta \text{ is a solution of } (\text{ROLS})_{z^o}^\beta\}$$

and the attainable set

$$V = \{u(q) : q \in Q_{ad}\} \quad .$$

Here and below we drop the index z^o for the notation of the solutions of $(\text{ROLS})_{z^o}^\beta$.

THEOREM 1. Let $\bar\beta > 0$ be chosen such that for a minimum norm solution q_m^o of $(\text{OLS})_{z^o}$

$$|q_m^o|^2 - \sup_{Q^\beta} |q^\beta|^2 < (K_1\, \tilde c_1(|q_m^o|))^{-2} \tag{5}$$

and define

$$\underline{\beta} = \text{dist}\ (z^o, V)^2\ [(K_1\, \tilde c_1(|q_m^o|))^{-2} - |q_m^o| + \sup_{Q^{\overline{\beta}}} |q^{\overline{\beta}}|]^{-1} \geq 0. \tag{6}$$

If $\underline{\beta} < \overline{\beta}$ then the diffusion coefficient q in (OLS) is ROLS-stable at z^o in Q_{ad} for $\beta \in (\underline{\beta}, \overline{\beta})$. In particular, if $z^o \in V$, then q is ROLS-stable in Q_{ad} for all $\beta \in (0, \overline{\beta})$.

Proof. The second assertion is a direct consequence of the first one which we verify by means of Proposition 1. In the notation of Proposition 1 we take $Y = \bigotimes_{i=1}^{n} C \times \mathbb{R}$, $K = \bigotimes_{i=1}^{n} C^+ \times \mathbb{R}^+$ and $W = L^2$. Here C^+ denotes the cone of nonnegative functions in C. Since the regular point condition and the necessary smoothness requirements are quite easy to verify we concentrate on the second order sufficiency condition. Let λ^* be a Lagrange multiplier in the dual cone of $\bigotimes_{i=1}^{n} C^+ \times \mathbb{R}^+$ and consider the corresponding Lagrange functional

$$F(q) = F(q, z^o) = |u(q) - z^o|^2 + \langle \lambda^*, g(q, z^o)\rangle + \beta |q|_Q^2 .$$

We compute for $q^\beta \in Q^\beta$ and $\beta \in (\underline{\beta}, \overline{\beta})$

$$\begin{aligned} F_{qq}(q^\beta)(h,h) &= 2|\eta|^2 + 2(u(q^\beta) - z^o, \xi) + \\ &\quad + \langle \lambda^*, g_{qq}(q^\beta)(h,h)\rangle + 2\beta |h|_Q^2 , \end{aligned} \tag{7}$$

where $\eta = u_q(q^\beta)(h)$ and $\xi = u_{qq}(q^\beta)(h,h)$. Using Lemma 1 and the fact that $g_{qq}(q^\beta)(h,h) \in K$ we find

$$\begin{aligned} F_{qq}(q^\beta)(h,h) &\geq 2|\eta|^2 - 4(u(q^\beta) - z^o,\ A^{-1}(q^\beta) A_1(h)\eta) + 2\beta|h|_Q^2 \\ &= 2|\eta|^2 - 4(A_1(h)A^{-1}(q^\beta)(u(q^\beta) - z^o), \eta) + 2\beta|h|_Q^2 \\ &\geq 2\beta|h|^2 - 2\ |A_1(h)A^{-1}(q^\beta)(u(q^\beta) - z^o)|^2 . \end{aligned} \tag{8}$$

By (2) and (3)

$$|A_1(h)A^{-1}(q^\beta)(u(q^\beta)-z^o)| \leq K_1 |h|_Q \tilde{c}_1(|q^\beta|) \ |u(q^\beta)-z^o| \quad .$$

This estimate is now used in (8) and we find

$$F_{qq}(q^\beta)(h,h) \geq 2|h|^2 \ [\beta - K_1^2 \tilde{c}_1(|q^\beta|)^2 \ |u(q^\beta)-z^o|^2] \quad .$$

On the other hand, we have for every minimum norm solution q_m^o of $(OLS)_{z^o}$ and every $q^\beta \in Q^\beta$

$$-|u(q^\beta)-z^o|^2 \geq \beta(|q^\beta|^2 - |q_m^o|^2) - \text{dist}\,(z^o,V) \ ,$$

see [5]. Therefore with $K_2 = K_1 \tilde{c}_1(|q^\beta|)$

$$F_{qq}(q^\beta)(h,h) \geq 2|h|^2 \ K_2^2 \ [\beta(K_2^{-2} + |q^\beta|^2 - |q_m^o|^2) - \text{dist}\,(z^o,V)^2]$$

$$\geq 2|h|^2 \ K_2^2 \ \text{dist}\,(z^o,V)^2 \ [\underline{\beta}^{-1} \ \beta-1] \ .$$

This implies $F_{qq}(q^\beta_{z^o})(h,h) \geq \text{const}\ |h|^2$ for every $\beta \in (\underline{\beta},\overline{\beta})$ and the proof is finished.

Theorem 1 gives results that guarantee continuous dependence of the solutions of $(ROLS)^\beta_{z^o}$ on the observation z^o. In the first part of the assertion there is no attainability assumption, but $\underline{\beta}$ may be greater than $\overline{\beta}$, in which case the assertion is void. In this case decreasing dist (z^o,V) either by a more accurate measurement z^o or an improvement of the model should lead to success. - Thus consider $z_n^o \to z^o$ in L^2 with $z^o \in V$. We denote the solutions of $(OLS)_{z_n^o}$ and $(ROLS)^\beta_{z_n^o}$ by $q_{z_n^o}$ and $q^\beta_{z_n^o}$. The following stability property can be obtained for z_n^o sufficiently close to $z^o \in V$. (Recall that stability is investigated with respect to the upper index in z_n^o.) By Q_z^o we denote the set of solutions of the unregularized problem $(OLS)_z$.

THEOREM 2. Let the assumptions of Theorem 1 hold. Choose a sequence of observations z_n^o with $z_n^o \to z^o$ in L^2 and $z^o \in V$. Then there exists

$\tilde{\beta} > 0$ with the following property: For all $\beta^* \in (0,\tilde{\beta})$ there exists an index $N(\beta^*) \in \mathbb{N}$ and a neighborhood $I(\beta^*)$ of β^* such that for all $n \geq N(\beta^*)$ the parameter q is ROLS-stable in Q_{ad} at z_n^o for all $\beta \in I(\beta^*)$.

For the proof we refer to [5].

DISCUSSION

1) We recall that G. Chavent [2,3] has introduced a stability concept for parameter estimation problems which is different from ours. A parameter is called output least squares identifiable (OLSI), if there exists a neighborhood $\tilde{V}$ of the attainable set V such that for every element $z \in \tilde{V}$ there exists a unique solution $q \in Q_{ad}$ of $(OLS)_z$ depending continuously on z. Chavent derives general sufficient conditions involving $\text{dist}(z, V)$, $\text{diam}(Q_{ad})$, lower and upper bounds on the first and second derivative of $q \to u(q)$ which imply OLSI. The main distinction between the stability concept introduced in this paper and OLSI is the fact that OLSI requires uniqueness of the solution of the output least squares problem whereas ROLS-stability (or also OLS-stability, see 4)) does not.

2) In a recent paper, Kravaris and Seinfeld [6] have studied the use of regularization for parameter estimation in parabolic partial differential equation. Their approach is based on a variant of Tikhonov's lemma which states that if a continuous function f between metric spaces X and Y is injective on a precompact subset $K \subset X$, then f is continuously invertible on $f(K)$. In applications to $(OLS)_{z^o}$ this requires that the regularization term in $(ROLS)^{\beta}_{z^o}$ is replaced by $\beta|q|_{Q_C}$, where Q_C is a compactly embedded subspace of Q with norm $|\cdot|_{Q_C}$. In computations this leads to some inconvenience, since it is more involved to implement the Q_C- than the Q-norm. Moreover the Tikhonov approach of [6] requires uniqueness of the solution of the unregularized problem, which is not needed in our analysis, where stability is checked at each solution of $(ROLS)_{z^o}$. If ROLS-stability can be guaranteed then the solutions of $(ROLS)_{z^o}$ are isolated.

3) The second order sufficient condition (see (7) and (8) in the proof of Theorem 1) can be used as a convenient tool to obtain some insight into the specific features of parameter estimation problems.

   (a) If $z^o \in V$, then in view of the term $\langle u(q^\beta) - z^o, \xi \rangle$ and the convergence of $u(q^\beta)$ to z^o as $\beta \to 0^+$ [5], the lower bound on $F_{qq}(q^\beta)(h,h)$ is easier to obtain than in case that $z^o \notin V$.

   (b) The advantage of a regularization term is obvious from (7),(8). It helps to achieve strict positivity of $F_{qq}(q^\beta)$ and the required second order sufficiency condition.

   (c) To explain the next observation, suppose that $z^o \in V$ and take $\beta = 0$. Then the second derivative of the Lagrange functional reduces to $|\eta| = |A^{-1}(q^\beta)\nabla(h\nabla u(q^\beta))|$ and it is apparent that continuous dependence of q in H^2 on z cannot hold, since $|A^{-1}(q^\beta)\nabla(h\nabla u(q^\beta))|$ can be bounded below by $|\nabla h\nabla(u(q^\beta))|_{H^{-2}}$ only and since, moreover, our conditions do not exclude the case meas $\{x : \nabla u(q^\beta)(x) = 0\} > 0$. Hence some kind of regularization is necessary.

4) If for $\beta = 0$ the same kind of stability of the solutions q_z of $(OLS)_z$ on z as required for ROLS-stability holds, then we call q OLS-stable. Special cases of OLS-stability are studied in [5].

REFERENCES

[1] W. Alt: Lipschitzian perturbations of infinite optimization problems; in: Mathematical Programming with Data Perturbations II, ed. A.V. Fiacco, Lecture Notes in Pure and Applied Mathematics 85, Marcel Dekker, New York, 1983, 7-21.

[2] G. Chavent: Local stability of the output least square parameter estimation technique, Matematica Applicada e Computacional, 2 (1983), 3-22.

[3] G. Chavent: On parameter identifiability, Proceedings of the 7-th IFAC Symposium on Identification and System Parameter Estimation, York, England, July 1985.

[4] F. Colonius and K. Kunisch: Stability for parameter estimation in two point boundary value problems, to appear in Journal Reine Angewandte Mathematik.

[5] F. Colonius and K. Kunisch: Output least squares stability in elliptic systems, submitted to Appl. Math. Optimization.

[6] C. Kravaris and J.H. Seinfeld: Identification of parameters in distributed systems by regularization, SIAM J. Control and Optimization 23 (1985), 217-241.

[7] O.A. Ladyzhenskaya and N.N. Ural'tseva: Linear and quasilinear elliptic equations, Academic Press 1986.

[8] D.L. Russell: Some remarks on numerical aspects of coefficient identification in elliptic systems, in: Optimal Control of Partial Differential Equations, ed. K.H. Hoffmann and W. Krabs, Birkhäuser 1984, 21o-228.

PERIODIC AND ALMOST PERIODIC OSCILLATIONS IN NONLINEAR SYSTEMS

C. Corduneanu

University of Texas at Arlington, Texas, USA

We are concerned in this paper with nonlinear dynamical systems which can be described by partial differential equations of the form

$$u_t = Lu + F(t,x,u) , \quad (t,x) \in R\times G , \tag{1}$$

or

$$u_{tt} = Lu + F(t,x,u) , \quad (t,x) \in R\times G , \tag{2}$$

where G stands for a bounded domain in R^n , L is a linear elliptic operator of order 2m , and F is a nonlinearity whose nature will be specified in the subsequent paragraphs.

If we regard a solution of (1) or (2) as a map from R into some function space , say $W_o^{2m,2}(G)$ or $L^2(G)$, it is quite natural to ask the problem of periodicity or almost periodicity of such solution with respect to the variable t , whenever periodicity or almost periodicity is assumed for the nonlinearity F (also with respect to t). Of course , a boundedness condition must hold for such solutions , in a convenient norm . We will find very convenient to assume that F is almost periodic in respect to t , in the sense of Stepanov. The almost periodicity of the solution will be obtained in the Bohr-Bochner sense , which is a more restrictive concept of almost periodicity. This kind of problem belongs obviously to the so-called Bohr-Neugebauer type of problem in regard to the almost periodicity of solutions of differential equations The method of approach will be based on certain results the author has recently obtained for *qualitative inequalities* [1],[3]. This paper follows , in general , the same pattern as the papers [1],[2], [4] in which ordinary differential equations , partial differential equations of parabolic or elliptic type , or even abstract differential equations have been dealt with . More precisely , this paper is a direct generalization of the results given in [2], to the case when the elliptic operator L is not necessarily a second order operator , and the coefficients can depend on x . In this respect , the *Gårding's inequality* for general elliptic operators turns out to be the right tool (as Poincare's inequality was in case of second order operators) .

Existence problems will not be discussed in the present paper . Under more stringent conditions than those necessary to discuss almost periodicity of bounded solutions , existence results can be found , for instance , in [7], where further references are included . In this paper , we will assume the existence of solutions whose periodicity or almost periodicity is to be established .

SOME AUXILIARY RESULTS

We shall state now the results on qualitative inequalities that are needed in the sequel . These results are very special cases of certain results to be found , with their complete proofs , in [3]. They can be also derived from Lemmas 3 and 4 of [1] , where somewhat more restrictive assumptions have been made (for instance , instead of local absolute continuity , the existence of a continuous derivative has been assumed) .

LEMMA A. Consider the differential inequalities on the entire real axis

$$y'(t) \geq ky(t) - f(t)\sqrt{y(t)} , \tag{3}$$

and

$$y''(t) \geq ky(t) - f(t)\sqrt{y(t)} , \tag{4}$$

in which $k > 0$ is a constant , and f is a nonnegative locally integrable integrable function such that

$$\sup \int_t^{t+1} f(s)ds = | f |_M < \infty , \quad t \in R . \tag{5}$$

If y is locally absolutely continuous , satisfies (3) a.e. , and is bounded on R , then necessarily

$$\sup y(t) \leq K(| f |_M)^2 , \quad t \in R , \tag{6}$$

where K is a positive constant depending on k only .

If y and y' are locally absolutely continuous , y is bounded on R and verifies (4) a.e. , then necessarily

$$\sup y(t) \leq K_1(|f|_M)^2 , \quad t \in R , \tag{7}$$

where K_1 is a positive constant depending on k only .

Let us state now Gårding's result related to elliptic operators given in a bounded domain $G \subset R^n$. We shall consider a differential operator of order 2m , say L , that can be represented as

$$L = \sum_{o}^{2m} a_j(x)D^j , \tag{8}$$

where $a_j(x)$ are continuous and bounded on G , together with their derivatives up to the order 2m . The ellipticity condition can be written as

$$(-1)^m \sum a_j(x)\xi^j \geq c_o|\xi|^{2m} , \quad |j| = 2m , \tag{9}$$

for any $x \in G$, and any $\xi \in R^n$, where $c_o > 0$ is a constant (strong ellipticity) .

LEMMA B. If the differential operator L given by (8) is strongly elliptic in G , then there exist real constants C > 0 and K , such that

$$\int_G uLudx + K\|u\|_o^2 \geq C\|u\|_m^2 , \tag{10}$$

for any $u \in W_o^{2m,2}(G)$.

The proof of Lemma B , known as Gårding's inequality , can be found under slightly varying assumptions in [5], [6]. It is usually stated for $u \in C_o^\infty(G)$, but it remains true under the assumptions of Lemma B due to the fact $C_o^\infty(G)$ is dense in $W_o^{2m,2}(G)$.

Remark. From inequality (10) one can see that the operator L + kI , with I the identity operator , satisfies the inequality

$$\int_G u(L+kI)udx \geq C \|u\|_m^2 , \tag{11}$$

which means that we can always reduce the general case in Gårding's inequality to the case corresponding to K = 0 . In order to achieve this , one has to add to the elliptic operator L the term Ku (and also subtract it from the other terms of the equation) . This remark will be very helpful in formulating our results in the next section .

THE MAIN RESULTS

Let us consider first the parabolic equation (1) in the domain $R\times G$, and assume that the elliptic operator L is satisfying Gårding's inequality with constant $K = 0$. As seen above , this property can be assured if we add to Lu the term Ku , and modify accordingly the nonlinear term $F(t,x,u)$.
In regard to the nonlinearity $F(t,x,u)$, we shall make the assumption

$$F(t,x,u(t,x)) \in L^2(G) , \tag{12}$$

for all $t \in R$, and for all u such that $u(t,\cdot) \in W_0^{2m,2}(G)$ for all $t \in R$. Moreover , we will assume that a condition of the form

$$\langle F(t,x,u) - F(t,x,v) , u - v \rangle_0 \geq \mu \| u - v \|_0^2 , \tag{13}$$

where μ is a constant related to C from (11) by

$$C + \mu > 0 . \tag{14}$$

The last assumption we shall make in regard to $F(t,x,u)$ is concerned with the almost periodicity of this function with respect to t . More precisely , we have in mind the Stepanov's type of almost periodicity which can be formulated as follows : for every $\varepsilon > 0$, there exists $\ell(\varepsilon) > 0$, with the property that any interval of length ℓ on the real axis contains at least one point T , such that

$$\sup \int_t^{t+1} \| F(s+T,x,u) - F(s,x,u) \|_0 ds < \varepsilon , \tag{15}$$

for all real u , the supremum being taken for all $t \in R$. The norm with the subscript 0 stand for the norm in $L^2(G)$, in accordance with the notation for norm in Sobolev spaces .

THEOREM 1. Let $u = u(t,x)$, $(t,x) \in R\,G$, be a solution of equation (1) , such that

$$u(t,\cdot) \in W_0^{2m,2}(G) , \tag{16}$$

for all real t , and

$$\int_G u^2(t,x)dx \leq M < \infty \text{ on } R. \tag{17}$$

Under above mentioned conditions on L and F , the solution $u(t,x)$ is

(Bohr-Bochner) almost periodic as a map from R into $L^2(G)$. In case F is periodic in t of period T , u is also periodic in t , with the same period T .

A similar result holds true in regard to the equation (2) . We can state this result as

THEOREM 2. Let u = u(t,x) , (t,x) R G , be a solution of equation (2) such that conditions (16) and (17) hold true . Assume further that the operator L , given by (8) , satisfies the condition (9) . If F(t,x,u) is as described above in this section , then the solution u(t,x) is (Bohr-Bochner) almost periodic as a map from R into $L^2(G)$. In case F is periodic in t , of period T , the same property holds for u(t,x) .

REMARK 1. Both conclusions in Theorems 1 and 2 , in the almost periodic case , could be improved in the sense that some of the derivatives of the solution are also almost periodic . More elaborate estimates are necessary in this regards , and we deliberately avoid getting into such details in this paper .

REMARK 2. While the derivatives in x of u(t,x) exist in accordance with the theory of Sobolev spaces , the derivative with respect to t is supposed to exist in the strong topology of the space $L^2(G)$.

PROOF OF THE MAIN RESULTS

We will provide first the proof of Theorem 1. Let us notice that in case of equation (1) it is not material whether the elliptic operator L is positive definite or negative definite . Indeed , for a solution defined on the entire real axis , the change of t into -t does not bring any qualitative changes . But such a change of sign reduces the case when L is negative definite to the case when L is negative definite , and viceversa . Therefore , we can assume (and cover both cases!) that L is given by (8) , and satisfies the inequality (9) . This assumption will assure the fact that Gårding's inequality holds true in the form (10) or (11) .

Let us denote now v(t,x) = u(t+T,x) - v(t,x) , where T is an arbitrarily chosen real number . In the periodic case , T can be chosen as the basic period of F with respect to t . It is obvious that v(t,x) satisfies the equation

$$v_t = Lv + F(t+T,x,u(t+T,x)) - F(t,x,u(t,x)) \ , \tag{18}$$

and $v(t,x) \in W_o^{2m,2}(G)$ for any $t \in R$. If we multiply both sides of the equation (18) by $v(t,x)$, integrate the result on G , and operate some elementary transformations , one obtains

$$\frac{1}{2}\frac{d}{dt}\|v\|_o^2 = \int_G vLvdx + \int_G v\{F(t+T,x,u(t+T,x)) - F(t+T,x,u(t,x))\}dx$$
$$+ \int_G v\{F(t+T,x,u(t,x)) - F(t,x,u(t,x))\}dx \ . \tag{19}$$

Taking now into account (11) , (13) , and applying Cauchy's inequality to the last term in the right hand side of (19) , one obtains the following differential inequality which must hold true on the entire real axis :

$$\frac{1}{2}\frac{d}{dt}\|v\|_o^2 \geq C\|v\|_m^2 + \mu\|v\|_o^2 - \|F(t+T,x,u(t,x)) - F(t,x,u(t,x))\|_o\|v\|_o . \tag{20}$$

Since by the definition of the norm in Sobolev spaces one has $\|v\|_m \geq \|v\|_o$, the inequality (20) yields

$$\frac{1}{2}\frac{d}{dt}\|v\|_o^2 \geq (C+\mu)\|v\|_o^2 - \|F(t+T,\ldots) - F(t,\ldots)\|_o\|v\|_o \ , \tag{21}$$

which has obviously the form (3) in Lemma A. Consequently , taking condition (14) into account , one can write the estimate

$$\|v\|_o \leq K \sup \int_t^{t+1} \|F(s+T,..) - F(s,..)\|ds \ , \tag{22}$$

for some K depending on $C+\mu$ only . The supremum in (22) is taken on the whole real axis . From the estimate (22) taking into account the definition of v , one derives the (Bohr-Bochner) almost periodicity of the solution u , regarded as a map from R into $L^2(G)$.
This ends the proof of Theorem 1.

The proof of Theorem 2 can be carried out following the same lines as in the proof of Theorem 1 . First of all , it is useful to notice that

$$vv_{tt} = \frac{1}{2}(v^2)_{tt} - (v_t)^2 \ ,$$

in order to write the inequality corresponding to (21) ;

$$\frac{1}{2}\frac{d^2}{dt^2}\|v\|_o^2 \geq (C+\mu)\|v\|_o^2 - \|F(t+T,..) - F(t,..)\|_o\|v\|_o . \qquad (23)$$

Based on our hypotheses one can easily find out that inequality (23) is of the type (4) in Lemma A . One obtains from the Lemma A the following estimate :

$$\|v\|_o = \|u(t+T,x) - u(t,x)\|_o \leq \sup\int_t^{t+1} \|F(s+T,..) - F(s,..)\|_o ds.$$

This inequality clearly shows that the solution u(t,x) of equation (2) is (Bohr-Bochner) almost periodic , as a map from R into $L^2(G)$. It is the right place to point out that the inequality above has the particular feature of emphasizing the connection between the almost periods of F , and those of the solution u .

REMARK. Results concerning the asymptotic almost periodicity of bounded solutions of the equations (1) and (2) can be established by means of the technique used in this paper .

R E F E R E N C E S

[1] Corduneanu , C. , "Some Almost Periodicity Criteria for Ordinary Differential Equations". *Libertas Mathematica*, III (1983) , 21 - 43.

[2] Corduneanu , C. , "Almost Periodic Solutions to Nonlinear Elliptic and Parabolic Equations". *Nonlinear Analysis* - TMA, 7 (1983) , 357 - 363.

[3] Corduneanu , C. ,"Two Qualitative Inequalities" . *Journal of Differential Equations*,(in print).

[4] Corduneanu , C. , and Goldstein , J. A. , "Almost Periodicity of Bounded Solutions to Nonlinear Abstract Equations". In *Differential Equations* , ed. I.W. Knowles and R.T. Lewis , Elsevier Science Publishers , 1984 , 115 - 121.

[5] Dunford , N. , and Scwartz , J. , *Linear Operators* , *Part II* . New York , John Wiley (Interscience) , 1958.

[6] Mizohata, Sigeru, *The Theory of Partial Differential Equations*. Cambridge University Press, Cambridge (U.K.), 1973.

[7] Pankov, A. A., "Bounded Solutions, Almost Periodic in Time, of a Class of Nonlinear Evolution Equations. *Mathematics USSR Sbornik*, 49 (1984), No. 1, 73 - 86.

DIFFERENTIABILITY OF A MIN MAX AND APPLICATION TO OPTIMAL CONTROL AND DESIGN PROBLEMS. Part I *

M.C.Delfour
Centre de recherches mathématiques
and Département de mathématiques et statistique,
Université de Montréal, C.P. 6128, Succ. A,
Montréal, Québec, Canada, H3C 3J7

J.-P. Zolésio
Laboratoire de Physique Mathématique,
U.S.T.L., Pl. Eugène Bataillon,
34060 - Montpellier Cédex,
France

ABSTRACT. We consider the Min Max of a functional which is parametrized by t. We show that, under appropriate conditions, the derivative of the Min Max with respect to t is the Min Max with respect to the points solution of the Min Max problem of the derivative of the original functional with respect to t. To illustrate the use of this theorem, we apply it to the control of an elliptic equation with a non-differentiable observation and to a shape optimal design problem.

1. INTRODUCTION.

In this paper we consider a functional $G(t, x, y)$ and the associated Min Max problem

$$g(t) = \{\text{Min}\,[\text{Max}\, G(t, x, y) \,:\, y \in Y] \,:\, x \in X\}$$

where t is real and X and Y are appropriate sets. We show that, under appropriate hypotheses, the derivative of $g(t)$ with respect to t is the Min Max of the partial derivative $\partial_t G(t, x, y)$ with respect to all points $(x, y) \in X \times Y$ solution of the Min Max problem. This is a generalization of the differentiation of a Min or a Max with respect to a parameter (cf. J.-P. Zolésio [3, Thm. 1.1, p. 1458].

This theorem and its eventual generalizations have many interesting applications. To illustrate that point we describe the associated techniques for two examples. The first one is a control or identification problem with a non-differentiable observation which depends on the state which is the solution of an elliptic equation which itself depends on the control function u. The second example is a shape optimal design problem where this technique makes it possible to completely by-pass the problem of the existence and interpretation of the Eulerian or material derivative of the state.

Those two simple examples are given for the purpose of illustration. However the techniques used here apply to the general linear case and some non linear situations.

The main theorem and its application to the first example have been announced in [4].

* This research was supported in part by the National Sciences and Engineering Council of Canada Operating Grant A-8730 and a FCAR Grant from the "Ministère de l'Education du Québec".

Notation.

R will denote the field of real numbers and R^n ($n \geq 1$, an integer) the n-fold Cartesian product of R. The inner product and norm in R^n will be defined as

$$x \bullet y = \Sigma_{i=1,n} x_i y_i \qquad |x| = (x \bullet x)^{1/2}.$$

The dual operator of a continuous linear operator $A : X \to Y$ will be denoted by A^*. The identity matrix in R^n will be written I_d. The composition of two applications f and g will be denoted by $f \circ g$.

2. DERIVATIVE OF A MIN MAX WITH RESPECT TO A PARAMETER.

Let $\mathcal{A} \subset X$ and $\mathcal{B} \subset Y$ be subsets of two Banach spaces X and Y. Given a map $G : R \times X \times Y \to R$, we consider the following functions :

$$H(t, x) = \quad \mathrm{Sup}\{G(t, x, y) : y \in \mathcal{B}\}, \ t \in R, x \in \mathcal{A} \tag{1}$$

$$g(t) = \mathrm{Inf}\{H(t, x) : x \in \mathcal{A}\}, \ t \in R. \tag{2}$$

As a result

$$g(t) = \mathrm{Inf}\{\mathrm{Sup}[G(t, x, y) : y \in \mathcal{B}] : x \in \mathcal{A}\}. \tag{3}$$

We wish to show that under appropriate hypotheses the function g is differentiable at $t = 0$ from the right

$$\lim_{\substack{t>0 \\ t \to 0}} (g(t) - g(0))/t \text{ exists.} \tag{4}$$

In order to better see the role of each hypothesis in the final result we proceed in a step by step fashion.

We first introduce hypotheses to ensure that the Sup and Inf problems have solutions.

(H1) $\exists \tau > 0, \ \forall t, \ 0 \leq t < \tau$, the set

$$A(t) = \{ x \in \mathcal{A} : g(t) = H(t, x) \} \tag{5}$$

is not empty.

(H2) $\exists \tau > 0, \ \forall t, \ 0 \leq t < \tau, \ \forall s, \ 0 \leq s < \tau, \ \forall x_t \in A(t)$, the set

$$B(s, x_t) = \{y \in \mathcal{B} : H(s, x_t) = G(s, x_t, y)\} \tag{6}$$

is not empty.

LEMMA 1. Under hypotheses (H1) and (H2) we have the following estimates : for all t, $0 \le t < \tau$

$$g(t) - g(0) \le G(t, x_0, y^*) - G(0, x_0, y^*), \ \forall x_0 \in A(0), \ \forall y^* \in B(t, x_0) \quad (7)$$

and

$$g(t) - g(0) \ge G(t, x_t, z^*) - G(0, x_t, z^*), \ \forall x_t \in A(t), \ \forall z^* \in B(0, x_t). \quad (8)$$

Proof. The proof uses standard arguments and will be omitted. ♦

In a second step we obtain upper and lower bounds on the differential quotient

$$(g(t) - g(0))/t. \quad (9)$$

We need the following additional hypothesis.

(H3) $\forall t, \ 0 \le t < \tau, \ \forall x_0 \in A(0), \forall x_t \in A(t), \forall z \in B(0, x_t), \forall y \in B(t, x_0)$

the fonctions

$$\alpha \rightarrow G(\alpha, x_0, y) \text{ and } \alpha \rightarrow G(\alpha, x_t, z) \quad (10)$$

are differentiable in a neighborhood of 0.

LEMMA 2. Under hypotheses (H1), (H2) and (H3), for each t, $0 < t < \tau$,

(i) there exists θ_1, $0 < \theta_1 < 1$, such that

$$(g(t)-g(0))/t \le \partial_t G(\theta_1 t, x_0, y^*), \ \forall x_0 \in A(0), \ \forall y^* \in B(t, x_0) \quad (11)$$

(ii) there exists θ_2, $0 < \theta_2 < 1$, such that

$$(g(t) - g(0))/t \ge \partial_t G(\theta_2 t, x_t, z^*), \ \forall x_t \in A(t), \ \forall z^* \in B(0, x_t). \quad (12)$$

Proof. (i) For $x_0 \in A(0)$ and $y^* \in B(t, x_0)$ define

$$\lambda(s) = G(st, x_0, y^*).$$

By hypothesis (H3) λ is differentiable in a neighborhood of 0. So by Taylor's Theorem : $\exists$ $\theta_1 \in]0,1[$ such that

$$\lambda(1) = \lambda(0) + d\lambda/ds\,(\theta_1)$$

and

$$G(t, x_0, y^*) - G(0, x_0, y^*) = t\,\partial_t G(\theta_1 t, x_0, y^*)$$

where $\partial_t G$ denotes the partial derivative of G with respect to the first argument. The proof of part (ii) is similar and will be omitted. ♦

In the next step we go to the limit in inequalities (11) and (12) as t goes to 0. So we introduce

$$\bar{d}\, g(0) = \limsup_{t\to 0^+} (g(t) - g(0))/t, \qquad \underline{d}\, g(0) = \liminf_{t\to 0^+} (g(t) - g(0))/t \tag{13}$$

They are the smallest upper and greatest lower bounds of the differential quotient in R. So there exist sequences $\{t_n\}$ and $\{t_n'\}$ of positive numbers in $]0, \tau]$ going to zero as n goes to $+\infty$ such that

$$\bar{d}\, g(0) = \lim_{n\to\infty} (g(t_n) - g(0))/t_n \tag{14}$$

$$\underline{d}\, g(0) = \lim_{n\to\infty} (g(t_n') - g(0))/t_n'. \tag{15}$$

We first consider the upper bound in (11). We use the following hypotheses of continuity.

(H4) (i) $\exists$ a topology τ_Y on Y and a compact subset K_Y of Y such that for all sequences $t_n \to 0$, $t_n > 0$, and all $x_0 \in A(0)$, $B(t_n, x_0) \cap K_Y \neq \emptyset$.

(ii) For all x_0 in $A(0)$ and all y in $\mathcal{B}$, the map $t \to G(t, x_0, y)$ is lower semi continuous.

(iii) For all x_0 in $A(0)$, the map $t, y \to G(t, x_0, y)$ is upper semi continuous in $R \times \tau_Y$.

(H5) For all $x_0 \in A(0)$ the map $s, y \to \partial_t G(s, x_0, y)$ is upper semi continuous in $R \times \tau_Y$.

PROPOSITION 1. Under hypotheses H1 to H5

$$\bar{d}\, g(0) \leq \operatorname*{Inf}_{x\in A(0)} \operatorname*{Sup}_{y\in B(0,x)} \partial_t G(0, x, y) . \tag{16}$$

Proof. Fix x_0 in $A(0)$ and let $\{t_n\}$ be the sequence in (14). Choose

$$y_n \in B(t_n, x_0) \cap K_Y.$$

By hypothesis H4, there exists a subsequence, still denoted $\{y_n\}$, such that

$$y_n \to y^* \quad \text{in} \quad \tau_Y\text{-topology}.$$

From inequality (11) we obtain using hypothesis H5

$$\bar{d}\, g(0) \leq \limsup_{n\to\infty} \partial_t G(\theta_n t_n, x_0, y_n) \leq \partial_t G(0, x_0, y^*) .$$

Hypotheses H4(ii) and H4(iii) imply that $y^* \in B(0, x_0)$. By definition of $B(t_n, x_0)$,

$$G(t_n, x_0, y_n) \geq G(t_n, x_0, y), \ \forall y \in \mathcal{B}$$

and

$$G(0, x_0, y^*) \geq \limsup_{n\to\infty} G(t_n, x_0, y_n)$$

$$\geq \limsup_{n\to\infty} G(t_n, x_0, y)$$

$$\geq G(0, x_0, y), \ \forall y \in \mathcal{B}.$$

As a result

$$\bar{d}\, g(0) \leq \sup\{\partial_t G(0, x_0, y) \ : \ y \in B(0, x_0)\}.$$

The last estimate is true for all x_0 in $A(0)$. This is sufficient to establish (16). ♦

We now turn to the lower bound (12). As before we need some compactness and continuity hypotheses.

(H6) There exists a topology τ_X on X and a compact subset K of X such that

$A(t) \cap K \neq \emptyset \ \ \forall\, 0 \leq t < \tau.$

This hypothesis implies that for a sequence $\{t_n'\}$ converging to 0, we can choose x_n in $A(t_n)$ and a subsequence $\{x_{n_k}\}$ of $\{x_n\}$ which converges in $K \subset X$.

(H7) There exists a topology $\tilde{\tau}_Y$ of Y for which the set-valued function $x \to B(0,x)$ is lower semi continuous in the sense of J.P. AUBIN [1, Déf. 9.4, p. 121] : for all convergent sequences $x_n \to x_0$ in X and all z^* in $B(0, x_0)$, there exists a sequence $z_n^* \in B(0, x_n)$ such that $z_n^* \to z^*$ in the $\tilde{\tau}_Y$-topology.

(H8) The map

$s, x, y \to \partial_t G(s, x, y)$

is lower semi continuous for the topology $R \times X \times \tilde{\tau}_Y$

(H9) (i) For all x in X the map

$t \to H(t, x)$

is upper semi continuous at $t = 0$; moreover

(ii) the map

$(t, x) \to H(t, x)$

is lower semi continuous on $R \times X$.

We state our final result

THEOREM 1. Under hypotheses (H1) to (H9), we have

$$\underline{d}\, g(0) = \bar{d}\, g(0) = \operatorname*{Inf}_{x \in A(0)} \ \operatorname*{Sup}_{y \in B(0,x)} \ \partial_t G(0, x, y)$$

and the function g is differentiable at 0 from the right :

$$\lim_{t \to 0^+} (g(t) - g(0))/t \qquad \text{exists}$$

Proof. We want to prove that

$$\underline{d}\, g(0) \quad \geq \quad \operatorname*{Inf}_{x \in A(0)} \quad \operatorname*{Sup}_{y \in B(0,x)} \quad \partial_t G(0, x, y)$$

Consider the converging sequence $t_n' \to 0^+$, $t_n' > 0$, and expression (15). It is always possible to choose a subsequence $\{x_n\}$ in X such that $x_n \in A(t_n')$. Under hypothesis (H6) this subsequence can be choosen in the compact subset K of X. So there exists another subsequence, still denoted $\{x_n\}$, such that

$$x_n \to x_0 \in X, \ \forall\, n, \ x_n \in A(t_n').$$

By definition of $A(t_n)$

$$H(t_n, x_n) \leq H(t_n, x), \ \forall\, x \in \mathcal{Q}$$

and

$$\limsup_{n \to \infty} H(t_n, x_n) \leq \limsup_{n \to \infty} H(t_n, x) \leq H(0, x)$$

by using hypothesis H9 (i). But

$$\liminf_{n \to \infty} \ H(t_n, x_n) \leq \limsup_{n \to \infty} \ H(t_n, x_n) \leq H(0, x)$$

and by hypothesis H9 (ii)

$$H(0, x_0) \leq H(0, x), \ \forall\, x \in \mathcal{Q}$$

As a result $x_0 \in A(0)$.

Fix an arbitrary element z^* in $B(0, x_0)$. By hypothesis (H7), there exists a sequence z_n^* in Y, $z_n^* \in B(t_n, x_n)$, such that

$$z_n^* \to z^* \text{ in } \tilde{\tau}_Y\text{-topology.}$$

We now use the lower bound (12) to establish the lower bound of $\underline{d}\, g(0)$:

$$\underline{d}\, g(0) \geq \liminf_{n \to \infty} \partial_t G(\theta_n t_n, x_n, z_n^*)$$

where x_n and z_n^* are as defined above.

Under hypothesis H8, we obtain

$\underline{d}\, g(0) \geq \partial_t G(0, x_0, z^*),\ \forall z^* \in B(0, x_0)$

for some x_0 in A(0). Finally

$$\underline{d}\, g(0) \geq \operatorname*{Sup}_{z \in B(0,x)} \partial_t G(0, x_0, z)$$

$$\geq \operatorname*{Inf}_{x \in A(0)} \operatorname*{Sup}_{z \in B(0,x)} \partial_t G(0, x, z) \ \blacklozenge$$

Remark 1. In order to obtain the lower bound on d g(0), we have used the lower semi continuity hypothesis H7 on the set-valued map B. Notice that this hypothesis is stronger than hypothesis H4 used to get the upper bound. Indeed hypothesis H7 implies that given a converging sequence $x_n \to x_0$, $x_n \in A(t_n)$, and any y in $B(0, x_0)$ there exists a sequence $\{y_n\}$, $y_n \in B(0, x_n)$, such that $y_n \to y$. In the case of hypothesis H4, there exists a converging subsequence, still denoted $\{y_n\}$, which converges to **some** y which belongs to the set of all limit points. So only those limit points can be approximated by a sequence $\{y_n\}$, $y_n \in B(0, x_n)$, and not all points y in $B(0, x_0)$. ♦

In view of the preceding Remark, hypothesis H7 can be weakened to the following hypothesis H7', but the upper and lower bounds on the differential quotient will no longer coincide.

(H7') (i) Given any convergent sequence $x_n \to x_0$ in X, there

exists a sequence $\{z_n\}$, $z_n \in B(0, x_n)$, a subsequence $\{z_{n}\}$ of $\{z_{n_k}\}$ and z^* in Y such that $z_{n_k} \to z^*$ for the τ_Y-topology

(ii) The map

$x \to G(0, x, z)$

is lower semi continuous on X, and the map

$x, z \to G(0, x, z)$

is upper semi continuous on $X \times \tilde{\tau}_Y$.

THEOREM 2. Under hypotheses H1 to H6, H7', H8 and H9

$$\operatorname*{Inf}_{x \in A(0)} \operatorname*{Inf}_{y \in B(0,x)} \partial_t G(0, x, y) \leq \underline{d}\, g(0)$$

$$d\, g(0) \leq \operatorname*{Inf}_{x \in A(0)} \operatorname*{Sup}_{y \in B(0,x)} \partial_t G(0, x, y) \ \blacklozenge$$

COROLLARY. If, in addition to the hypotheses of Theorem 2, the set $B(0, x)$ is a singleton for each x in $A(0)$,

$$\forall x \in A(0),\ B(0, x) = \{y_x\},$$

then g is differentiable at 0 from the right and

$$d\,g(0) = \operatorname{Inf}\,\{\partial_t G(0, x, y_x) \ :\ x \in A(0)\}. \ \blacklozenge$$

Remark 2. The Corollary can also be proved directly by two consecutive applications of the theorem on the differentiability of a Min. ♦

3. DERIVATIVE OF A NON-DIFFERENTIABLE OBSERVATION FUNCTIONAL WITH RESPECT TO THE CONTROL VARIABLE.

Let Ω be a bounded domain in R^n with smooth boundary Γ, $f \in L^2(\Omega)$ and u be a function in the interior U of $L_+^\infty(\Omega)$, that is

$$\forall u \in U,\ \exists\, \alpha > 0 \text{ such that } u(x) \geq \alpha \text{ a.e. in } \Omega. \tag{1}$$

Consider the solution $y = y(u)$ in $H_0^1(\Omega)$ of the variational problem

$$-\operatorname{div}(u \nabla y) = f \text{ in } \Omega,\ y = 0 \text{ on } \Gamma. \tag{2}$$

Associate with u and y the cost function

$$J(u) = \int_\Omega |y - y_d|\, dx,\ y_d \in L^1(\Omega). \tag{3}$$

We want to compute the derivative of $J(u)$ with respect to u subject to the constraint (2).

We consider the state equation (2) as a constraint and remove it by introducing a Min Sup. It is easy to check that

$$J(u) = \operatorname{Min}\{\operatorname{Sup}[\int_\Omega \mu(\varphi - y_d)\, dx + d\,E(u, \varphi; 0, p) : (p, \mu) \in H_0^1(\Omega) \times M\}] : \varphi \in H_0^1(\Omega)\} \tag{4}$$

where $d\,E(u, \varphi; 0, p)$ is the right Gateaux derivative of

$$E(u, \varphi) = 1/2 \int_\Omega [u\, |\nabla\varphi|^2 - 2 f\varphi]\, dx \tag{5}$$

at (u, φ) in the direction $(0, p)$ and

$$M = \{\mu \in L^\infty(\Omega) \ :\ |\mu(x)| \leq 1, \text{ a.e. in } \Omega\}. \tag{6}$$

In this form, it is not directly possible to apply Theorem 1 in section 2. It is necessary to introduce a perturbed functional indexed by a parameter $r > 0$ (which is not necessarily infinitesimally small)

$$-G_r(u, (p, \mu), \varphi) = \int_\Omega \mu(\varphi - y_d)\, dx + d\,E(u, \varphi; 0, p) + r\,\{\, E(u, \varphi) - e(u)\,\} \tag{7}$$

where

$$e(u) = \operatorname{Inf}\{\, E(u, \varphi) \ :\ \varphi \in H_0^1(\Omega)\,\}. \tag{8}$$

Define

$$J_r(u) = \underset{\varphi \in H_0^1(\Omega)}{\text{Min}} \quad \underset{(p,\mu) \in H_0^1(\Omega) \times M]}{\text{Sup}} \quad -G_r(u, (p, \mu), \varphi) \tag{9}$$

and the dual functional

$$\text{(10)} \quad J_r^*(u) = - \underset{(p,\mu) \in H_0^1(\Omega) \times M}{\text{Inf}} \quad \underset{\varphi \in H_0^1(\Omega)}{\text{Max}} \quad G_r(u, (p, \mu), \varphi) .$$

PROPOSITION 2. For each u in U and r, $0 < r < 2$, the functional $G_r(u, \bullet, \bullet)$ has saddle points and

$$J_r^*(u) = J_r(u) = - \underset{(p,\mu) \in H_0^1(\Omega) \times M}{\text{Min}} \quad \underset{\varphi \in H_0^1(\Omega)}{\text{Max}} \quad G_r(u, (p, \mu), \varphi) \tag{11}$$

Proof . The first identity (11) follows from Ekeland and Temam [2 , Prop. 2.4, p. 177] applied to the functional

$$F_r(u, p, \varphi) = \text{Sup}\{-G_r(u, (p, \mu), \varphi) \; : \; \mu \in M\} \tag{12}$$

which is equal to

$$\int_\Omega \{|\varphi - y_d| \, dx + d\, E(u, \varphi; 0, p) + r[E(u, \varphi) - e(u)]. \tag{13}$$

It suffices to check the following two conditions

$$\exists\, p \in H_0^1(\Omega) \quad \text{such that} \quad \lim_{\|\varphi\| \to \infty} F_r(u, p, \varphi) = +\infty \tag{14}$$

$$\lim_{\|p\| \to \infty} \quad \underset{\varphi \in H_0^1(\Omega)}{\text{Inf}} \quad F_r(u, p, \varphi) = -\infty \tag{15}$$

The first condition is verified for $p = 0$. For the second condition, we fix p and choose $\varphi = -p$

$$\text{Inf}\{F_r(u, p, \varphi) \; : \; \varphi \in H_0^1(\Omega)\} \le F_r(u, p, -p)$$

and show that the upper bound goes to $-\infty$ as $\|p\|$ goes to $+\infty$:

$$F_r(u, p, -p) = \int_\Omega \{|-p - y_d| - u|\nabla p|^2 - fp + r/2\, (u|\nabla p|^2 + 2fp)\}\, dx - re(u). \tag{16}$$

The L^2-norm of ∇p goes to $+\infty$ since it is equivalent to the $H_0^1(\Omega)$-norm. So for r, $0 < r < 2$, the right -hand-side of (16) goes to $-\infty$ and (15) is verified. This shows the existence of a saddle-point for $F_r(u, \bullet, \bullet)$:

$$\underset{\varphi}{\text{Min}} \; \underset{p}{\text{Sup}} \; F_r(u, p, \varphi) = \underset{p}{\text{Max}} \; \underset{\varphi}{\text{Inf}} \; F_r(u, p, \varphi) . \tag{17}$$

The next step is to show that for a fixed p,

$$\operatorname*{Inf}_{\varphi} \ \operatorname*{Sup}_{\mu \in M} \ -G_r(u, p, \mu, \varphi) = \operatorname*{Max}_{\mu \in M} \ \operatorname*{Inf}_{\varphi} \ -G_r(u, p, \mu, \varphi) . \tag{18}$$

In view of the properties of $-G_r$ and the fact that M is bounded, this is a consequence of Remark 2.3 and Proposition 2.3 in Ekeland and Temam [2, p.162]. By combining (17) and (18)

$$\operatorname*{Min}_{\varphi} \ \operatorname*{Sup}_{(p, \mu)} \ -G_r = \operatorname*{Max}_{(p, \mu)} \ \operatorname*{Inf}_{\varphi} \ -G_r$$

and by Proposition 1.2 in Ekeland and Temam [2, p. 155], $-G_r(u, (\bullet, \bullet), \bullet)$ has saddle points. In view of (10), this is sufficient to establish (11). ♦

It is now important to notice that for all $r \geq 0$

$$J_r(u) = J_0(u) = J(u). \tag{19}$$

For $0 < r < 2$, $G_r(u, \bullet, \bullet)$ has saddle points and

$$J_r(u) = J_r^*(u). \tag{20}$$

We now apply Theorem 1 in section 2 to $J_r^*(u)$.

For $0 < r < 2$, $u \in U$ and $v \in L^\infty(\Omega)$, there exists $\tau > 0$ small enough such that $u + \tau v \in U$. Define for t in $[0, \tau]$

$$G(t, q, \varphi) = G_r(u + tv, q, \varphi) \tag{21}$$

for $q = (\mu, p) \in X = M \times H_0^1(\Omega)$ and $\varphi \in Y = H_0^1(\Omega)$. In view of the above proposition, the saddle points of $G(t, \bullet, \bullet)$ are completely characterized by the following set of equations (cf. Ekeland and Temam [2, Prop. 1.6, p. 157]) :

$$-\mathrm{div}[(u + t\, v)\, \nabla y_t] = f \ \text{ in } \Omega, \ y_t = 0 \ \text{ on } \Gamma \tag{22}$$

$$-\mathrm{div}[(u + t\, v)\, \nabla p_t] + \mu_t = 0 \ \text{ in } \Omega, \ p_t = 0 \ \text{ on } \Gamma. \tag{23}$$

$$\mu_t \in M_{dt} = \{\mathrm{sgn}(y_t - Y_d) - \alpha \chi_{\Omega_{dt}} : \alpha \in M\}, \tag{24}$$

where

$$\Omega_{dt} = \{ x \in \Omega : y_t(x) = y_d(x) \} \tag{25}$$

is a measurable set. The technique with the term in r could have been completely by-passed by noticing that the system of equations (22)-(25) has solutions and applying Proposition 1.6 in Ekeland and Temam [2, p 157] to show that they are saddle points of G_r.

Introduce the constants

$$\beta = 1/2\ \|u\|_{L^\infty(\Omega)} \qquad \tau = \beta / \|V\|_{L^\infty(\Omega)}. \tag{26}$$

The sets $\mathcal{A}$, $\mathcal{B}$ are

$$\mathcal{A} = H_0^1(\Omega) \times M, \ \mathcal{B} = H_0^1(\Omega) \tag{27}$$

and the sets A(t), $0 \le t \le \tau$, and B(s, q), $0 \le s \le \tau$, are characterized by the following lemma.

LEMMA 3. Given r, $0 < r < 2$, then for all t, $0 \le t \le \tau$, and all s, $0 \le s \le \tau$,

$$A(t) = \{ (\mu_t(\alpha), p_t(\alpha)) : \mu_t(\alpha) = \operatorname{sgn}(y_t - y_d) - \alpha \chi_{\Omega_{dt}}\ \alpha \in M \} \tag{28}$$

$$B(s, q) = B(s) = \{y_s\} \tag{29}$$

where $p_t(\alpha)$ is the solution of (21) with $\mu_t = \mu_t(\alpha)$ and y_s is the solution of (20) with $t = s$.

Proof. In fact (28) and (29) mean that the sets A(t) and B(t, x_t) for $x_t \in A(t)$ reduce to the saddle points $((\mu_t, p_t), y_t)$. This result is not a priori true. In general, only the following inclusion is true for all $r \ge 0$.

LEMMA 4. For all $r \ge 0$

(i) $A(t) \supset \{ (\mu, p) : ((\mu, p), y)$ is a saddle point of $-G_r$ for some $y\}$

(ii) $\forall\ q_t = (\mu_t, p_t) \in A(t)$,

$B(t, q_t) \supset \{y_t : (q_t, y_t)$ is a saddle point for some $q_t\}$. ♦

The proof of Lemma 4 uses standard arguments and will be omitted.

Proof of Lemma 3. It suffices to prove the converse of the two inclusions in Lemma 4 (for $0 < r < 2$). Given $q = (p, \mu)$ we introduce

$$H(u, q) = \underset{\varphi}{\operatorname{Max}} - G_r(u, q, \varphi).$$

This maximum φ is achieved at a unique point y. Then the map

$$q \to H(u, q)$$

is Gateaux differentiable and

$$dH(u, q\,; 0, g) = -\,dG_r(u, q, y\,; 0, g, 0).$$

Moreover the map $q \to H(u, q)$ is convex and lower semi continuous. Thus $q \in A(u)$ if and only if

$$-\,dG_r(u, q, y\,; 0, g, 0) \ge 0 \text{ for all admissible g.}$$

So in view of the characterization of y, (q, y) is a saddle point of $-G_r$. ♦

For $0 \le t \le \tau$, $A(t) \neq \emptyset$ and H1 is verified. For each s in $[0, \tau]$, the set $B(s, q)$ reduces to a singleton which is independent of q. So it is non-empty and H2 is verified. Hypothesis (H3) is clear. For H4(i) we choose for τ_Y the weak topology on $Y = H_0^1(\Omega)$ and for K_Y the weakly compact ball of radius R since for all t in $[0, \tau]$ and $\{y_t\} = B(t)$

$$\beta \int_\Omega |\nabla y_t|^2 dx \le \int_\Omega (u + t\, v)\, |\nabla y_t|^2 dx = \int_\Omega f\, y_t\, dx \le c\|f\|_{L^2} \|\nabla y_t\|_{L^2}$$

$$\Rightarrow \|\nabla y_t\|_{L^2} \le R = c/\beta\, \|f\|_{L^2}, \quad \forall t \in [0, \tau].$$

H4(ii), H4(iii) and H5 are also obvious. For H6 we choose for τ_X the weak topology on $X = M \times H_0^1(\Omega)$ and for K the ball of radius $R' = c'/\beta$ (c' as defined below). Indeed for all t in $[0, \tau]$

$$\begin{aligned} \beta\|\nabla p\|^2 &\le \int_\Omega (u + t\, v)\, \nabla p \bullet \nabla p\, dx = -\int_\Omega [\operatorname{sgn}(y_t - y_d) - \alpha\, \chi_\Omega\ ]\, p\, dx \\ &\le c\|p\| \le c'\, \|\nabla p\| \\ \Rightarrow \|\nabla p\| &\le R' = c'/\beta, \quad \forall t \in [0, \tau]. \end{aligned}$$

For H7, the map

$$q \to B(0, q) = B(0) = \{y_0\}$$

is constant and single-valued. For H8

$$\partial_t G(s, q, \varphi) = -\int_\Omega v\, \nabla p \bullet \nabla \varphi\, dx + r[d\, E(u + sv, \varphi; v, 0) - de(u + sv; v)].$$

From [3, Thm. 1.1, p. 1458], it is known that

$$de(u + sv; v) = d\, E(u + sv, y_s; v, 0)$$

where y_s is the solution of (20) with $t = s$. With $\tilde{\tau}_Y$ being the strong topology on $Y = H_0^1(\Omega)$, $\partial_t G$ is jointly continuous with respect to its arguments.

Hypothesis H9 consists of two conditions on the functional

$$H(t, q) = \sup_{\varphi \in H_0^1(\Omega)} G(t, q, \varphi)$$

where

$$G(t, q, \varphi) = -\{\int_\Omega \mu\, (\varphi - y_d)\, dx + d\, E(u + tv, \varphi; 0, p) + r[E(u + tv, \varphi) - e(u + tv)]\}.$$

Since G is lower semi continuous in the variables (t, q, φ), the functional

$$(t, q) \to H(t, q)$$

is lower semi-continuous and H9(ii) is verified. The second hypothesis essentially requires that $t \to H(t, q)$ be continuous at 0. This follows from the continuity with respect to t of the minimizing element φ_t of $-G(t, q, \varphi)$ with respect to t.

We summarize our results in the next proposition.

PROPOSITION 3. For all u in U, v in $L^\infty(\Omega)$ and $0 < r < 2$, there exists $\tau > 0$ such that hypotheses H1 to H9 on the function $G(t, q, \varphi)$ in (19) be verified (recall that $q = (\mu, p) \in X = M \times H_0^1(\Omega)$ and that $\varphi \in Y = H_0^1(\Omega)$). For t in $[0, \tau]$, the set $\mathcal{A}$, $\mathcal{B}$, A(t) and B(t, q) are given by (27) to (29). ♦

THEOREM 3. For all u in U and v in $L^\infty(\Omega)$, the functional J(u) is Gateaux differentiable from the right at u in the direction v and

$$J'(u; v) = \mathrm{Sup}\{\int_\Omega v\, \nabla p(\alpha) \bullet \nabla y\, dx \ :\ \alpha \in M\} \tag{30}$$

where y and $p(\alpha)$ are the respective solutions of

$$-\mathrm{div}(u\, \nabla y) = f \text{ in } \Omega,\ y = 0 \text{ on } \Gamma \tag{31}$$

$$-\mathrm{div}(u\, \nabla p(\alpha)) + \mathrm{sgn}(y - y_d) - \alpha \chi_{\Omega_d} = 0 \text{ in } \Omega,\ p(\alpha) = 0 \text{ on } \Gamma \tag{32}$$

$$\Omega_d = \{x \in \Omega : y(x) = Y_d(x)\}. \tag{33}$$

Proof. Recall that for $0 < r < 2$ and $0 \le t \le \tau$

$$J_r(u + t\, v) = J(u + t\, v).$$

Computing the derivative of J is equivalent to computing the derivative of J_r for some fixed r, $0 < r < 2$. The results of section 2 are now available. It is sufficient to notice that the integral in (30) is $\partial_t G(0, (\mu(\alpha), p(\alpha)), y)$, where

$$\mu(\alpha) = \mathrm{sgn}(y - y_d) - \alpha\chi_{\Omega_d}, \quad \alpha \in M.$$

Expression (30) then follows from Proposition 3 and Theorem 1 in section 2. ♦

Remark 3. Notice that the map $\alpha \to p(\alpha)$ is affine and continuous. In (30) the Sup occurs at extremal points of M. By defining

$$M_d = \{\alpha \in L^\infty(\Omega) : \alpha(x) = \pm 1 \text{ in } \Omega_d \text{ and } \alpha(x) = 0 \text{ elsewhere}\}, \tag{34}$$

we obtain

$$J'(u; v) = \mathrm{Sup}\{\int_\Omega v\, \nabla p(\alpha) \bullet \nabla y\, dx \ :\ \alpha \in M_d\}. \tag{35}$$ ♦

Remark 4. Throughout our analysis, the parameter r, $0 < r < 2$, is fixed but arbitrary and the saddle points of G_r are independent of r. ♦

Remark 5. The interest behind this method for cost functionals of the form

$$J(u) = F(u, y(u)) \tag{36}$$

is to justify the differentiation of $J(u)$ at u in the direction v without using the intermediate step of differentiating the state $y(u)$ of u in the direction v. ♦

The above results formally extend to the class of linear variational problems. For instance, let

$$E : U \times X \to R, \; E(u, x) = (1/2)\, a(u, x, x) - L(u, x) \tag{37}$$

for some open set U, a Hilbert space X, a continuous symmetrical bilinear form $a(u, \bullet, \bullet)$ and a continuous linear form $L(u, \bullet)$. For each u in U, define

$$e(u) = \operatorname{Inf}\{E(u, x) \; : \; x \in X\} \tag{38}$$

and assume that the minimizing element y is unique and completely characterized by the variational equation

$$d\,E(u, y; 0, \psi) = 0, \;\; \forall\, \psi \in X. \tag{39}$$

We can readily extend this method to cost functionals of the form

$$J(u) = F(u, y(u)) \tag{40}$$

for some map

$$F : U \times X \to R \tag{41}$$

where $y(u)$ is the solution of (37).

Then the associated functional G is given by

$$G(u, p, x) = F(u, x) + d\,E(u, x; 0, p) \tag{42}$$

for $u \in U$, $p \in X$ and $x \in X$. Let $P(u)$ is the set of solutions of the adjoint inequation

$$d\,F(u, y(u); 0, \psi) + d^2 E(u, y(u); 0, p; 0, \psi) \le 0, \; \forall\, \psi \in X \tag{43}$$

where $y(u)$ is the solution of (39).

THEOREM 4. (i) Assume that for all x in X, the map

$$u \to E(u, x) \tag{44}$$

is differentiable and that its "derivative" is continuous with respect to its arguments (u, x); for each u, the map

$$x \to F(u, x) \tag{45}$$

is convex, lower semi continuous and the map

$$u \to F(u, x) \tag{46}$$

is right Gateaux differentiable.

(ii) Further assume that the set

$$P(u) \neq \emptyset. \tag{47}$$

Then $J(u)$ defined by (40) is right Gateaux differentiable and

$$dJ(u; p) = \operatorname{Sup}\,\{dF(u, y(u); v, 0) + dE(u, y(u); 0, p; v, 0) : p \in P(u)\}. \; ♦ \tag{48}$$

The proof of this theorem will be given after a short discussion of the fundamental hypothesis (47).

Remark 4. If the map

$$v \to dF(u, y; 0, v) \tag{49}$$

is linear and continuous, the adjoint problem (33) is variational and P(u) reduces to the usual solution of the associated variational problem. ♦

When (49) is non-differentiable and non-linear we can use the augmented Lagrangian technique previously developed.

PROPOSITION 4. Assume the existence of a number r, $0 < r < 2$, such that

$$\left.\begin{aligned} &F(u, x) + r/2\ a(u, x, x) \to +\infty \\ &-F(u, -x) + (1 - r/2)\ a(u, x, x) \to +\infty \end{aligned}\right\} \text{ as } \|x\| \to \infty \tag{50}$$

Then P(u) is not empty. ♦

Proof of Theorem 4. First notice that

$$J(u) = \operatorname*{Inf}_{\varphi \in X} \ \operatorname*{Sup}_{\psi \in X} \ G(u, \varphi, \psi) \tag{51}$$

where G is defined by (42). Now the solution (y, p) of equation (43) and

$$dE(u, y; 0, \varphi) = 0, \forall \varphi \in X \tag{52}$$

is also a solution of the system of inequations

$$dG(u, y, p; 0, \varphi - y, 0) \geq 0, \forall \varphi \in X \tag{53}$$

$$dG(u, y, p; 0, 0, \psi - p) \leq 0, \forall \psi \in X\ . \tag{54}$$

But any solution of the system (53) - (54) is a saddle point of G(u, •, •) (cf. Ekeland and Temam [2, Prop. 1.6]). Now (54) reduces to (39) and y is unique. The final expression (48) is obtained by applying Theorem 1. ♦

Proof of Proposition 4. As in our illustrative example, we construct the augmented Lagrangian

$$G_r(u, x, p) = G(u, x, p) + r\,[E(u, x) - e(u)]\ .$$

It is easy to show that the functional G_r is convex and lower semicontinuous in x and concave and upper semicontinuous in p. From hypotheses (50),

$$G_r(u, x, 0) \to +\infty \text{ as } \|x\| \to \infty$$

$$\operatorname*{Inf}_{x \in X}\ G_r(u, x, p) \leq G_r(u, -p, p) \to -\infty \text{ as } \|p\| \to \infty\ .$$

Again by Proposition 2.4 in Ekeland and Temam[2, p. 164], G_r(u, •, •) has saddle points in X x X. They are completely characterized by the system (53) - (54) which is equivalent to (39) and (43). This is sufficient to establish that P (u) is not empty. ♦

REFERENCES.

[1] J.P. AUBIN, L'analyse non linéaire et ses motivations économiques, Masson, Paris, New-York, 1984.

[2] I. EKELAND and R. TEMAM, Analyse convexe et problèmes variationnels, Dunod, Gauthiers-Villars, Paris, Bruxelles, Montréal, 1974.

[3] J.P. ZOLESIO, Semi-derivative of repeated eigenvalues, in "Optimization of distributed parameter structures", E.J. Haug and J. Céa, eds., pp. 1457-1473, Sijthoff and Noordhoff, Alphen aan den Rijn, Netherlands 1980.

[4] M.C. DELFOUR and J.P. ZOLESIO, Dérivation d'un Min Max et application à la dérivation par rapport au contrôle d'une observation non différentiable de l'état, C.R. Acad. Sc. Paris, to appear.

DIFFERENTIABILITY OF A MIN MAX AND APPLICATION TO OPTIMAL CONTROL AND DESIGN PROBLEMS. Part II *

M.C.Delfour
Centre de recherches mathématiques
and Département de mathématiques et statistique,
Université de Montréal, C.P. 6128, Succ. A,
Montréal, Québec, Canada, H3C 3J7

J.-P. Zolésio
Laboratoire de Physique Mathématique,
U.S.T.L., Pl. Eugène Bataillon,
34060 - Montpellier Cédex,
France

ABSTRACT. We consider the Min Max of a functional which is parametrized by t. We show that, under appropriate conditions, the derivative of the Min Max with respect to t is the Min Max with respect to the points solution of the Min Max problem of the derivative of the original functional with respect to t. To illustrate the use of this theorem, we apply it to the control of an elliptic equation with a non-differentiable observation and to a shape optimal design problem.

4. SHAPE DERIVATIVE OF A FUNCTIONAL: A SIMPLE EXAMPLE.

4.1 Shape optimization problem.

Consider the following simple example. Let Ω be a bounded open domain in R^n with a smooth boundary Γ. Let $y = y(\Omega)$ be the solution of the variational problem

$$\text{Inf}\{E(\Omega, \varphi) \; : \; \varphi \in H^1(\Omega)\} \tag{1}$$

where

$$E(\Omega, \varphi) = 1/2 \int_\Omega [|\nabla\varphi|^2 + |\varphi|^2 - 2 f \varphi] \, dx \tag{2}$$

for some fixed function f in $H^1(R^n)$. We associate with y a cost function

$$J(\Omega) = F(\Omega, y(\Omega)). \tag{3}$$

For instance we can choose the standard cost function

$$F(\Omega, y) = 1/2 \int_\Omega (y - Y_d)^2 \, dx, \;\; Y_d \in H^1(R^n). \tag{4}$$

4.2. The Velocity Field Method.

We briefly recall the notion of a shape derivative. Let $V(t, x)$, $t \geq 0$, $x \in R^n$, be a **velocity field of deformation.** Under the action of V, the points of Ω are transported onto a new domain $\Omega_t = T_t(\Omega)$, where the transformation $T_t : R^n \to R^n$ is generated by the solutions of the equation

$$(\partial/\partial t) \, T_t(x) = V(t, T_t(x)), \;\; t \geq 0, \;\; T_0(x) = x \tag{5}$$

(cf. J.P. ZOLESIO [3]). Let y_t be the solution of problem (1) on the transformed domain Ω_t

$$\text{Inf}\{E(\Omega_t, \varphi) \; : \; \varphi \in H^1(\Omega_t)\} \tag{1'}$$

and associate with y_t the cost function

$$J(\Omega_t) = F(\Omega_t, y_t). \tag{3'}$$

4.3. *The Inf Sup formulation of the perturbed problem.*

In general our objective is the minimization of the cost function J with respect to Ω. In particular we want to compute the shape derivative of J at Ω in the direction of the velocity field of deformations V. To do this we transform the problem (1') - (3') into an Inf Sup problems. This approach is widespread in the engineering and mathematical literature.

The solution of (1') is completely characterized by the variational equation

$$dE(\Omega_t, y_t; \varphi) = 0, \ \forall\, \varphi \in H^1(\Omega_t) \tag{6}$$

where

$$dE(\Omega_t, \psi; \varphi) = \int_{\Omega_t} [\nabla\psi \bullet \nabla\varphi + \psi\,\varphi - f\,\varphi]\, dx, \quad \varphi, \psi \in H^1(\Omega_t). \tag{7}$$

Define for $r \geq 0$

$$-G_r(t, \varphi, p) = F(\Omega_t, \varphi) + dE(\Omega_t, \varphi; p) + r[E(\Omega_t, \varphi) - e(t)] \tag{8}$$

where

$$e(t) = \mathrm{Min}\{E(\Omega_t, \varphi) \ : \ \varphi \in H^1(\Omega_t)\}. \tag{9}$$

But it is readily seen that

$$\mathrm{Sup}\{-G_r(t, \varphi, p) \ : \ p \in H^1(R^n) = \begin{cases} F(\Omega_t, \varphi), & \text{if } \varphi \text{ is solution of (6)} \\ +\infty & \text{, otherwise} \end{cases} \tag{10}$$

As a result for all $r \geq 0$

$$J(\Omega_t) = \mathrm{Inf}\{\mathrm{Sup}[-G_r(t, \varphi, p) \ : \ p \in H^1(R^n)] \ : \ \varphi \in H^1(R^n)\}. \tag{11}$$

Note that now the spaces involved are fixed and independent of the parameter $t \geq 0$.

4.4. *Perturbed dual functional J_r^* and existence of saddle points.*

Our objective is to show the existence of saddle points for $r > 0$ and use the results of section 2 together with identity (11).

For $r \geq 0$, define the functionals

$$J_r(\Omega_t) = -\mathrm{Sup}\{\mathrm{Inf}[G_r(t, \varphi, p) \ : \ p \in H^1(R^n)] \ : \ \varphi \in H^1(R^n)\} \tag{12}$$

and

$$J_r^*(\Omega_t) = -\mathrm{Inf}\{\mathrm{Sup}[G_r(t, \varphi, p) \ : \ \varphi \in H^1(R^n)] \ : \ p \in H^1(R^n)\}. \tag{13}$$

Recall that in view of (11)

$$J_r(\Omega_t) = J_0(\Omega_t) = J(\Omega_t), \ \forall \ r \geq 0. \tag{14}$$

In general

$$J_r^*(\Omega_t) \leq J_r(\Omega_t) \tag{15}$$

since J_r^* is the dual functional associated with the perturbed functional G_r.

We have made the above construction in order to apply Theorem 1 to the dual problem for $r > 0$; for $r = 0$ certain hypotheses would not be verified.

To show the existence of a saddle point of G_r for all $r \geq 0$ we can use Theorem 4 directly.

PROPOSITION 5. (i) Given $\tau > 0$ small enough, then for all r, $0 \leq r$, and t, $0 \leq t \leq \tau$, $G_r(t, \bullet, \bullet)$ has saddle points (Y_r, P_r) in $H^1(R^n) \times H^1(R^n)$ and

$$J_r^*(\Omega_t) = J_r(\Omega_t). \tag{16}$$

(ii) The restriction of each saddle point to Ω_t

$$(y_t^r, p_t^r) = (Y_r|_{\Omega_t}, P_r|_{\Omega_t}) \tag{17}$$

coincide with the unique pair (y_t, p_t) solution of the system

$$dE(\Omega_t, y_t; \varphi) = 0, \ \forall \varphi \in H^1(\Omega_t) \tag{18}$$

$$dF(\Omega_t, y_t; \psi) + d^2 E(\Omega_t, y_t; p_t; \psi) = 0, \ \forall \psi \in H^1(\Omega_t) \tag{19}$$

where

$$dF(\Omega_t, y_t; \psi) = \int_{\Omega_t} (y_t - Y_d) \ \psi \, dx \tag{20}$$

$$d^2 E(\Omega_t, y_t; p_t; \psi) = \int_{\Omega_t} [\nabla p_t \bullet \nabla \psi + p_t \psi] \, dx. \tag{21}$$

Proof. The conditions characterizing a saddle point (Y_r, P_r) of G_r are precisely (18) - (19). Both equations are elliptic with a unique solution in $H^1(\Omega_t)$ which is independent of $r \geq 0$. ♦

For $r > 0$, the existence of saddle points could also be proved in a way similar to the technique used in Proposition 4.

PROPOSITION 6. For each $t \geq 0$ and $r > 0$, the functional $G_r(t, \bullet, \bullet)$ has saddle points.

Proof. From Ekeland and Temam [2, Prop. 2.4, p. 177]. It suffices to check the following two conditions

$$\exists\, p \in H^1(R^n) \text{ such that } \lim_{\|\varphi\| \to \infty} -G_r(t, \varphi, p) = +\infty. \tag{22}$$

$$\lim_{\|p\| \to \infty} \operatorname{Inf}\{-G_r(t, \varphi, p) \; : \; \varphi\} = -\infty. \tag{23}$$

The first condition is verified with $p = 0$. For the second condition we fix p and choose $\varphi = -p$:

$$\operatorname{Inf}\{-G_r(t, \varphi, p) \; : \; \varphi\} \le -G_r(t, -p, p)$$

and show that the upper bound goes to $-\infty$ as $\|p\|$ goes to $+\infty$:

$$\begin{aligned} -G_r(t, -p, p) &= F(\Omega_t, -p) + dE(\Omega_t, -p; p) + r[E(\Omega_t, -p) - e(t)] \\ &= 1/2 \int_{\Omega_t} (-p - Y_d)^2\, dx - \int_{\Omega_t} [|\nabla p|^2 + |p|^2 + fp]\, dx \\ &\quad + r/2 \{ \int_{\Omega_t} [|\nabla p|^2 + |p|^2 + 2 f p]\, dx - 2\, e(t)\}. \end{aligned} \tag{24}$$

For $0 < r < 1$ the quadratic terms are

$$-(1 - r/2) \int_{\Omega_t} |\nabla p|^2\, dx - (1/2 - r/2) \int_{\Omega_t} |p|^2\, dx$$

and (23) is necessarily true. ♦

4.5. Application of Theorem 1.

The next step is the application of Theorem 1 from section 2 to the function $g(t) = J_r(\Omega_t) = J_r^*(\Omega_t)$. We first study the differentiability of the functional $G_r(t, \varphi, \psi)$ as defined by (8) with respect to t for all φ and ψ in the space $X = H^1(R^n)$:

$$\begin{aligned} &- G_r(\Omega_t, \varphi, \psi) \\ &= 1/2 \int_{\Omega_t} (\varphi - Y_d)^2\, dx + \int_{\Omega_t} [\nabla\psi \bullet \nabla\varphi + \psi\, \varphi - f\, \varphi]\, dx + r/2 \int_{\Omega_t} [|\nabla\varphi|^2 + |\varphi|^2 - 2 f\, \varphi]\, dx. \end{aligned} \tag{25}$$

If ψ and φ were smoother (e.g. in $H^2(R^n)$) to make sure that the traces of $|\nabla\varphi|^2$ and $\nabla\psi \bullet \nabla\varphi$ exist, the standard expression for the derivative would be given by a boundary integral (cf. J.P. ZOLESIO [3]). Unfortunately this is not the case. Expression (25) transported from Ω_t onto Ω is given by

$$\begin{aligned} &\int_\Omega A(t)\; [\nabla (\psi \circ T_t) + r/2\; (\nabla\varphi \circ T_t)] \bullet \nabla(\varphi \circ T_t)\, dx \\ &+ \int_\Omega \{1/2\, (\varphi \circ T_t - Y_d \circ T_t)^2 + [\psi \circ T_t - f \circ T_t) + r/2\, (\varphi \circ T_t - 2 f \circ T_t)\,](\varphi \circ T_t)\}\, J(t)]dx \end{aligned} \tag{26}$$

where $D T_t$ is the Jacobian matrix of the transformation T_t and

$$J(t) = \det (D T_t), \quad A(t) = J(t)\, ((D T_t)^{-1})^* (D T_t)^{-1} \tag{27}$$

(* indicates the transposed matrix). Again to differentiate the above expression with respect to t would require that φ and ψ be in $H^1(R^n)$. To get around this difficulty we need the special technique which is described below.

Given the smooth velocity field V, define the transformation

$$T_t = T_t(V) \ : \ R^n \to R^n \tag{28}$$

which transports R^n onto R^n, $\Omega_0 = \Omega$ onto Ω_t and $\Gamma_0 = \Gamma$ onto Γ_t. The space $H^1(\Omega_t)$ is transported in a similar way. As the functions φ and ψ fill in the whole space X, so do the functions $\psi \circ T_t^{-1}$ and $\varphi \circ T_t^{-1}$. As a result

$$g(t) = J_r(\Omega_t) = J_r^*(\Omega_t) = -\operatorname*{Inf}_{\varphi \in X} \ \operatorname*{Sup}_{\psi \in X} -\tilde{G}_r(t, \varphi, \psi) \tag{29}$$

where

$$\tilde{G}_r(t, \varphi, \psi) = G_r(t, \varphi \circ T_t^{-1}, \psi \circ T_t^{-1}). \tag{30}$$

By introducing the quantities

$$J(t) = \det (D T_t), \quad A(t) = J(t)\, ((D T_t)^{-1})^* (D T_t)^{-1} \tag{31}$$

($D T_t$ is the Jacobian matrix of the transformation T_t) and the change of variable $x' = T_t(x)$, we obtain

$$-\tilde{G}_r(t, \varphi, \psi) = -\tilde{G}_0(t, \varphi, \psi) + r[E(\Omega, \varphi \circ T_t) - e(t)] \tag{32}$$

where

$$e(t) = \operatorname{Inf}\{E(\Omega_t, \varphi) \ : \ \varphi \in H^1(R^n)\} = \operatorname{Inf}\{E(\Omega, \varphi \circ T_t) \ : \ \varphi \in H^1(R^n)\} \tag{33}$$

and

$$\begin{aligned} -\tilde{G}_0(t, \varphi, \psi) = &\int_\Omega [\,(A(t)\ \nabla \psi) \bullet \nabla \varphi + \psi\, \varphi\, J(t)]\, dx \\ &+ \int_\Omega [1/2\, (\varphi - Y_d \circ T_t)^2 - (f \circ T_t)\psi]\, J(t)\, dx \end{aligned} \tag{34}$$

$$E(\Omega, \varphi \circ T_t) = \int_\Omega [1/2\, (A(t)\ \nabla \varphi) \bullet \nabla \varphi + (f \circ T_t)\, \varphi\, J(t)]\, dx. \tag{35}$$

The right Gateaux derivative with respect to t can easily be obtained for every terms in $\tilde{G}_r(t, \bullet, \bullet)$ except possibly $e(t)$. Fortunately we know from J.P. ZOLESIO [3, Thm. 1.1, p.

1458] that

$$d_t\, e(0) = \text{Inf}\{\partial_t\, E(\Omega_0, \varphi) \ :\ \varphi \in X,\ E(\Omega_0, \varphi) = e(0)\}$$

$$= \partial_t\, E(\Omega_0, Y) \tag{36}$$

where Y is the solution

$$E(\Omega_0, Y) = \text{Inf}\{E(\Omega_0, \varphi) \ :\ \varphi \in H^1(R^n)\}$$

or equivalently

$$Y \in H^1(R^n),\ d\,E(\Omega, Y; \varphi) = 0,\ \forall \varphi \in H^1(R^n). \tag{37}$$

We finally obtain the following intermediate result prior to the application of Theorem 1 in section 2.

PROPOSITION 7. For all r, $0 < r < 1$, t, $0 \le t \le \tau$, $\tilde{G}_r(t, \varphi, p)$ is Gateaux differentiable with respect to t at 0 and

$$-\partial_t \tilde{G}_r(0, \varphi, p) = \int_\Omega [(A'(0)\, \nabla p) \bullet \nabla\varphi + p\,\varphi\, \text{div}\, V(0)]\, dx$$

$$- \int_\Omega [1/2\, (\varphi - Y_d)\, \nabla Y_d + p\, \nabla f] \bullet V(0)\, dx$$

$$+ \int_\Omega [1/2\, (\varphi - Y_d)^2 - f\, p]\, \text{div}\, V(0)\, dx \tag{38}$$

$$+ r[\partial_t\, E(\Omega_0, \varphi) - \partial_t\, E(\Omega_0, E)]$$

where

$$A'(0) = [\text{div}\, V(0)]\, I_d - [D\, V(0) + (D\, V(0))^*\,]. \tag{39}$$

($(D\, V(0))^*$ is the transposed matrix of $D\, V(0)$). ♦

All the hypotheses of Theorem 1 in section 2 are now verified.

THEOREM 5. For all r, $0 < r < 1$,

$$d\, J_r(\Omega; V(0)) = -\, \partial_t \tilde{G}_r(0, p, y) \tag{40}$$

where y and p are the solutions of

$$d\, E(\Omega, y; \varphi) = 0,\ \forall \varphi \in H^1(\Omega). \tag{41}$$

and

$$d\, F(\Omega, y; \psi) + d^2\, E(\Omega, y; p; \psi) = 0,\ \forall \psi \in H^1(\Omega). \quad ♦ \tag{42}$$

Remark 6. The derivative of J_r was obtained without ever considering the problem of the differentiability of y. ♦

Finally recall identity (14) to obtain the desired result.

THEOREM 6. (i)The function g is differentiable at $t = 0$ and

$$dJ(\Omega; V(0)) = \int_\Omega [(A'(0)\, \nabla p) \bullet \nabla y\ -\ [1/2\, (y - Y_d)\, \nabla\, Y_d + p\, \nabla\, f] \bullet V(0)]\, dx$$

$$+ \int_\Omega [1/2\, (y - Y_d)^2 + p\, y - f\, p]\, \text{div}\, V(0)\, dx \tag{43}$$

where y and p are the solutions of (41) and (42).

(ii) If, in addition, p and y belong to $H^{3/2+\rho}(\Omega)$, $\rho > 0$, then

$$d\,J(\Omega; V(0)) = \int_\Gamma [\nabla y \bullet \nabla p + y\,p - f\,p + 1/2\,(y - Y_d)^2]\,V(0) \bullet n\,d\Gamma. \tag{44}$$

Proof. It suffices to notice that for $\varphi = y$ the term which contains the r in identity (38) is identically zero. When y and p are sufficiently smooth, expression (43) is equivalent to the standard boundary integral formulation in Shape Optimization. ♦

Remark 7. The above simple example contains several techniques which will turn out to be quite fundamental in different problems. For instance the introduction of the functional

$$\tilde{G}_r(t, \varphi, \psi) = G_r(t, \varphi \circ T_t^{-1}, \psi \circ T_t^{-1})$$

followed by the transport of the resulting expression from the domain Ω_t onto Ω makes it possible to keep the tests functions in $H^1(\Omega)$ instead of going to the larger space $H^1(R^n)$. For instance this is extremely important for the homogeneous Dirichlet problem in $H_0^1(\Omega)$ where it would not be possible to substitute $H^1(R^n)$. ♦

5. SHAPE DERIVATIVE OF A FUNCTIONAL : OTHER EXAMPLES.

In this last section we describe two other examples to further illustrate the applicability of Theorem 1 and our associated techniques.

The first one goes over the discussion at the end of section 4 in Remark 7. Key details are provided to show how problems with Dirichlet boundary conditions can be handled. In fact the suggested construction could also have been used right from the beginning in section 4, but we preferred to do it in a different way to better appreciate its importance.

The second example shows that we can handle problems where the smoothness of the solution of the saddle point equations is minimal. Other techniques based on Implicit Function Theorems would require more smoothness.

5.1. Dirichlet Boundary Condition.

We go back to the problem (1) to (4) in section 4 but with $H_0^1(\Omega)$ instead of $H^1(\Omega)$. Let $y = y\,(\Omega)$ in $H_0^1(\Omega)$ be the solution of the variational problerm

$$\operatorname{Inf}\,\{E(\Omega,\varphi) : \varphi \in H_0^1(\Omega)\} \tag{1}$$

where

$$E(\Omega,\varphi) = 1/2 \int_\Omega [\,|\nabla\varphi|^2 + |\varphi|^2 - 2\;f\varphi]\;dx \tag{2}$$

for some fixed function f in $H^1(R^n)$. We associate with y a cost function

$$J(\Omega) = F(\Omega, y(\Omega)). \tag{3}$$

Again for simplicity we assume that it is of the form

$$F(\Omega,\varphi) = 1/2 \int_\Omega (\varphi - Y_d)^2 \, dx, \quad \varphi \in H_0^1(\Omega), \; Y_d \in H^1(R^n). \tag{4}$$

Assume that V is a smooth vector field which transports Ω onto Ω_t, its boundary Γ onto Γ_t and the Sobolev space $H^1(\Omega)$ onto $H^1(\Omega_t)$ at time $t \geq 0$. As a result it also transports functions in $H_0^1(\Omega)$ onto functions in $H_0^1(\Omega_t)$ and

$$H_0^1(\Omega_t) = \{\varphi \circ T_t^{-1} : \varphi \in H_0^1(\Omega)\} \tag{5}$$

Here we use techniques described at the end of section 4 in Remark 7. Introduce the new functional

$$\varphi \to \tilde{E}(t, \varphi) = E(\Omega_t, \varphi \circ T_t^{-1}) : H_0^1(\Omega) \to R \tag{6}$$

and notice that

$$\operatorname*{Inf}_{\varphi \in H_0^1(\Omega)} \tilde{E}(t, \varphi) = \operatorname*{Inf}_{\psi \in H_0^1(\Omega_t)} E(\Omega_t, \psi) . \tag{7}$$

Denote by y^t and y_t the respective minimizing unique solutions of $\tilde{E}(t,\varphi)$ in $H_0^1(\Omega)$ and $E(\Omega_t, \psi)$ in $H_0^1(\Omega_t)$, respectively. Then in view of (5)

$$y_t = y^t \circ T_t^{-1} . \tag{8}$$

The two formulations are equivalent, but the differentiation of $\tilde{E}(t,\varphi)$ with respect to t does not require that the function φ smoother than $H^1(\Omega)$ since

$$\tilde{E}(t, \varphi) = 1/2 \int_{\Omega_t} [|\nabla (\varphi \circ T_t^{-1})|^2 + |\varphi \circ T_t^{-1}|^2 - 2 \, f (\varphi \circ T_t^{-1})] \, dx$$

and

$$\tilde{E}(t, \varphi) = 1/2 \int_\Omega \{(A(t) \nabla\varphi) \bullet \nabla\varphi + [|\varphi|^2 - 2 (f \circ T_t) \varphi] J(t)\} \, dx, \tag{9}$$

where DT_t is the Jacobian matrix associated with the transformation T_t,

$$J(t) = \det (DT_t) , \quad A(t) = J(t) \, ((DT_t)^{-1})^* \, (DT_t)^{-1} \tag{10}$$

and * denote the transposed matrix.

If we want to work with $\tilde{E}(t, \Omega)$ and y^t, we must also transform the functional F into a new functional

$$\varphi \to \tilde{F}(t, \varphi) = F(\Omega_t, \varphi \circ T_t^{-1}) : H_0^1(\Omega) \to R \tag{11}$$

As a result the cost function

$$J(\Omega_t) = F(\Omega_t, y_t) = F(\Omega_t, y^t \circ T_t^{-1}) = F(t, y^t) \tag{12}$$

and again the differentiability of $F(t, \varphi)$ with respect to t does not require that the function φ be smoother than $H^1(\Omega)$:

$$\tilde{F}(t, \varphi) = 1/2 \int_{\Omega_t} (\varphi \circ T_t^{-1} - Y_d)^2 \, dx$$

and

$$F(t, \varphi) = 1/2 \int_\Omega (\varphi - Y_d \circ T_t)^2 J(t) \, dx \; . \tag{13}$$

Thus we are led to the construction of the functional

$$(\varphi, \psi) \rightarrow \tilde{G}_r(t, \varphi, \psi) = G_r(\Omega_t, \varphi \circ T_t^{-1}, \psi \circ T_t^{-1}) : H_0^1(\Omega) \times H_0^1(\Omega) \rightarrow R \tag{14}$$

and the technique used in section 4.

We do not repeat the details here since the results are the same as those in Theorem 6 except that the functions y and p are the solutions of the variational equations

$$y \in H_0^1(\Omega) \,, \; dE(\Omega, y; \varphi) = 0 \,, \forall \, \varphi \in H_0^1(\Omega) \tag{15}$$

$$p \in H_0^1(\Omega) \,, \; dF(\Omega, y; \psi) + d^2E(\Omega, y; p, \psi) = 0 \,, \forall \, \psi \in H_0^1(\Omega). \tag{16}$$

So formally it suffices to substitute $H_0^1(\Omega)$ for $H^1(\Omega)$ in Theorem 6.

5.2. An example with less smoothness.

In the two previous examples, the solutions (y, p) of the optimality system (42) - (43) or (15) - (16) are smoother than anticipated and belong to $H^2(\Omega)$. So it would be possible to argue that all the results can also be obtain by application of some form of Implicit Function Theorem.

It is not difficult to slightly modify the example of section 4 to prevent this situation from happening. Firstly change the functionals E and F to

$$E(\Omega, \varphi) = 1/2 \int_\Omega [|\nabla \varphi|^2 + |\varphi|^2 - 2 \, f \bullet \nabla \varphi] \, dx, \;\; \varphi \in H^1(\Omega). \tag{17}$$

where $f \in (L^2(\Omega))^n$

$$F(\Omega, \varphi) = \int_\Omega |\nabla \varphi| \, dx, \;\; \varphi \in H^1(\Omega). \tag{18}$$

The minimization problem

$$e(\Omega) = \operatorname{Inf} \{E(\Omega, \varphi) : \varphi \in H^1(\Omega)\} \tag{19}$$

still has a unique solution y in $H^1(\Omega)$ which coincides with the solution of the boundary value problem

$$-\operatorname{div} \nabla y + y - \operatorname{div} f = 0 \text{ in } \Omega, \;\; (\partial y/\partial n) = 0 \text{ on } \Gamma. \tag{20}$$

However f is only a vector of L^2-functions and the above equation only holds in a distributional sense. So its solution belongs to $H^1(\Omega)$ but not much more.

As in section 5.1 we introduce the new functionals

$$\tilde{E}(t, \varphi) = E(\Omega_t, \varphi \circ T_t^{-1}), \quad \tilde{F}(t, \varphi) = F(\Omega_t, \varphi \circ T_t^{-1}) \quad J(t) = J(\Omega_t) \tag{21}$$

and transport all the integrals from Ω_t to Ω. We are now back to the set-up at the end of section 3 and Theorem 4 and Proposition 4 apply with $u = t$.

As a result

$$dJ(\Omega; V) = (d/dt)\, J(t)\, |_{t=0}$$

$$= \operatorname{Sup}\{d\tilde{F}(0, y; 1, 0) + d\tilde{E}(0, y; 0, p; 1, 0) : p \in \tilde{P}(0)\} \tag{22}$$

where y is the solution of

$$d\tilde{E}(0, y; 0, \psi) = 0, \quad \forall \psi \in H^1(\Omega) \tag{23}$$

and $\tilde{P}(0)$ is the set of solutions of the adjoint inequation

$$d\tilde{F}(0, y; 0, \psi) + d^2\, \tilde{E}(0, y; 0, p; 0, \psi) \leq 0, \quad \forall \psi \in H^1(\Omega)\,. \tag{24}$$

Note that the set $\tilde{P}(0)$ is not empty since the hypotheses of Proposition 4 are verified. However the elements of $P(0)$ belongs to $H^1(\Omega)$ but again not much more. In fact (24) reduces to

$$\int_{\Omega_+} (\nabla y/|\nabla y|) \bullet \nabla\psi\, dx + \int_{\Omega_0} |\nabla\psi|\, dx + \int_\Omega [\nabla p \bullet \nabla\psi + p\psi]\, dx \leq 0$$

$$p \in H^1(\Omega), \quad \forall \psi \in H^1(\Omega) \tag{25}$$

where

$$\Omega_+ = \{x \in \Omega : \nabla y(x) \neq 0\}, \quad \Omega_0 = \{x \in \Omega : \nabla y(x) = 0\}. \tag{26}$$

This is an example where all techniques based on an H^2-smoothness of the solution (y, p) of the saddle point equation and inequation would fail.

REFERENCES.

[1] J.P. AUBIN, L'analyse non linéaire et ses motivations économiques, Masson, Paris, New-York, 1984.

[2] I. EKELAND and R. TEMAM, Analyse convexe et problèmes variationnels, Dunod, Gauthiers-Villars, Paris, Bruxelles, Montréal, 1974.

[3] J.P. ZOLESIO, Semi-derivative of repeated eigenvalues, in "Optimization of distributed parameter structures", E.J. Haug and J. Céa, eds., pp. 1457-1473, Sijthoff and Noordhoff, Alphen aan den Rijn, Netherlands 1980.

[4] M.C. DELFOUR and J.P. ZOLESIO, Dérivation d'un Min Max et application à la dérivation par rapport au contrôle d'une observation non différentiable de l'état, C.R. Acad. Sc. Paris, to appear.

OPTIMAL CONTROL OF NONLINEAR SYSTEMS: CONVERGENCE OF SUBOPTIMAL CONTROLS. II

H.O. Fattorini

University of California, Los Angeles, USA

§1. INTRODUCTION. Optimal control problems for general nonlinear input-output systems have been studied in [10], [11], [12]; the main tool in these papers is Ekeland's variational principle ([6]) and the final result is a version of Pontryagin's maximum principle that applies equally well to systems described by nonlinear ordinary differential equations, nonlinear partial differential equations (either with distributed or boundary control) or nonlinear functional differential equations. Using arguments of a similar sort, strong convergence results for sequences of suboptimal control can be obtained; this has been done in [16] for a particular nonlinear hyperbolic equation, and in [13] for general systems. We obtain here a new version of the *convergence principle* in [13] that (although restricted to the time optimal problem) applies to quasilinear equations where the resolvent of the principal part is compact, but where the solution operator may not be compact, so that the results are especially suited to quasilinear controlled hyperbolic equations. In contrast, the results in [13] require compactness of the solution operator of the principal part, and the applications there are to quasilinear controlled parabolic equations. (In the parabolic case, convergence results have been obtained by different methods; see [4], [5]). The results in this paper, as well as those in [13], refer to the set target problem, rather than the point target problem, which shall be treated in a forthcoming paper. Compared with those in [13], the applications we consider here suffer from the predictable limitations stemming from finite velocity of propagation of disturbances, inherent to the hyperbolic case, thus the convergence results are somewhat incomplete (see §6).

§2. *SYSTEMS.* We denote by E, F arbitrary Banach spaces; U is a subset of F called the *control set*. Given $k \geq 0$, $T > 0$ the *control space* is the set of all (equivalence classes of) strongly measurable F-valued functions $u(\hat{t})$ defined in $-k \leq t \leq T$ such that $u(t)$ belongs to U almost everywhere. The control space (which will be denoted by $W(-k,T;U)$) is a metric space equipped with

$$d(u(\hat{t}),v(\hat{t})) = \text{meas}\,\{t;\ u(t) \neq v(t)\}\ . \tag{2.1}$$

(Here and below $u(\hat{t})$ indicates the function $t \to u(t)$, whereas $u(t)$ denotes its value at t; the same convention applies to other functions). The *output space* $C(0,T;E)$ consists of all E-valued continuous functions $y(\hat{t})$ defined in $0 \leq t \leq T$.

Consider a map $X : W(-k,T;U) \to C(0,T;E)$. We shall call X a *system* if (a) X is *causal* (that is, if the *trajectory* $y(t,u) = (Xu)(t)$ does not change in $0 \leq t \leq \bar{t}$ if $u(t)$ is modified in $\bar{t} \leq t \leq T$), (b) X is (pointwise) *continuous* in the sense that the map $u(\hat{t}) \to y(\bar{t},u)$ from $W(-k,\bar{t};U)$ into E is continuous for $\bar{t}$ fixed, (c) X is *differentiable with respect to needle perturbations* in the sense that, for each $u(\hat{t})$ in $W(-k,\bar{t};U)$ there exists a set $e = e(u)$ of full measure in $0 \leq t \leq \bar{t}$ such that

$$\xi(\bar{t},s,u,v,u(s)) = \lim_{r \to 0} \frac{1}{r}(y(\bar{t},u_{s,r,v}) - y(\bar{t},u)) \tag{2.2}$$

exists for all $v \in U$ and all $s \in e$; the *needle perturbation* $u_{s,r,v}(t)$ of $u(t)$ is defined as usual by $u_{r,s,v}(t) = v$ in $s - r \leq t < s$, $u_{s,r,v}(t) = u(t)$ elsewhere. The function ξ in (2.2) is assumed defined for $0 \leq s \leq t$, u in $W(-k,\bar{t};U)$ and v, w in U. Systems are meant to represent input-output relationships generated by ordinary differential equations, partial differential equations (both with boundary or distributed control), functional differential equations, etc., thus results obtained at this level of generality (such as the convergence principle in §4) will apply to all these situations. See [12] for further details, in particular on how assumptions (a), (b) and (c) can be relaxed. Note that (c) in this paper is less stringent than the corresponding condition in [12] (continuity of ξ is not postulated) due to the different nature of the results.

Finally, we point out that the constant k in the definition of control space is meant to account for delays in control action: for examples where $k > 0$ see [12, Section 3]. In the systems in §5 and §6 there is no delay in control action, so that $k = 0$.

§3. THE TIME OPTIMAL PROBLEM. We assume from now on that E and F are Hilbert spaces (although some of the material extends to the general setting). Let Y be a subset of E, called the *target set*. The *target condition* of the problem is

$$y(t,u) \in Y \, , \tag{3.1}$$

and the time optimal problem is, as customary, that of finding the optimal control(s) that satisfy (3.1) in minimum time $\bar{t}$. We shall consider chiefly the case

$$Y = Y_\delta = \{y; \|y - y_0\| \le \delta\} \, , \quad \delta > 0 \, . \tag{3.2}$$

A control $u(\hat{t}) \in W(-k,t;U)$ is called $(\tilde{t},\varepsilon)$-*suboptimal* if (3.1) holds for $t = \tilde{t}$ and

$$\tilde{t} \le \bar{t} + \varepsilon \, . \tag{3.3}$$

In this definition the target condition $y(t,u) \in Y_\delta$ may be replaced by an approximate target condition $y(t,u) \in Y_{\delta+\varepsilon}$.

Let $\{u^n\}$ be a sequence of controls, each in a different space $W(-k,t_n;U)$. Assume that $t_n \to t_0$. We say that $\{u^n\}$ *converges weakly* to $u \in W(-k,t_0;U)$ if u^n, extended to $t \ge t_n$ (if $t_n < t_0$) by setting $u^n(t) = 0$ there, or chopped off at t_0 (if $t_n > t_0$) converges to u weakly in $L^2(0,t_0;U)$. A similar meaning will be given to expressions like (3.5) below.

We add to (a), (b) and (c) the following assumption:

(d) *Let* $\{u^n\}$ *be a sequence of controls,* $u^n \in W(-k,t_n;U)$, *with* $t_n \to t_0$. *Then there exists a subsequence of* $\{u^n\}$ *(which we denote by the same symbol) and a* $\bar{u} \in W(-k,t_0;U)$ *such that*

$$u^n(t) \to \bar{u}(t) \quad \text{weakly in } L^2(-k,t_0;F) \, , \tag{3.4}$$

$$y(t,u^n) \to y(t,\bar{u}) \quad \text{strongly in } L^2(0,t_0;E) \, , \tag{3.5}$$

$$y(t_n,u^n) \to y(t_0,\bar{u}) \quad \text{weakly in } E \, . \tag{3.6}$$

As we shall see in the next section, Assumption (d) goes well with quasilinear hyperbolic equations in bounded domains. Of course, (3.6) alone implies existence of optimal controls; in this result, Y can be considerably more general than the set in (3.2).

Theorem 3.1 Suppose the system X *satisfies* (3.6) *of Assumption* (d) *and that the target set* Y *is weakly closed. Assume there exists a control* $u \in W(-k,t;U)$ *satisfying the target condition* (3.1). *Then an optimal control* $\bar{u}(\hat{t})$ *exists.*

The proof is classical and we omit it. The following result (which will not be used in the sequel) also uses only (3.6).

Theorem 3.2 Let the target set be (3.2). *Assume* X *satisfies* (3.6) *of Assumption* (d) *and let* $\{u^n\}$ *be a sequence of* (t_n,ε_n)-*suboptimal controls with* $\varepsilon_n \to 0$. *Then there exists a subsequence of* $\{u^n\}$ *(which we denote by the same symbol) such that*

$$t_n \to \bar{t} \quad (\bar{t} \text{ the optimal time}), \tag{3.7}$$

$$u^n \to \bar{u} \in W(-k,\bar{t};U) \quad \text{weakly in } L^2(-k,\bar{t};F)\ , \tag{3.8}$$

$$y(t_n,u^n) \to \bar{y} = y(\bar{t},\bar{u}) \in Y \quad \text{strongly in } E\ . \tag{3.9}$$

Proof: We can obviously achieve (3.7) and (3.8) by taking a subsequence. Let

$$\delta_0 = \lim\inf \|y(t_n,u^n) - y_0\|\ , \quad \delta_1 = \|\bar{y} - y_0\|\ .$$

Plainly, $\delta_1 \le \delta_0$. If $\delta_1 < \delta$, we apply Theorem 3.1 and obtain an alleged optimal trajectory with final point $\bar{y} = y(\bar{t},\bar{u})$ interior in Y_δ; this is a contradiction since we can hit Y_δ earlier by simply stopping at the boundary. Hence, $\delta_1 = \delta_0 = \delta$, and passing again if necessary to a subsequence we may assume that $\|y(t_n,u^n) - y_0\| \to \|\bar{y} - y_0\|$. Since $y(t_n,u^n) - y_0 \to \bar{y} - y_0$ weakly, the strong convergence claim (3.9) follows and the proof is complete.

Remark 3.3 If $\{u^n\}$ is a sequence of (t_n,ε_n)-suboptimal controls with $\varepsilon_n \to 0$, it follows that $t_n \to \bar{t}$, where $\bar{t}$ is the optimal time; otherwise, applying an obvious weak convergence argument based on (3.6), we could construct a trajectory hitting Y at a time $t < \bar{t}$, which is a contradiction.

§4. *THE CONVERGENCE PRINCIPLE.* We assume henceforth that the system X satisfies (d) and that the target set Y is the sphere Y_δ in (3.2).

Theorem 4.1 Let $\{u^n\}$ *be a sequence of* (t_n, ε_n)*-suboptimal controls with* $\varepsilon_n \to 0$. *Then there exists a subsequence of* $\{u^n\}$ *(which we denote by the same symbol), a sequence* $\{\tilde{t}_n\}$ *such that*

$$\tilde{t}_n < \bar{t}, \quad \tilde{t}_n \to \bar{t} \quad , \tag{4.1}$$

an optimal control $\bar{u} \in W(-k,\bar{t};U)$ *with*

$$u^n(t) \to \bar{u}(t) \quad \text{weakly in } L^2(-k,\bar{t};F) \ , \tag{4.2}$$

$$y(\tilde{t}_n,u^n) \to \bar{y} = y(\bar{t},\bar{u}) \quad \text{strongly in } E \ , \tag{4.3}$$

a second sequence of controls $\{\tilde{u}^n\}$ *with* $\tilde{u}^n \in W(-k,\tilde{t}_n;U)$ *such that*

$$d_n(u^n,\tilde{u}^n) \to 0 \ , \tag{4.4}$$

$(d_n$ *the distance in* $W(-k,\tilde{t}_n;U))$, *a sequence* $\{y^n\}$ *in* E *such that*

$$\|y^n\| = 1, \quad y^n \to y \quad \text{strongly in } E \ , \tag{4.5}$$

and a set e *of full measure in* $0 \le s \le \bar{t}$ *such that*

$$(y^n, \ \xi(\tilde{t}_n,s,\tilde{u}^n,v,\tilde{u}^n(s))) \ge -\delta_n \to 0$$

$$(v \in U, \ 0 \le s \le \tilde{t}_n, \ s \in e) \ . \tag{4.6}$$

Proof: Since L^2 convergence implies convergence almost everywhere (of a subsequence) using (3.5), taking subsequences repeatedly and finally selecting a diagonal subsequence we can construct a sequence $\{\tau_n\}$ in $0 \le t \le \bar{t}$, $\tau_n \to \bar{t}$ such that, for each n,

$$y(\tau_n,u^m) \to y(\tau_n,\bar{u}) \quad \text{strongly in } E$$

as $m \to \infty$. Combining this relation with the continuity of the trajectory $y(\hat{t},\bar{u})$ it is clear that a sequence $\{\tilde{t}_n\}$ satisfying (4.1) and (4.3) can be constructed (as a subsequence of $\{\tau_n\}$).

Let θ be an angle such that $\pi/4 < \theta < \pi/2$, and let b be a positive number such that the cone $\Gamma(\bar{y},z,\theta,b)$ of summit $\bar{y}$, aperture θ, height b and generator $z = \delta^{-1}(y_0 - \bar{y})$ defined by

$$\|y - \bar{y}\| \cos\theta \leq (y - \bar{y}, z) \leq b \tag{4.7}$$

is entirely contained in Y_δ. Let $\gamma > 0$,

$$\kappa_n = \|y(\tilde{t}_n,u^n) - \bar{y}\|/\gamma . \tag{4.8}$$

Since $\kappa_n \to 0$, we may assume that $\kappa_n < b/3$. Define

$$\bar{y}_n = \bar{y} + \kappa_n z \tag{4.9}$$

and consider the function

$$F_n(u) = \|y(\tilde{t}_n,u) - \bar{y}_n\| . \tag{4.10}$$

in the metric space $W(-k,\tilde{t}_n;U)$ (which is complete with respect to the distance d_n). Obviously, $F_n(u) > 0$ (otherwise, we could hit the target Y_δ at time $\tilde{t}_n < \bar{t}$). Moreover, in view of postulate (b) for systems, F_n is continuous. The control u^n is an approximate minimum of F_n in the sense that

$$F_n(u^n) = \|y(\tilde{t}_n,u^n) - \bar{y}_n\| \leq$$

$$\leq \|y(\tilde{t}_n,u^n) - \bar{y}\| + \|\bar{y} - \bar{y}_n\| \leq (1 + \gamma)\kappa_n . \tag{4.11}$$

Applying Ekeland's variational principle [6, Theorem 1] we obtain a control $\tilde{u}^n \in W(-k,\tilde{t}_n;U)$ such that

$$F_n(\tilde{u}^n) = \|y(\tilde{t}_n,\tilde{u}^n) - \bar{y}_n\| \leq F_n(u^n) \leq (1 + \gamma)\kappa_n , \tag{4.12}$$

$$d_n(u^n,\tilde{u}^n) \leq (1 + \gamma)^{1/2}\kappa_n^{1/2} , \tag{4.13}$$

$$F_n(w) \geq F(\tilde{u}^n) - (1 + \gamma)^{1/2}\kappa_n^{1/2} d_n(w,\tilde{u}^n)$$
$$(w \in W(-k,\tilde{t}_n;U)) . \tag{4.14}$$

We use (4.14) for needle perturbations $\tilde{u}^n_{s,r,v}(\hat{t})$ of $\tilde{u}^n(\hat{t})$, taking advantage of postulate (c) in the definition of system; here the fact that $F_n(u) > 0$ is essential (see [11], [12] for further details).

The result is

$$(y^n,\ \xi(\tilde{t}_n,s,\tilde{u}^n,v,\tilde{u}^n(s))) \geq -\ (1+\gamma)^{1/2}\kappa_n^{1/2}$$

$$(v \in U,\ s \in e_n)\ , \tag{4.15}$$

where e_n is the set in $0 \leq t \leq t_n$ postulated in (c) corresponding to $\tilde{u}^n$ and

$$y^n = (y(\tilde{t}_n,\tilde{u}^n) - \bar{y}_n)/\|y(\tilde{t}_n,\tilde{u}^n) - \bar{y}_n\|\ . \tag{4.16}$$

Applying resolution of triangles in the two dimensional subspace of E spanned by $y(\tilde{t}_n,\tilde{u}^n) - \bar{y}_n$ and the generator z of $\Gamma(\bar{y},z,\theta,b)$ we deduce that $\{y^n\}$ must belong to the cone $\Gamma(0,-z,\theta_\gamma,\infty)$ of equation

$$(y,\ -z) \geq \|y\| \cos\ \theta_\gamma\ , \tag{4.17}$$

with

$$\theta_\gamma = \pi - \theta - \text{arc sin}\ ((1+\gamma)^{-1}\sin\ \theta)\ . \tag{4.18}$$

(See Figure 1). Since $\theta_\gamma \to 0$ as $\theta \to \pi/2$ and $\gamma \to 0$, it is clear that $\{y^n\}$ belongs to the cone $\Gamma(0,-z,\theta',\infty)$ where θ' is as small as we wish. Using this for a sequence $\theta'_n \to 0$ and selecting at the end a diagonal sequence, we conclude that

$$y^n \to -z = \delta^{-1}(\bar{y} - y_0)\ ,$$

finally proving all the claims in Theorem 4.1 for the set $e = \lim\sup e_n$.

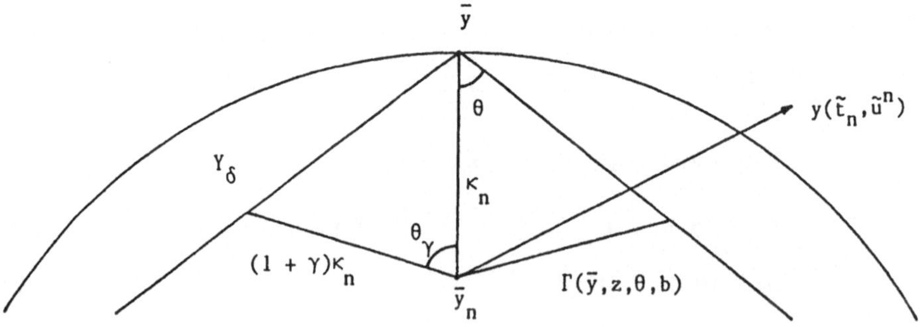

Figure 1

Remark 4.2 Theorem 4.1 should be compared with its counterpart, Theorem 4.1 in [13]. The only difference in the conclusions is that convergence of y^n in [13] is only weak (to a nonzero limit). This is due to the fact that the target sets in [13] are more general.

§5. *APPLICATIONS.* We consider here and in next section the initial value problem

$$y'(t) = Ay(t) + f(t,y(t),u(t)) \quad (0 \leq t \leq T) , \tag{5.1}$$

$$y(0) = y_0 , \tag{5.2}$$

where A is the infinitesimal generator of a strongly continuous semigroup $S(t)$ $(t \geq 0)$ in the Hilbert space E and f maps $[0,T] \times E \times U$ into E; here U is a bounded subset of a Banach space F. A *solution* of (5.1)-(5.2) is, by definition, a solution of the integrated version

$$y(t) = S(t)y_0 + \int_0^t S(t - \sigma)f(\sigma,y(\sigma),u(\sigma))\, d\sigma \quad (0 \leq t \leq T). \tag{5.3}$$

We assume that $f(t,y,u)$ has a Fréchet derivative $\partial_y f(t,y,u)$ with respect to y and that f (resp. $\partial_y f$) is continuous (resp. strongly continuous) and bounded on bounded subsets of $[0,T] \times E \times U$. These conditions being satisfied, (5.3) can be uniquely solved by succesive approximations in some interval $0 \leq t \leq T_0 \leq T$. To construct solutions in $0 \leq t \leq T$ we need *a priori* bounds on the solutions (to prevent them from blowing up in time $t < T$). When (5.1)-(5.2) models a controlled hyperbolic equation (as in next section), these estimates are best obtained using partial differential (rather than operator) equation methods. Here we *assume* that

$$\|y(t)\| \leq C \quad (0 \leq t \leq T_0) \tag{5.4}$$

for any solution in any interval $[0,T_0]$, where C does not depend on the interval. Under (5.4) and the rest of the assumptions on f we can prove that the map defined by

$$(Xu)(t) = y(t,u) = y(t) , \tag{5.5}$$

where $y(t)$ is the only solution of (5.3), satisfies postulates (a), (b) and (c) in §2: the function $\xi(t,s,u,v,w)$ in (2.3) is

$$\xi(t,s,u,v,w) = S(t,s;u)\{f(s,y(s,u),v) - f(s,y(s,u),w)\} \qquad (5.6)$$

where $S(t,s;u)$ is the solution operator of the *linearized equation* $z'(t) = (A + B(t))z(t)$, $B(t) = \partial_y f(t,y(t,u),u(t))$, that is, the only strongly continuous solution of the operator equation

$$S(t,s;u)y = S(t - s)y + \int_s^t S(t - \sigma)B(\sigma)S(\sigma,s;u)y \, d\sigma \qquad (5.7)$$

in $0 \leq s \leq t \leq \bar{t}$. For proofs and additional details, see [12].

Lemma 5.1 Let $S(t)$ *be a strongly continuous semigroup in a Hilbert space* E. *Assume that* $R(\mu,A) = (\mu I - A)^{-1}$ *is compact for some* $\mu \in \rho(A)$. *Then the operator*

$$(\Lambda u)(t) = \int_0^t S(t - \sigma)u(\sigma) \, d\sigma \qquad (5.8)$$

from $L^2(0,T;E)$ *(equipped with the weak topology) into* $C(0,T;E)$ *(equipped with the* $L^2(0,T;E)$ *norm) is continuous.*

Proof: Set $S(t) = 0$ for $t < 0$ and extend $u(t)$ in the same way to $t < 0$ and $t > T$. We can then write (5.8) as the convolution $S*u$; moreover, translating A if necessary, we may assume as well that $\|S(t)\| \leq Ce^{-ct}$ $(c > 0)$ in $t \geq 0$. Accordingly, if we denote by Φ the Fourier transform, we have

$$(\Phi\Lambda u)(\sigma) = R(-i\sigma,A)\Phi u(\sigma) \ . \qquad (5.9)$$

Let $\{u_n(\hat{t})\}$ be a sequence converging weakly to $u(t)$. Then

$$(u,\Phi u_n(\sigma)) = \int_0^T (e^{i\sigma t}u,u_n(t)) \, dt \to \int_0^T (e^{i\sigma t}u,u(t)) \, dt = (u,\Phi u(\sigma))$$

for $u \in E$ and $-\infty \leq \sigma \leq \infty$ (actually, convergence is uniform on compacts of $-\infty < \sigma < \infty$ since, by the Arzelà - Ascoli theorem the set $\{e^{i\sigma t}u;\ |\sigma| \leq a\}$ is precompact in $L^2(0,T;E)$ for any a). On account of the fact that each $R(-i\sigma,A)$ is compact (consequence of the second resolvent equation) and of the weak convergence of $\{u_n(\sigma)\}$ we deduce that $R(-i\sigma,A)\Phi u_n(\sigma) \to R(-i\sigma,A)\Phi u(\sigma)$ strongly. We use then the bound $\|R(-i\sigma,A)\| \leq C$, the dominated convergence theorem and Plancherel's theorem and deduce that $\Lambda u_n(\hat{t}) \to \Lambda u(\hat{t})$ in $L^2(0,T;E)$. This completes the proof.

We assume in the sequel that the nonlinearity in (5.1) satisfies

$$f(t,y,u) = f(t,y) + Bu \ , \tag{5.10}$$

where $B : F \to E$ is a linear bounded operator.

Theorem 5.2 Let the semigroup $S(t)$ *in* (5.3) *satisfy the assumptions of Lemma 5.1, and let* U *be closed, bounded and convex. Then the system* X *defined by (5.5) satisfies Assumption* (d).

Proof: Let $\{u^n(\hat{t})\}$ be a sequence of controls, $u^n \in W(0,t_n;U)$ with $t_n \to t_0$. We achieve (3.4) passing to a subsequence; due to the hypoteses on U, $\bar{u} \in W(0,t_0;U)$.

The solution $y(t,u^n)$ of (5.3) corresponding to $u = u^n$ is defined as the limit of the sequence $\{y_m(t,u^n)\}$ defined by $y_0(t,u^n) = y_0$,

$$y_{m+1}(t,u^n) = S(t)y_0 +$$

$$\int_0^t S(t-\sigma)f(\sigma,y_m(\sigma,u^n))\,d\sigma + \int_0^t S(t-\sigma)Bu^n(\sigma)\,d\sigma \tag{5.11}$$

The sequence $\{y_m(t,u^n);\ m = 1,2,\ldots\}$ converges absolutely and uniformly (with respect to σ and n) as $m \to \infty$ in some interval $[0,T_0]$, where T_0 does not depend on n. We use (5.11) combined with Lemma 5.1: passing to a subsequence, we deduce that $y_1(t,u^n) \to y_1(t,\bar{u})$ in L^2 and almost everywhere in $0 \le t \le T_0$. Using then the dominated convergence theorem in the first term of (5.11) and passing again to a subsequence, we obtain that $y_2(t,u^n) \to y_2(t,\bar{u})$ in L^2 and almost everywhere in $0 \le t \le T_0$ as well. Proceeding inductively in the same fashion and selecting a diagonal subsequence at the end, we show that $y_m(t,u^n) \to y(t,\bar{u})$ in L^2 and almost everywhere in $[0,T_0]$ for all m. Writing

$$y(t,u^n) - y(t,\bar{u}) = (y(t,u^n) - y_m(t,u^n)) +$$

$$(y_m(t,u^n) - y_m(t,\bar{u})) + (y_m(t,\bar{u}) - y(t,\bar{u})) \tag{5.12}$$

and using the convergence properties of $\{y_m(t,u^n)\}$ precised above, (3.5) follows, although *only in* $[0,T_0]$. To extend the result to the entire interval $[0,t_0]$ we argue as follows. If necessary shifting T_0 a little to the left we may assume that $y(T_0,u^n)$ is convergent and proceed to solve by succesive approximations in an interval $[T_0,T_1]$, where we can apply the same arguments as in $[0,T_0]$; using then the same reasoning in intervals $[T_1,T_2]$, $[T_2,T_3],\ldots$ whose lenght does not tend to zero because of (5.4), we obtain (3.5) in full.

The proof of (3.6) is similar; we begin by showing that

$$(\Lambda Bu^n)(t_n) \to (\Lambda Bu^n)(t_0) \quad \text{weakly in } E \ , \tag{5.13}$$

(compactness of $R(\mu,A)$ is not needed for (5.13)) and operate inductively with (5.11) using (5.12) at the end, relying on convergence of $y_m(t,u^n)$ at each step. Extension to the whole interval $[0,t_0]$ is handled as above: we omit the details.

Theorem 5.3 Let the semigroup $S(t)$ *and the control set* U *satisfy the assumptions of Theorem 5.1 and let* $\{u^n\}$ *be a sequence of controls,* $u^n \in W(0,t_n;U)$ *with* $t_n \to t_0$*. Then there exists a subsequence of* $\{u^n\}$ *(denoted by the same symbol) and a* $\bar{u} \in W(0,t_0;U)$ *such that*

$$u^n(t) \to \bar{u}(t) \quad \text{weakly in } L^2(0,t_0;F) \ , \tag{5.14}$$

$$S(t_n,s;u^n)^*y \to S(t_0,s;\bar{u})^*y \ , \tag{5.15}$$

for $0 < s < t_0$*, where* $S(t,s;u)$ *is the operator defined by* (5.7).

Proof: Taking adjoints in (5.7) we obtain the following integral equation for $S(t,s;u)^*$:

$$S(t,s;u)^*y = S(t-s)^*y + \int_s^t S(\sigma,s;u)^*B(\sigma)^*S(t-\sigma)^*y \, d\sigma \ . \tag{5.16}$$

Obviously, (5.14) can be achieved taking a subsequence: (5.15) is easily obtained using the uniform boundedness of $y(t,u^n)$, the L^2 convergence of (a subsequence of) $y(t,u^n)$, the assumptions on $\partial_y f(t,u)$ and Gronwall's inequality (for a somewhat similar argument, see [13]).

We apply the convergence principle (Theorem 4.1) to the system defined by (5.5). We obtain

$$(B^*S(\tilde{t}_n,s;u^n)^*y^n, \ v - u^n(s)) \geq -\delta_n \to 0 \tag{5.17}$$

for $v \in U$, $0 < s < t_n$, $s \in e$. In view of the fact that $y^n \to y$ and of Lemma 5.3 we may transform (5.17) into

$$(B^*S(\bar{t},s;\bar{u})^*y,\ v - u^n(s)) \geq -\delta_n \to 0$$
$$(v \in U,\ 0 \leq s \leq \bar{t},\ s \in e)\ , \tag{5.18}$$

modifying the δ_n if necessary.

Let $z \neq 0$. Denote by $U(z,\delta)$ the set of all $u \in U$ such that $(z,\ v - u) \geq -\delta$ (all $v \in U$), where $\delta > 0$ (See Figure 2)

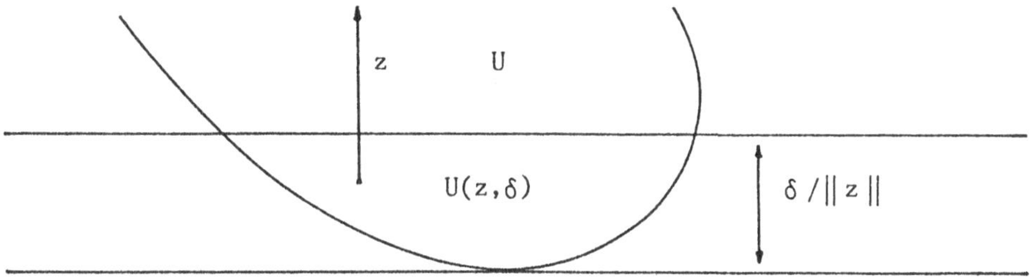

Figure 2

Theorem 5.4 Let X *be the system defined by* (5.5). *Assume that* $R(\mu,A)$ *is compact for some* μ *and that the control set* U *is closed, convex and bounded and satisfies*

$$\operatorname{diam} U(z,\delta) \to 0 \quad \text{as} \quad \delta \to 0 \tag{5.19}$$

for every $z \neq 0$. *Then, if the optimal control* $\bar{u}(t)$ *is unique and* c *is the subset of* $0 \leq s \leq \bar{t}$ *(*$\bar{t}$ *the optimal time) where*

$$B^*S(\bar{t},s;\bar{u})^*y \neq 0\ , \tag{5.20}$$

any sequence $\{u^n\}$ *of suboptimal controls converges to* $\bar{u}$ *in* $L^p(c;F)$, $1 \leq p < \infty$.

The proof is essentially contained in (5.18). The fact that the *entire* sequence $\{u^n\}$ converges to $\bar{u}$ is a consequence of the uniqueness of $\bar{u}$; for a similar argument see [13]. If $\bar{u}$ is not known to be unique, only convergence of subsequence can be guaranteed.

§6. *QUASILINEAR HYPERBOLIC DISTRIBUTED PARAMETER SYSTEMS.* We apply the results to controlled wave equations. The first example is

$$D_t^2 y(t,x) = \sum\sum D^j(a_{jk}(x)D^k y(t,x)) - g(y(t,x)) + u(t,x) \tag{6.1}$$

in a bounded domain Ω of class $C^{(2)}$ with boundary Γ in m-dimensional Euclidean space R^m; here $D_t = \partial/\partial t$, $D^j = \partial/\partial x_j$, $x = (x_1, x_2, \ldots, x_m)$. The solution $y(t,x)$ of (6.1) is expected to satisfy the Dirichlet boundary condition

$$y(t,x) = 0 \quad (x \in \Gamma,\ 0 \leq t \leq T) . \tag{6.2}$$

We assume that the a_{jk} are continuously differentiable and satisfy

$$\sum\sum a_{jk}(x)\eta_j\eta_k \geq \kappa|\eta|^2 \quad (x \in \Omega,\ \eta \in R^m) \tag{6.3}$$

for some $\kappa > 0$. We assume of course that $a_{jk} = a_{kj}$. We reduce (6.1) to a first order system in the customary way:

$$D_t y(t,x) = y_1(t,x) , \tag{6.4}$$

$$D_t y_1(t,x) = \sum\sum D^j(a_{jk}(x)D^k y(t,x)) - g(y(t,x)) + u(t,x) , \tag{6.5}$$

in the space $E = H_0^1(\Omega) \times L^2(\Omega)$; this casts the equation in the form (5.1), with

$$A = \begin{pmatrix} 0 & I \\ \sum\sum D^j(a_{jk}(x)D^k) & 0 \end{pmatrix}$$

(domain $D(A) = (H^2(\Omega) \cap H_0^1(\Omega)) \times H_0^1(\Omega)$),

$$f(t,(y,y_1),u) = \begin{pmatrix} 0 \\ g(y) + u \end{pmatrix}$$

The differentiability assumptions required in §5 will be satisfied if the map

$$(\phi(y(\hat{x})))(x) = g(y(x)) \tag{6.6}$$

from $H_0^1(\Omega)$ into $L^2(\Omega)$ has a Fréchet derivative $\partial\phi$ and ϕ (resp. $\partial\phi$) is everywhere continuous (resp. strongly continuous). Using Sobolev's imbedding theorem ([1, p. 97]) we check easily that, for $m = 1$, this will be the case if g is continuously differentiable;

for $m > 1$, we require also the estimate

$$|g'(y)| \leq C(1 + |y|^{\alpha}) , \tag{6.7}$$

for α arbitrary when $m = 2$; for $m > 2$ we take $\alpha = 2/(m - 2)$. This conditions guarantee that ϕ will satisfy the smoothness conditions in §5; we have

$$(\partial\phi(y(\hat{x}))h(\hat{x}))(x) = g'(y(x))h(x) \tag{6.8}$$

It follows that the arguments in §5 can be applied to construct local solutions of (6.1)-(6.2) given initial conditions

$$y(0,x) = y_0(x) , \quad y_t(0,x) = y_1(x) \quad (x \in \Omega) . \tag{6.9}$$

In order to establish the *a priori* bound (5.4) we require that

$$yg(y) \geq 0 \quad (- \infty < y < \infty) \tag{6.10}$$

If $y(t,x)$ is a smooth solution of (6.1)-(6.2) (5.4) can be obtained easily enough multiplying (6.1) by y_t, integrating in the cylinder $[0,T_0] \times \Omega$, applying the divergence theorem and Gronwall's lemma. However, the solutions of (5.1)-(5.2) constructed through the integral equation (5.3) may not be smooth, thus a far subtler analysis must be used. For the necessary details, see [14] .

We define a system through (5.5) with $E = H_0^1(\Omega) \times L^2(\Omega)$, $F = L^2(\Omega)$, U the unit sphere in $L^\infty(\Omega)$, so that $W(0,T;U)$ is the unit sphere of $L^\infty([0,T] \times \Omega)$. Obviously, U is bounded, closed and convex in $F = L^2(\Omega)$; compactness of the resolvent $R(\mu,A)$ is a classical result (see [15]). Consequently, the results in §5 apply: the operator $S(t,s;u)^*$ is the solution operator of the system

$$D_t z(t,x) = \sum\sum D^j(a_{jk}(x)D^k z_1(t,x)) - g'(y(t,x,u))z_1(t,x) , \tag{6.11}$$

$$D_t z_1(t,x) = z(t,x) , \tag{6.12}$$

with boundary condition (6.3). The treatment of (6.11)-(6.12) is essentially the same as that of (6.4)-(6.5), but simpler; details are omitted. The operator $B : L^2(\Omega) \to H_0^1(\Omega) \times L^2(\Omega)$ is given by $Bu = (0,u)$, so that $B^* : H_0^1(\Omega) \times L^2(\Omega) \to L^2(\Omega)$ is $B^*(z,z_1) = z_1$.

Assume that there exists an infinite set c in $0 \leq t \leq \bar{t}$ where $B\,(z(t),z_1(t)) = 0$, where $(z(t),z_1(t)) = (z(t,x),z_1(t,x))$ is a smooth solution of (6.11)-(6.12). It follows from the second equation that if t is an accumulation point of c then $(z(t),z_1(t)) = 0$ so that, by uniqueness, $(z(t),z_1(t))$ is the null solution of (6.11)-(6.12). This is as well true for an arbitrary (not necessarily smooth) solution; the argument is the same used in [9, p.169]. Consequently, it follows that the set d in Theorem 5.4 where (5.20) occurs is the complement of a finite set in $0 \leq t \leq \bar{t}$. Nevertheless Theorem 5.4 cannot be applied directly since the control set U does not satisfy (5.19). However, nontrivial information can be obtained. We apply directly (5.18), deducing existence of a continuous $L^2(\Omega)$-valued function $z(s,\hat{x})$ such that $z(s,\hat{x}) = 0$ (as an element of $L^2(\Omega)$) only in a finite set of points in $0 \leq s \leq \bar{t}$ satisfying

$$\int_\Omega z(s,x)(v(s,x) - u^n(s,x))\,dx \geq -\delta_n \to 0 \qquad (6.13)$$

for all $v(s,x)$ in $W(0,\bar{t};U) = L^2([0,\bar{t}]\times\Omega)$. Assume the optimal control $\bar{u}(t) = \bar{u}(s,x)$ is unique; then we deduce integrating (6.13) in $0 \leq s \leq \bar{t}$ that $\{u^n(s,x)\}$ converges in measure (thus in $L^2(\Xi)$) to the optimal control

$$\bar{u}(s,x) = -\operatorname{sign} z(s,x)\ , \qquad (6.14)$$

where Ξ is the subset of $[0,\bar{t}]\times\Omega$ where

$$z(s,x) \neq 0\ .$$

However, we do not obtain any information outside of Ξ.

The case where U is the unit sphere of $L^2(\Omega)$ is much easier to deal with (Theorem 5.4 can be applied directly). Also, the final convergence result is much more satisfactory, since we obtain convergence of the sequence $\{u^n(s,x)\}$ of suboptimal controls in $L^2([0,\bar{t}]\times\Omega)$ (in fact, L^p convergence for any $p < \infty$), always under the uniqueness condition for $\bar{u}$; if that fails, convergence of a subsequence follows. However we note that in this case, convergence results are available as well in the more demanding *point target* case, where the target condition (3.1) becomes

$$y(t,u) = \bar{y}\ .$$

The arguments are somewhat different, depending on controllability properties of the linearized system.

The results in this paper, being independent of controllability assumptions, can be applied even if the control space F is "very small", although, when Theorem 5.2 is applied, approximate controllability enters in the identification of the set d. For instance, consider the linear version of (6.1) ($g = 0$) with 1-dimensional control $u(t,x) = b(x)u(t)$, $b(x) \in L^2(\Omega)$. Here $B : R^1 \to H_0^1(\Omega) \times L^2(\Omega)$ is $Bu = (0, b(x)u)$, so that $B^*(z,z_1) = \int b(x)z_1(x)dx$. In space dimension $m = 1$, if $b(x)$ has nonzero scalar product with all eigenvalues of the problem, then the linearized system will be approximately controllable in time $t >$ a certain t_{min}, so that d cannot exclude an interval of length $> t_{min}$. This is not much in comparison to what we can do in the parabolic case, where analyticity arguments may be used to show that d has full measure in $[0,\bar{t}]$ in any space dimension (see [13]).

We note finally that, in any case, suboptimal controls can be computed by the ε-method in [3] (see again [13]) and that the boundary condition (6.2) can be replaced by $D^\nu = \gamma(x)u(t,x)$, where D^ν is the conormal derivative at Γ with respect to the principal part of (6.1).

REFERENCES.

[1] R.A. ADAMS, *Sobolev Spaces*. Academic Press, New York, 1975.

[2] J.P. AUBIN and I. EKELAND, *Applied Nonlinear Analysis*. Wiley, New York, 1984.

[3] A.V. BALAKRISHNAN, On a new computing technique in optimal control, SIAM J. Control 6 (1968) 149-173.

[4] V. BARBU, *Optimal Control of Variational Inequalities*. Research Notes in Math. 100, Pitman, London 1984.

[5] V. BARBU, The time optimal problem for a class of nonlinear distributed systems. Preprint.

[6] I. EKELAND, Nonconvex minimization problems. Bull. Amer. Math. Soc. 1(NS) (1979) 443-474.

[7] H.O. FATTORINI, The time optimal control problem in Banach spaces. Appl. Math. Optimization 1 (1974) 163-188.

[8] H.O. FATTORINI, Local controllability of a nonlinear wave equation. Math. Systems Theory 9 (1975) 30-44.

[9] H.O. FATTORINI, The time optimal problem for distributed control of systems described by the wave equation. Control Theory of Systems Described by Partial Differential Equations, Academic Press, New York (1977) 151-175.

[10] H.O. FATTORINI, The maximum principle for nonlinear nonconvex systems in infinite dimensional spaces. Proceedings of the 2nd. International Conference on Control Theory of Distributed Parameter Systems, Vorau (1984).
[11] H.O. FATTORINI, The maximum principle for nonlinear nonconvex systems with set targets. Proceedings of the 24th. IEEE Conference on Decision and Control, Fort Lauderdale (1985) 1999-2004.
[12] H.O. FATTORINI, A unified theory of necessary conditions for nonlinear nonconvex control systems. To appear.
[13] H.O. FATTORINI, Optimal control of nonlinear systems: convergence of suboptimal controls, I. To appear.
[14] K. JÖRGENS, Das Anfangswertproblem in Grossen für eine Klasse nichtlinearer Wellengleichungen, Math. Z. 77 (1961) 295-308.
[15] V.P. MIHAILOV, *Partial Differential Equations*. Mir, Moscow, 1978.
[16] V.I. PLOTNIKOV and M.I. SUMIN, The construction of minimizing sequences in problems of control of systems with distributed parameter systems, Zh. Vycisl. Mat. mat. fiz. 22 (1982) 49-56.

This work was supported in part by the National Science Foundation under grant DMS 82-00645

COATING REFLECTIVITY

W.W. Hager[1] and R. Rostamian[2]

1. Introduction

Recently, we have studied the following problem: Design a viscoelastic coating to reduce the reflection of sound from a wall. In [5] we analyze the case where the incident wave is normal to the reflective surface. A formula is derived which gives the strength of the reflected wave relative to the strength of the incident wave. Utilizing this formula for the reflectivity, efficient techniques are presented in [4] and [5] to minimize the maximum amplitude of reflected sound waves corresponding to waves with frequencies contained in some given interval. We now develop a formula which gives the coating reflectivity for waves which strike the coating at an oblique angle. This formula can be combined with the algorithms of [4] and [5] to solve the minimax problem.

2. Reflection at an interface

In the framework of linear elasticity, the equation of motion for an isotropic elastic material is (see Gurtin's treatise[3])

$$\rho\frac{\partial^2 \mathbf{v}}{\partial t^2} = \operatorname{div}\left[\,\mu(\nabla\mathbf{v} + (\nabla\mathbf{v})^T) + \lambda(\operatorname{div}\,\mathbf{v})\mathbf{I}\,\right] \tag{2.1}$$

[1]Department of Mathematics, Pennsylvania State University, University Park, Pennsylvania, 16802. This author was partly supported by the National Science Foundation grant DMS-8401758.

[2]Department of Mathematics, University of Maryland Baltimore County, Catonsville, Maryland, 21228. This author was partly supported by the Institute for Mathematics and Its Applications, Minneapolis, Minnesota.

where $\mathbf{v} = \mathbf{v}(\mathbf{x},t)$ is the displacement vector, ρ is the density, and the coefficients μ and λ are the Lamé moduli. If the mechanical properties ρ, μ, and λ vary with position, we say that the material is *inhomogeneous*. When ρ, μ, and λ are constants in some region, we say that the material is homogeneous in that region. In either case, the material has a *strongly elliptic* elasticity tensor at a point if the inequalities

$$\mu > 0 \quad \text{and} \quad 2\mu + \lambda > 0 \tag{2.2}$$

hold at that point.

A homogeneous, isotropic linearly elastic material with a strongly elliptic elasticity tensor admits exactly two types of wave propagation mechanisms which can be described as follows: Given any three unit vectors $\mathbf{d}$, $\mathbf{s}$, and $\mathbf{p}$ where $\mathbf{s}\cdot\mathbf{p} = 0$, it can be verified that the traveling waves

$$\mathbf{v}(\mathbf{x},t) = \mathbf{d}f(t-D\mathbf{x}\cdot\mathbf{d}) \tag{2.3}$$

and

$$\mathbf{v}(\mathbf{x},t) = \mathbf{p}g(t-S\mathbf{x}\cdot\mathbf{s}) \tag{2.4}$$

formally satisfy the equation of motion (2.1) for any choice of the wave profiles f and g where D and S are defined by

$$D = \sqrt{\frac{\rho}{2\mu + \lambda}} \quad \text{and} \quad S = \sqrt{\frac{\rho}{\mu}}\,. \tag{2.5}$$

Of course, (2.3) represents a plane wave which travels in direction $\mathbf{d}$ with speed $c_d = 1/D$ while (2.4) represents a plane wave which travels in direction $\mathbf{s}$ with speed $c_s = 1/S$. Since the motion in (2.3) is along the direction of propagation $\mathbf{d}$, this wave is called *longitudinal* or *dilatational*. Since the motion in (2.4) is perpendicular to the direction of propagation $\mathbf{s}$, this wave is called *transverse* or *shear*. Thus c_d is the dilatational wave speed and c_s is the shear wave speed. Their reciprocals, D and S, are sometimes called the *dilatational slowness* and the *shear slowness* respectively (see [6]).

Let us consider two half-spaces of homogeneous, isotropic elastic materials with distinct mechanical properties and with a common plane interface. A dilatational wave striking the interface typically produces a reflected dilatational wave, a reflected shear wave, a transmitted dilatational wave, and a transmitted shear wave. Similarly, a shear wave striking the interface typically produces a reflected dilatational wave, a reflected shear wave, a transmitted dilatational wave, and a transmitted shear wave. Therefore, when a combination of dilatational and shear waves strikes the interface, eight different waves are generated altogether. The propagation directions of the eight outgoing waves can be determined from the directions of the incident waves by substituting (2.3) and (2.4) into (2.1). The resulting relations, known as Snell's Law, can be stated as follows:

> The propagation vectors $\mathbf{s}_r$ and $\mathbf{d}_r$ of the reflected wave and the propagation vectors $\mathbf{s}_t$ and $\mathbf{d}_t$ of the transmitted wave lie in the plane formed by the propagation vectors $\mathbf{d}$ or $\mathbf{s}$ of the incident wave and the normal to the interface. Moreover, if $\mathbf{m}$ is a unit vector at the intersection between this plane and the interface (see Figure 1), then for an incident dilatational wave, we have
>
> $$D\mathbf{d}\cdot\mathbf{m} = D\mathbf{d}_r\cdot\mathbf{m} = S\mathbf{s}_r\cdot\mathbf{m} = D_t\mathbf{d}_t\cdot\mathbf{m} = S_t\mathbf{s}_t\cdot\mathbf{m} \tag{2.6}$$
>
> and for an incident shear wave, we have
>
> $$S\mathbf{s}\cdot\mathbf{m} = D\mathbf{d}_r\cdot\mathbf{m} = S\mathbf{s}_r\cdot\mathbf{m} = D_t\mathbf{d}_t\cdot\mathbf{m} = S_t\mathbf{s}_t\cdot\mathbf{m}\,. \tag{2.7}$$

Throughout this paper, the subscript t is attached to parameters associated with the transmitted wave while the subscript r is attached to parameters associated with the reflected wave. Given the unit propagation vector $\mathbf{d}$ of the incident dilatational wave, (2.6) determines the propagation vectors of the four scattered waves. Similarly, given the unit propagation vector $\mathbf{s}$ of the incident shear wave, (2.7) determines the propagation vectors of the four scattered waves.

By (2.6) and (2.7), we conclude that if the incident dilatational and shear waves propagate in directions $\mathbf{d}$ and $\mathbf{s}$ which satisfy

$$D\mathbf{d}\cdot\mathbf{m} = S\mathbf{s}\cdot\mathbf{m} \tag{2.8}$$

where $\mathbf{d}$, $\mathbf{s}$, and $\mathbf{m}$ all lie in the same plane, then the propagation directions of the two reflected dilatational waves and the two transmitted dilatational

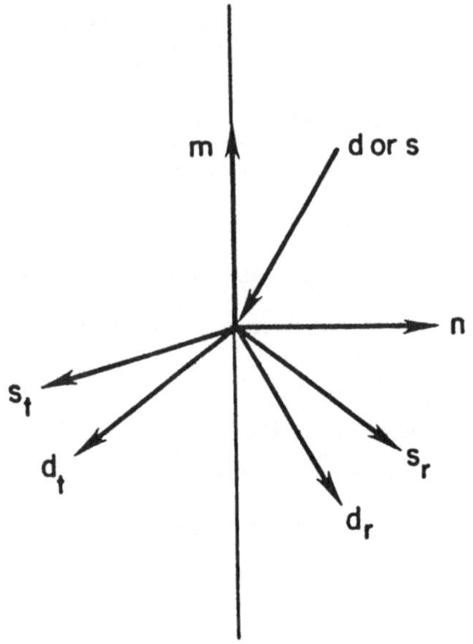

Figure 1. Incident, reflected, and transmitted waves.

waves are identical. Similarly, the propagation directions of the two reflected shear waves and the two transmitted shear waves are identical. Thus the two incident waves will produce just four distinct scattered waves as opposed to the usual eight scattered waves. We call a pair $(\mathbf{d}, \mathbf{s})$ of dilatational and shear waves which satisfy (2.8) a *conjugate pair* of waves. Note that both the reflected pair $(\mathbf{d}_r, \mathbf{s}_r)$ and the transmitted pair $(\mathbf{d}_t, \mathbf{s}_t)$ corresponding to an incident dilatational wave or an incident shear wave are always conjugate.

3. Transmission and reflection matrices

Let δ denote the amplitude of an incident sinusoidal dilatational wave and let σ denote the amplitude of an incident sinusoidal shear wave. A conjugate pair of incident waves produces a conjugate pair of reflected waves and a conjugate pair of transmitted waves. It can be shown that the corresponding amplitudes δ_r and σ_r of the reflected waves and the corresponding amplitudes

δ_t and σ_t of the transmitted waves depend linearly upon the amplitudes of the incident waves. In other words, there is a 2×2 matrix **R** called the *reflection matrix* and a 2×2 matrix **T** called the *transmission matrix* such that

$$\begin{bmatrix} \delta_r \\ \sigma_r \end{bmatrix} = \begin{bmatrix} R_{11} & R_{12} \\ R_{21} & R_{22} \end{bmatrix} \begin{bmatrix} \delta \\ \sigma \end{bmatrix} \tag{3.1a}$$

and

$$\begin{bmatrix} \delta_t \\ \sigma_t \end{bmatrix} = \begin{bmatrix} T_{11} & T_{12} \\ T_{21} & T_{22} \end{bmatrix} \begin{bmatrix} \delta \\ \sigma \end{bmatrix}. \tag{3.1b}$$

The matrices **R** and **T** depend on the mechanical properties of the material through which the waves propagate as well as the propagation vectors **d** and **s**. To evaluate **R** and **T**, equate displacements and stresses at the interface. Since this computation is quite lengthy, we just state the final results. Figure 2 depicts a conjugate pair (**d**, **s**) scattering at an interface. The indicated angles are related by Snell's law which takes the form

$$D\sin\alpha = S\sin\beta = D_t\sin\alpha_t = S_t\sin\beta_t\,. \tag{3.2}$$

Let us introduce the 2×2 matrices **A** and **B** defined by

$$\mathbf{A} = \begin{bmatrix} \cos\alpha & \sin\beta \\ -\sin\alpha & \cos\beta \end{bmatrix} \tag{3.3a}$$

and

$$\mathbf{B} = \begin{bmatrix} \rho c_d \cos 2\beta & \rho c_s \sin 2\beta \\ -2\rho c_s \sin\beta\cos\alpha & \rho c_s \cos 2\beta \end{bmatrix}. \tag{3.3b}$$

Analogous matrices $\mathbf{A}_t$ and $\mathbf{B}_t$ are defined for the region which contains the transmitted wave. These matrices are obtained by adding a "t" subscript to each variable in (3.3). It can be shown that the reflection and transmission matrices are solutions to the system

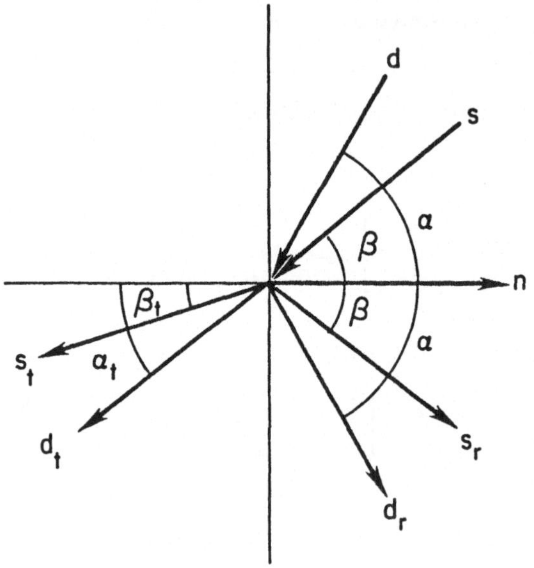

Figure 2. The angles associated with a scattering conjugate pair.

$$\begin{bmatrix} \mathbf{PA} & \mathbf{A}_t \\ -\mathbf{PB} & \mathbf{B}_t \end{bmatrix} \begin{bmatrix} \mathbf{R} \\ \mathbf{T} \end{bmatrix} = \begin{bmatrix} \mathbf{A} \\ \mathbf{B} \end{bmatrix} \tag{3.4}$$

where

$$\mathbf{P} = \begin{bmatrix} 1 & 0 \\ 0 & -1 \end{bmatrix}.$$

In summary, the amplitude of the reflected and the transmitted wave is given by (3.1) where $\mathbf{R}$ and $\mathbf{T}$ satisfy (3.4).

Remark 1. If the incident wave is normal to the interface, then $\alpha = \beta = 0$ and by (3.2), we have $\alpha_t = \beta_t = 0$. The solution to (3.4) is

$$\mathbf{R} = \begin{bmatrix} \dfrac{p_t - p}{p_t + p} & 0 \\ 0 & \dfrac{q_t - q}{q_t + q} \end{bmatrix} \tag{3.5a}$$

and

$$\mathbf{T} = \begin{bmatrix} \dfrac{2p}{p_t + p} & 0 \\ 0 & \dfrac{2q}{q_t + q} \end{bmatrix} \tag{3.5b}$$

where $p = \rho c_d$ is the dilatational *compliance* and $q = \rho c_s$ is the shear *compliance*. Since **R** and **T** are diagonal matrices, the effects of the dilatational and shear waves are decoupled when the incident wave is normal to the interface. The case of normal incidence is thoroughly analyzed in our paper[5].

The expressions in (3.5) bring out the explicit dependence of the reflection and transmission matrices on the mechanical properties associated with each side of the interface. Motivated by this relation, we introduce the *compliance matrix* $\mathbf{H} = \mathbf{B}\mathbf{A}^{-1}$ for the right half-space and the compliance matrix $\mathbf{H}_t = \mathbf{B}_t\mathbf{A}_t^{-1}$ for the left half-space. Solving (3.4), we obtain

$$\mathbf{R} = \mathbf{A}^{-1}(\mathbf{H}_t\mathbf{P} + \mathbf{P}\mathbf{H})^{-1}(\mathbf{H}_t - \mathbf{H})\mathbf{A} \tag{3.6}$$

and

$$\mathbf{T} = \mathbf{A}_t^{-1}(\mathbf{H}\mathbf{P} + \mathbf{P}\mathbf{H}_t)^{-1}(\mathbf{P}\mathbf{H} + \mathbf{H}\mathbf{P})\mathbf{A}\ , \tag{3.7}$$

which are the analogues of (3.5) for oblique incidence. Using (3.6), we can solve for $\mathbf{H}_t$ in terms of the reflection matrix and material properties for the right half-space:

$$\mathbf{H}_t = (\mathbf{H}\mathbf{A} + \mathbf{P}\mathbf{H}\mathbf{A}\mathbf{R})(\mathbf{A} - \mathbf{P}\mathbf{A}\mathbf{R})^{-1}\ . \tag{3.8}$$

Also, (3.3) can be combined with Snell's law to express the compliance matrix **H** in terms of the reflection angles and the dilatational and shear compliances:

$$\mathbf{H} = \frac{1}{\cos(\alpha-\beta)} \begin{bmatrix} p\cos\beta & -q\sin(\alpha-2\beta) \\ q\sin(\alpha-2\beta) & q\cos\alpha \end{bmatrix}. \tag{3.9}$$

4. Inhomogeneous media

In this section we develop a method based on invariant embedding (see [1], [2], or [5]) to compute the reflection and transmission matrices for an inhomogeneous media. Let us consider the slab in Figure 3 sandwiched between the two half-spaces $x_1 \geq T$ and $x_1 \leq 0$. The left and right half-spaces are assumed to be homogeneous while the slab of thickness T may have mechanical properties (such as density and Lamé moduli) which vary along the x_1 axis. The vertical axis in Figure 3 may represent density for example. Henceforth, x_1 is abbreviated x. Our goal in this section is to determine the

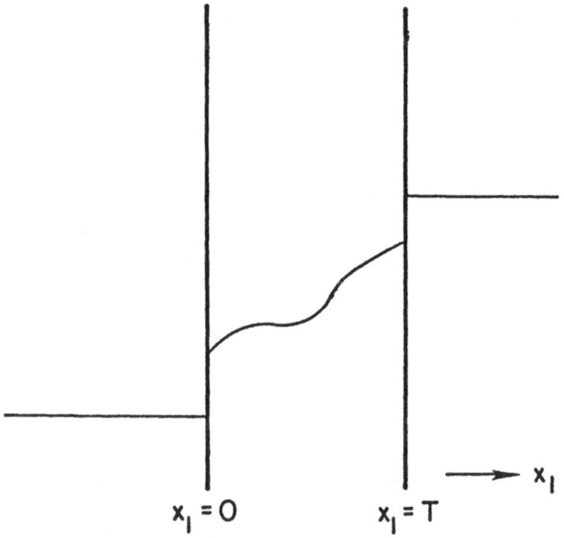

Figure 3. The slab cross-section.

reflection and the transmission matrices corresponding to the interface $x = T$. To help motivate the analysis which follows, let us review the procedure developed in [5] to compute the reflectivity (ratio between the amplitude of the reflected wave and the amplitude of the incident wave) for normal incidence.

Let μ_1, λ_1, and ρ_1 denote the Lamé moduli and the density corresponding to the region $x > T$ and let μ_2, λ_2, and ρ_2 denote the Lamé moduli and the density corresponding to the region $x < 0$. Defining the parameter $\kappa_j = 2\mu_j + \lambda_j$, it follows from our analysis in [5] that for normal incidence, the reflectivity is given by

$$r = \frac{G(T) - p_1}{G(T) + p_1} \tag{4.1}$$

where $p_1 = \sqrt{\kappa_1 \rho_1}$ is the scalar compliance for the right half-space and G is the solution to the differential equation

$$G'(a) = -i\omega\left(\rho(a) - \frac{1}{\kappa(a)}G(a)^2\right) \tag{4.2}$$

with the initial condition $G(0) = p_2 = \sqrt{\kappa_2 \rho_2}$. Of course, the parameter $\rho(a)$ in (4.2) is the density at $x = a$ and $\kappa(a)$ denotes the quantity $2\mu(a) + \lambda(a)$ where $\mu(a)$ and $\lambda(a)$ are the Lamé moduli at $x = a$. Notice that (4.1) expresses the reflectivity in terms of the compliance p_1 of the region $x > T$ and in terms of a parameter $G(a)$ which depends on the material properties for the region $x < a$. Comparing (4.1) to (3.5), we see that $G(T)$ acts as a "generalized compliance" for the half-space $x < T$. For this reason, we refer to $G(a)$ as the compliance of the material in the region $x < a$.

Extending this strategy for evaluating the reflectivity to the oblique incidence case, we define the 2×2 compliance matrix $\mathbf{G}(a)$ for the material in the region $x < a$ using equation (3.8), but with $\mathbf{H}_t$ replaced by $\mathbf{G}(a)$:

$$\mathbf{G}(a) = (\mathbf{H}_1\mathbf{A}_1 + \mathbf{P}\mathbf{H}_1\mathbf{A}_1\mathbf{R}(a))(\mathbf{A}_1 - \mathbf{P}\mathbf{A}_1\mathbf{R}(a))^{-1}\,. \tag{4.3}$$

Again, the subscript 1 refers to material in the region $x > a$. If $\mathbf{G}$ can be evaluated, then the reflection matrix $\mathbf{R}(T)$ corresponding to the interface at $x = T$ in Figure 3 can be determined using (4.3):

$$\mathbf{R}(T) = \mathbf{A}_1^{-1}(\mathbf{G}(T)\mathbf{P} + \mathbf{P}\mathbf{H}_1)^{-1}(\mathbf{G}(T) - \mathbf{H}_1)\mathbf{A}_1\,.$$

One of our main results is that the matrix $\mathbf{G}$ defined in (4.3) is a physical property of the material in the region $x < a$. Moreover, $\mathbf{G}(a)$ can be evaluated by integrating a matrix Riccati equation over the region $0 \leq x \leq a$. The derivation of a differential equation for $\mathbf{G}$ is quite technical so we just sketch the argument.

We first consider an auxiliary problem where the slab in Figure 3 is homogeneous with thickness a. Using results from § 3, the reflectivity and hence the compliance $\mathbf{G}$ corresponding to the interface $x = a$ can be determined. This expression for $\mathbf{G}$ has the form:

$$\mathbf{G} = (\mathbf{H}_1 + \mathbf{P}\mathbf{H}_1\mathbf{L})(\mathbf{I} - \mathbf{P}\mathbf{L})^{-1} \tag{4.4}$$

where

$$\mathbf{L} = (\mathbf{P}\mathbf{H}_1+\mathbf{H}_3\mathbf{P})^{-1}(\mathbf{H}_3-\mathbf{H}_1) + (\mathbf{P}\mathbf{H}_1+\mathbf{H}_3\mathbf{P})^{-1}(\mathbf{H}_3\mathbf{P}+\mathbf{P}\mathbf{H}_3)\mathbf{M}(\mathbf{H}_1\mathbf{P}+\mathbf{P}\mathbf{H}_3)^{-1}(\mathbf{H}_1\mathbf{P}+\mathbf{P}\mathbf{H}_1)$$

and

$$\mathbf{M}=[\mathbf{A}_3\mathbf{D}_3^{-1}\mathbf{A}_3^{-1}(\mathbf{H}_2-\mathbf{H}_3)^{-1}(\mathbf{P}\mathbf{H}_3+\mathbf{H}_2\mathbf{P})\mathbf{A}_3\mathbf{D}_3^{-1}\mathbf{A}_3^{-1}-(\mathbf{H}_1\mathbf{P}+\mathbf{P}\mathbf{H}_3)^{-1}(\mathbf{H}_1-\mathbf{H}_3)]^{-1}.$$

Here $\mathbf{H}_1$ is the compliance for the half-space $x > a$, $\mathbf{H}_2$ is the compliance for the half-space $x < 0$, and $\mathbf{H}_3$ is the compliance for the homogeneous slab in the region $0 \leq x \leq a$. Similarly, $\mathbf{A}_3$ is the matrix $\mathbf{A}$ defined in (3.3) corresponding to a wave propagating in the slab. The phase delay matrix $\mathbf{D}$ above is given by

$$\mathbf{D} = \begin{bmatrix} e^{i\omega aD\cos\alpha} & 0 \\ 0 & e^{i\omega aS\cos\beta} \end{bmatrix}.$$

Of course, $\mathbf{D}_3$ will be the phase delay matrix corresponding to a wave propagating in the slab. Note that the dependence of $\mathbf{G}$ on the slab thickness a only enters through the matrix $\mathbf{D}$.

It can be shown that the formula (4.4) for $\mathbf{G}(a)$ is independent of $\mathbf{H}_1$ and after some work, we obtain

$$\mathbf{G}(a) = \mathbf{H}_3\mathbf{Z}(a)(\mathbf{Z}(a) - \mathbf{I} - \mathbf{H}_3^{-1}\mathbf{P}\mathbf{H}_3)^{-1}$$

where

$$\mathbf{Z}(a) = \mathbf{H}_3^{-1}\mathbf{P}\mathbf{H}_3 + \mathbf{A}_3\mathbf{D}_3(a)^{-1}\mathbf{A}_3^{-1}(\mathbf{H}_2 - \mathbf{H}_3)^{-1}(\mathbf{P}\mathbf{H}_3 + \mathbf{H}_2\mathbf{P})\mathbf{A}_3\mathbf{D}_3(a)^{-1}\mathbf{A}_3^{-1}.$$

Differentiating $\mathbf{G}(a)$ with respect to a and letting a approach zero, we obtain an expression for $\mathbf{G}'(0)$ in terms of $\mathbf{G}(0)$ which eventually leads us to a matrix Riccati equation for $\mathbf{G}(a)$:

$$\mathbf{G}' = i\omega[(\mathbf{GP}+\mathbf{PH})\mathbf{F}(\mathbf{G}-\mathbf{H}) + (\mathbf{G}-\mathbf{H})\mathbf{FP}(\mathbf{GP}+\mathbf{PH})\mathbf{P}] \qquad (4.5)$$

where $\mathbf{H}$ is defined in (3.9) and

$$\mathbf{F} = \frac{\rho}{2}\begin{bmatrix} \dfrac{1}{p^2} & -\dfrac{1}{pq}\sin(\alpha-\beta) \\ \dfrac{1}{pq}\sin(\alpha-\beta) & -\dfrac{1}{q^2} \end{bmatrix}.$$

Here $p = \rho c_d$ and $q = \rho c_s$ are the scalar compliances. The variable $\mathbf{G}$ as well as the material matrices $\mathbf{F}$ and $\mathbf{H}$ in (4.5) are evaluated at $x = a$. The angles which appear in the definitions of $\mathbf{F}$ and $\mathbf{H}$ are determined from Snell's law:

$$D_1 \sin\alpha_1 = S_1 \sin\beta_1 = D(a)\sin\alpha(a) = S(a)\sin\beta(a)$$

where α_1 and β_1 denote the angles of an incident conjugate pair of waves. The starting condition for equation (4.5) is $\mathbf{G}(0) = \mathbf{H}_2$, the compliance for the half-space $x < 0$.

Remark 2. For normal incidence, $\alpha = \beta = 0$ and both $\mathbf{H}_2$ and the material matrices appearing in (4.5) are diagonal. Hence, $\mathbf{G}(a)$ is diagonal for all $a > 0$ and equation (4.5) uncouples into the form:

$$G'_{11} = -i\omega\left(\rho - \frac{1}{2\mu+\lambda}G_{11}^2\right),$$

$$G'_{22} = -i\omega\left(\rho - \frac{1}{\mu}G_{22}^2\right).$$

REFERENCES

[1] Bellman, R. *Methods of Nonlinear Analysis.* Vol. II. New York: Academic Press, 1973.

[2] Bellman, R., and R. Kalaba. "Functional Equations, Wave Propagation and Invariant Imbedding." *J. Math. Mech.* 8 (1959): 683–704.

[3] Gurtin, M.E. "Linear Theory of Elasticity." *Handbuch der Physik* VIa/2 (1972): 1–295.

[4] Hager, W.W., and D.L. Presler. "Dual Techniques for Minimax." *SIAM J. Control Optim.* Forthcoming.

[5] Hager, W.W., and R. Rostamian. "Optimal Coatings, Bang-Bang Controls, and Gradient Techniques." *Optimal Control: Applications and Methods.* Forthcoming.

[6] Kennett, B.L.N. *Seismic Wave Propagation in Stratified Media.* Cambridge: Cambridge University Press, 1983.

BOUNDARY CONTROLLABILITY OF MAXWELL'S EQUATIONS IN A SPHERICAL REGION

Katherine A. Kime and David L. Russell

University of Wisconsin, Madison, Wisconsin, USA

We consider Maxwell's equations

$$\begin{aligned} \nabla \circ \underline{H} &= 0 \\ \nabla \circ \underline{E} &= 0 \\ \nabla \times \underline{E} &= -\partial \underline{H}/\partial t \\ \nabla \times \underline{H} &= \partial \underline{E}/\partial t \end{aligned} \qquad (1)$$

in Ω the unit ball in $\mathbb{R}^3$, assuming no internal charges or currents. Here $\underline{E}$ and $\underline{H}$ are 3-dimensional vectors representing the electric and magnetic fields, respectively.

The question of influencing the behavior of $\underline{E}$ and $\underline{H}$ inside Ω by means of external forces arises from the need to stabilize plasma confinement in attempts to achieve controlled nuclear fusion. In this case, Maxwell's equations contain terms representing internal charge and current densities, and are coupled to equations describing the plasma evolution. Until recently, [1], however, controllability questions for the simpler system (1) had not been addressed to much extent.

We are interested here in the possibility of controlling the fields $\underline{E}$, $\underline{H}$ inside Ω by means of a current $\underline{J}(\circ,t)$ flowing tangentially on $\partial\Omega$, the effect of which is described by the boundary condition

$$\underline{n} \times \underline{H} = \underline{J} \quad \text{on } \partial\Omega \quad (\underline{n} \text{ the unit outward normal}) \qquad (2)$$

Thus we state the

Control Problem: Given $T > 0$ and prescribed initial data, find a control current $\underline{J}(\circ,t)$ defined on $\partial\Omega$ such that the solutions $\underline{E}$, $\underline{H}$ of (1), (2) with this initial data also satisfy the terminal condition

$$\underline{E}(\circ,T) = \underline{H}(\circ,T) = 0.$$

In [1], the control problem for Ω a circular or rectangular cylinder was treated, under assumption of no dependence of the fields in the axial direction.

If $\underline{E}$, $\underline{H}$ are smooth solutions of (1), (2), then there exists a vector $\underline{W}$ (see e.g, [2]) with

$$\underline{H} = \nabla \times \underline{W} \qquad \underline{E} = -\partial\underline{W}/\partial t$$

which satisfies

$$\left.\begin{aligned} \Box\circ\underline{W} &= 0 \\ \nabla\circ\underline{W} &= 0 \end{aligned}\right\} \text{ in } \Omega \tag{3}$$

$$\underline{n} \times (\nabla \times \underline{W}) = \underline{J} \quad \text{on } \partial\Omega \tag{4}$$

We have the following definitions [3]:

$\mathfrak{L}^2(\Omega)$ is the real Hilbert space of vectors $\underline{u} = (u_1,u_2,u_3)$, $u_i \in L^2(\Omega)$.

$\mathcal{H}^1(\Omega)$ is the real Hilbert space of vectors which belong, along with their first-order derivatives, to $\mathfrak{L}^2(\Omega)$.

$J(\Omega)$ is the closure in $\mathfrak{L}^2(\Omega)$ of $\{\underline{u}: \underline{u} \in C^\infty(\Omega),\ \nabla\circ\underline{u} = 0\}$

$\hat{J}(\Omega)$ is the closure in $\mathfrak{L}^2(\Omega)$ of $\{\underline{u}: \underline{u} \in C_0^\infty(\Omega),\ \nabla\circ\underline{u} = 0\}$.

$J^1(\Omega) = J(\Omega) \cap \mathcal{H}^1(\Omega)$

$J_n^1(\Omega) = \{\underline{u}: \underline{u} \in J^1(\Omega),\ \underline{u}\circ\underline{n}|_{\partial\Omega} = 0\}$; $J_n^1(\Omega)^*$ denotes the dual of $J_n^1(\Omega)$.

Using a coercive estimate, [3], [4] for the form

$$a(\underline{\varphi},\underline{\psi}) = \int_\Omega (\nabla \times \underline{\varphi})\circ(\nabla \times \underline{\psi})\, dx \qquad \underline{\varphi},\ \underline{\psi} \in J_n^1(\Omega)$$

and transposition of adjoints, [5], we obtain the following existence and uniqueness result:

<u>THEOREM 1</u> Given $T > 0$, $\underline{J} \in L^2[0,T;\ \mathfrak{L}^2(\partial\Omega)]$, $\begin{bmatrix} w_0 \\ w_1 \end{bmatrix} \in \hat{J}(\Omega) \times J_n^1(\Omega)^*$, there exists a unique $\underline{W} \in L^2(0,T;\ \hat{J}(\Omega))$ which satisfies

$$\int_0^T\!\!\int_\Omega \underline{W}\circ\Box\underline{\varphi}\, dxdt = \int_0^T\!\!\int_{\partial\Omega} -\underline{J}\circ\underline{\varphi}\, dsdt + \int_\Omega w_1\circ\underline{\varphi}(\circ,0) - w_0\circ\frac{\partial\underline{\varphi}}{\partial t}(\circ,0) \tag{5}$$

for every $\underline{\varphi}$ belonging to a suitably defined test function space $X \subset \hat{J}(\Omega)$. Since any classical solution $W^\ddagger$ of (3), (4) would satisfy (5), with $w_0 = W^{\ddagger}(\circ,0)$, $w_1 = \frac{\partial W^{\ddagger}}{\partial t}(\circ,0)$, we take W to be the weak solution of (3), (4).

Use of divergence-free eigenfunctions of the vector Laplacian (the "multipole fields" [2], [6]) shows that a $\underline{J}$ which "drives $\underline{W}$ to zero" exists if, for every pair nm, n = 0,1,2..., m = 0,1,2...2n, there is a solution $\sigma_{nm}(t)$ to the moment problem

$$\int_0^T e^{i\pi\gamma_{nl}t}\sigma_{nm}(t)\,dt = a_{nml}$$

$$l = 1,2,3\ldots \qquad (6)$$

$$\int_0^T e^{-i\pi\gamma_{nl}t}\sigma_{nm}(t)\,dt = b_{nml}$$

and a solution $\pi_{nm}(t)$ to the moment problem

$$\int_0^T e^{i\pi\beta_{nl}t}\pi_{nm}(t)\,dt = c_{nml}$$

$$l = 1,2,3\ldots \qquad (7)$$

$$\int_0^T e^{-i\pi\beta_{nl}t}\pi_{nm}(t)\,dt = d_{nml}$$

Here the a's, b's, c's and d's depend on w_0, w_1, and

$$\beta_{nl} \text{ is the } l^{th} \text{ root of } j_n(\pi\beta) = 0, \quad l = 1,2\ldots$$

$$\gamma_{nl} \text{ is the } l^{th} \text{ root of } \frac{d}{d\gamma}[\gamma j_n(\pi\gamma)] = 0, \quad l = 1,2\ldots,$$

where j_n is the nth spherical Bessel function. We have

<u>THEOREM 2</u> Let $n \geqq 1$, $0 \leqq m \leqq 2n$. Let the sequences $\{a_{nml}\}$, $\{b_{nml}\}$, $\{c_{nml}\}$, and $\{d_{nml}\}$ be square summable.

Then, if T > 2, the moment problems (6), (7) have solutions $\sigma_{nm}(t), \pi_{nm}(t)$ in $L^2[0,T]$.

The proof of Theorem 2 proceeds by first showing, using [7], that there exists a lower bound on the spacings of the coefficients γ_{nl} and β_{nl}. From this, a result of Ingham shows that when T > 2, hypotheses of a theorem of Boaz are satisfied, as in [8], and the existence of the solutions $\sigma_{nm}(t)$, $\pi_{nm}(t)$ follows.

Sufficient conditions on the initial data $\begin{bmatrix} w_0 \\ w_1 \end{bmatrix}$ which ensure satisfaction of the summability assumption are discussed in [9].

Note: This is a summary of results from [9], which are in preparation for publication in detailed form.

References

1. Russell, D.L. "The Dirichlet-Neumann Boundary Control Problem Associated with Maxwell's Equations in a Cylindrical Region." *SIAM J. Control and Optimization* 24, no. 2 (1986): 199-229.

2. Jackson, J.D. *Classical Electrodynamics*. Second Edition, New York: John Wiley and Sons, 1975.

3. Ladyzhenskaya, O.A, and V.A. Solonikov. "The Linearization Principle and Invariant Manifolds for Problems of Magnetohydronamics." *Journal of Soviet Math* 8 (1977): 384-422.

4. Bykhovskii, E.B. and N.V. Smirnov. "On the orthogonal decomposition of the space of vector functions square summable in a given domain and the operators of vector analysis." *Tr. Mat. Inst. Steklov* 59 (1960): 6-36.

5. Lions, J.L. *Optimal Control of Systems Governed by Partial Differential Equations*. New York: Springer-Verlag, 1971.

6. Morse, P.M. and Feshbach, H. *Methods of Theoretical Physics*. New York: McGraw-Hill Book Company, Inc., 1953.

7. Graham, K.D. "Separation of Eigenvalues of the Wave Equation for the Unit Ball in R^n." *Studies in Applied Mathematics* LII (1973): 329-343.

8. Graham, K.D. and Russell, D.L. "Boundary Value Control of the Wave Equation in a Spherical Region." *SIAM J. Control* 13, no. 1 (1975): 174-196.

9. Kime, K.A. "Boundary Controllability of Maxwell's Equations". Ph.D. thesis, University of Wisconsin, 1986.

NUMERICAL SOLUTION OF TIME-MINIMAL CONTROL PROBLEMS

W. Krabs and U. Lamp

Technical University of Darmstadt, The Federal Republic of Germany

1. THE PROBLEM AND BASIC RESULTS

As in [2] and [3] the following abstract process of vibrations is investigated: Let $y : [0,T] \to H$, for any $T > 0$, be a function that describes the deviation of a vibrating medium from the position of rest as a function of time t with values in a (finite- or infinite-dimensional) Hilbert space H. We assume y to satisfy an abstract wave equation of the form

$$\ddot{y}(t) - Ay(t) = f(t), \quad t \in (0,T), \tag{1.1}$$

where $\ddot{y}$ denotes the derivative with respect to t, A is a self adjoint positive definite linear operator defined on a dense domain D(A) in H and $f(t) \in H$ for almost all $t \in [0,T]$, $\|f(\cdot)\|_H$ is measurable and satisfies $\int_0^T \|f(t)\|_H^2 \, dt < \infty$ where $\|\cdot\|_H$ denotes the norm in H. The space of all (classes of) such functions is called $L^2([0,T], H)$. Let $N = \dim H$. In addition to the above requirements we assume that A has a complete sequence $(\varphi_j)_{j=1,\dots,N}$ of orthonormal eigenelements $\varphi_j \in D(A)$ and a corresponding sequence $(\lambda_j)_{j=1,\dots,N}$ of eigenvalues λ_j of finite multiplicity with $0 < \lambda_1 \le \lambda_2 \le \dots$ and $\lim_{j\to\infty} \lambda_j = \infty$, if $N = \infty$.

Then it follows that

$$D(A) = \{v \in H \mid \sum_{j=1}^{N} \lambda_j^2 |\langle v,\varphi_j\rangle_H|^2 < \infty\}$$

and

$$Av = \sum_{j=1}^{N} \lambda_j \langle v,\varphi_j\rangle_H \, \varphi_j \quad \text{for all } v \in D(A).$$

Furthermore there exists a unique "square root" $A^{1/2}$ of A with domain

$$D(A^{1/2}) = \{v \in H \mid \sum_{j=1}^{N} \lambda_j \langle v, \varphi_j \rangle_H^2 < \infty\}$$

which is defined by

$$A^{1/2} v = \sum_{j=1}^{N} \lambda_j^{1/2} \langle v, \varphi_j \rangle_H \varphi_j \quad \text{for all } v \in D(A^{1/2}).$$

Let $y_0 \in D(A^{1/2})$ and $\dot{y}_0 \in H$ be given. Then we require initial conditions for $t = 0$ to be given by

$$y(0) = y_0 \quad \text{and} \quad \dot{y}(0) = \dot{y}_0. \tag{1.2}$$

We put $V = D(A^{1/2})$ provided with the scalar product

$$\langle v, w \rangle_V = \sum_{j=1}^{N} \lambda_j \langle v, \varphi_j \rangle_H \langle w, \varphi_j \rangle_H, \quad v, w \in V.$$

Then V becomes a separable Hilbert space which is continuously and densely imbedded into H because of $D(A) \subset V \subset H$ and

$$\|v\|_V \geq \lambda_1^{1/2} \|v\|_H \quad \text{for all } v \in V.$$

The dual space V^* of V consists of all linear functionals $v^* : V \to \mathbb{R}$ such that

$$\sum_{j=1}^{n} \frac{1}{\lambda_j} v^*(\varphi_j)^2 < \infty$$

and

$$v^*(v) = \sum_{j=1}^{N} \langle v, \varphi_j \rangle_H v^*(\varphi_j), \quad v \in V.$$

If we identify H with its dual space, we obtain the following chain of continuous and dense imbeddings: $V \subset H \subset V^*$.
If we define a linear mapping $\tilde{A} : V \to V^*$ by

$$(\tilde{A}v)(w) = \langle A^{1/2} v, A^{1/2} w \rangle_H = \langle v, w \rangle_V$$

for all $v, w \in V$, then we have

$$(\tilde{A}v)(v) = \|v\|_V^2 \quad \text{for all } v \in V$$

and all the assumptions of Theorem 1.1 in Chapter IV of [5] are satisfied. This in connection with Remark 1.3 loc. cit. implies the existence of exactly one weak solution $y : [0,T] \to H$ of (1.1), (1.2) for every choice of $f \in L^2([0,T], H)$, $y_0 \in V = D(A^{1/2})$ and $\dot{y}_0 \in H$ in the following sense:

1) $y \in C([0,T], V)$, $\dot{y}(t)$ exists for all $t \in [0,T]$ in the strong sense and $\dot{y} \in C([0,T], H)$ where, for an arbitrary Hilbert space Z, $C([0,T], Z)$ is the space of continous functions from $[0,T]$ into Z.

2) $\lim_{t \to 0+} \|y(t) - y(0)\|_V = 0$ and $\lim_{t \to 0+} \|\dot{y}(t) - \dot{y}_0\|_H = 0$.

3) The second derivative y in the sense of distributions can be identified with a function in $L^2([0,T], V^*)$ and

$$\ddot{y}(t)(v) + \tilde{A}y(t)(v) = \langle f(t), v \rangle_H \tag{1.3}$$

is satisfied for all $v \in V$ and almost all $t \in (0,T)$. This weak solution can be explicitly expressed with the aid of sine and cosine operators (see [2] and [3]).

The main concern of this paper is the

PROBLEM OF TIME-MINIMAL NULL-CONTROLLABILITY

Given $(y_0, \dot{y}_0) \in V \times H$ and some $M > 0$.

a) Does there exist a time $T > 0$ and a control function $f \in L^2([0,T], H)$ with

$$\|f\|_{2,T} = \left(\int_0^T \|f(t)\|_H^2 \, dt\right)^{1/2} \leq M \tag{1.4}$$

such that the weak solution $y : [0,T] \to H$ of (1.1), (1.2) satisfies

$$y(T) = \Theta_V \quad \text{and} \quad \dot{y}(T) = \Theta_H \tag{1.5}$$

with Θ_V and Θ_H being the zero element of V and H, respectively? (We assume, of course, that $(y_0, \dot{y}_0) \neq (\Theta_V, \Theta_H)$).

b) If a) is possible, then the infimum $T(M)$ of all such times $T > 0$ is well defined and the question is whether there exists a time-minimal control function $f \in L^2([0,T(M)], H)$ with (1.4) for $T = T(M)$ such that the corresponding weak solution $y : [0,T(M)] \to H$ satisfies (1.5) for $T = T(M)$.

A further question is the uniqueness of time-minimal control functions and how they can be characterized.

In [3] we have proved the following results:

1) For every $T > 0$ there exists exactly one control function $f_T \in L([0,T], H)$ with least norm such that the corresponding weak solution $y : [0,T] \to H$ of (1.1), (1.2) satisfies (1.5). Moreover, the norm of f_T can be estimated in the form

$$\|f_T\|_{2,T}^2 \le \frac{8\sqrt{\lambda_1}}{T\sqrt{\lambda_1} - 1} (\|y_0\|_V^2 + \|y_0\|_H^2), \tag{1.6}$$

if $T > 1/\sqrt{\lambda_1}$. This estimate implies

$$\lim_{T\to\infty} \|f_T\|_{2,T} = 0$$

so that part a) of the problem of time-minimal null-controllability has a solution.

2) For every $M > 0$ the least norm control $f_{T(M)} \in L^2([0,T(M)], H)$ in 1) is the unique time-minimal control function and is characterized by

$$\|f_T\|_{2,T} = M \iff T = T(M). \tag{1.7}$$

2. GALERKIN'S METHOD

In [4] we have demonstrated how the problem of time-minimal null-controll= ability can be solved by using Galerkin's method. In this note we will present a modification of the method being described in [4] which starts with a more general situation, leads to a further simplification and facilitates the convergence proof (which will be published elsewhere).

Instead of an orthonormal basis (as in [4]) we start with linearly independent basis $(v_j)_{j=1,...,N}$ of V (observe, $N = \dim H$). In order to describe Galerkin's method in its basic form we define, for every $n \in \{1,...,N\}$,

$$Y_n = \{Y_n = \sum_{j=1}^{n} y_j^n v_j \mid y_j^n \in H^2[0,T],\ j = 1,...,n\}$$

which is an n-dimensional linear subspace of $H^2([0,T], V) \subset C([0,T], V)$ and we replace the weak equation (1.3) by

$$\langle \ddot{y}_n(t), v^n\rangle_H + \langle A^{1/2} y_n(t), A^{1/2} v^n\rangle_H = \langle f(t), v^n\rangle_H \qquad (1.3)_n$$

for almost all $t \in (0,T)$ and all

$$v^n \in V_n = \{\sum_{j=1}^{n} c_j v_j \mid c_j \in \mathbb{R}, \quad j = 1,\dots,n\}$$

which is equivalent to

$$B^n \ddot{y}^n(t) + A^n y^n(t) = f^n(t) \qquad (2.1)$$

for almost all $t \in (0,T)$

where $A^n = (a^n_{jk})$, $B^n = (b^n_{jk})$, $f^n(t) = (f^n_1(t),\dots,f^n_n(t))^T$ with

$$a^n_{jk} = \langle A^{1/2} v_j, A^{1/2} v_k\rangle_H,$$

$$b_{jk} = \langle v_j, v_k\rangle_H \quad \text{for all } j,k = 1,\dots,n \text{ and}$$

$$f^n_j(t) = \langle f(t), v_j\rangle_H \quad \text{for all } j = 1,\dots,n.$$

The initial conditions (1.2) are replaced by

$$y^n(0) = y^n_0 \quad \text{and} \quad \dot{y}^n(0) = y^n_0 \qquad (2.2)$$

where

$$A^n y^n_0 = (\langle A^{1/2} v_1, A^{1/2} y_0\rangle_H,\dots,\langle A^{1/2} v_n, A^{1/2} y_0\rangle_H)^T$$

and

$$B^n \dot{y}^n_0 = (\langle v_1, \dot{y}_0\rangle_H,\dots,\langle v_n, \dot{y}_0\rangle_H)^T.$$

These linear systems are uniquely solvable for y^n_0 and $\dot{y}^n_0 \in \mathbb{R}^n$, since A^n and B^n are symmetric and positive definite $n \times n$-matrices. It is well known (see, for instance, [1]) that the generalized eigenvalue problem

$$A^n \varphi = \lambda B^n \varphi, \ \varphi \in \mathbb{R}^n, \ \lambda \in \mathbb{R},$$

has n eigenvalues $\lambda^n_1,\dots,\lambda^n_n \in \mathbb{R}$ with $0 < \lambda^n_1 \le \lambda^n_2 \le \dots \le \lambda^n_n$ and corresponding eigenvectors $\varphi^n_1,\dots,\varphi^n_n \in \mathbb{R}$ such that

$$\langle \varphi^n_j, B^n \varphi^n_k\rangle_{\mathbb{R}^n} = \delta_{jk} \quad (= \text{Kronecker's symbol}) \qquad (2.3)$$

for $j,k = 1,\dots,n$.

In order to define the modification of Galerkin's method as mentioned above we put

$$v_j^n = \sum_{k=1}^{n} \varphi_{jk}^n v_k \quad \text{for } j = 1,\dots,n.$$

Then, by virtue of (2.3)

$$\langle v_j^n, v_k^n \rangle_H = \langle \varphi_j^n, B^n \varphi_k^n \rangle_{\mathbb{R}^n} = \delta_{jk}$$

and

$$\langle A^{1/2} v_j^n, A^{1/2} v_k^n \rangle_H = \langle \varphi_j^n, A^n \varphi_k^n \rangle_{\mathbb{R}^n} = \lambda_j^n \delta_{jk}$$

for $j,k = 1,\dots,n$.

If we replace the basis $\{v_1,\dots,v_n\}$ of V_n by $\{v_1^n,\dots,v_n^n\}$, then we obtain an orthonormal basis of V_n and $(1.3)_n$ turns out to be equivalent to

$$\ddot{y}_j^n(t) + \lambda_j^n y_j^n(t) = f_j^n(t) \quad \text{for } j = 1,\dots,n \text{ and almost all } t \in (0,T) \qquad (1.1)_n$$

where

$$y_n(t) = \sum_{j=1}^{n} y_j^n(t) v_j^n.$$

The initial conditions (2.2) can be equivalently rewritten in the form

$$y_j^n(0) = \frac{1}{\lambda_j^n} \langle A^{1/2} v_j^n, A^{1/2} y_0 \rangle_H \quad \text{and} \quad \dot{y}_j^n(0) = \langle v_j^n, \dot{y}_0 \rangle_H \qquad (1.2)_n$$

for $j = 1,\dots n$.

The unique solution $y_j \in H^2(0,T)$, for every $j = 1,\dots,n$, of $(1.1)_n$, $(1.2)_n$ is given by

$$y_j^n(t) = y_{0j}^n \cos\sqrt{\lambda_j^n}\,t + \frac{\dot{y}_{0j}^n}{\sqrt{\lambda_j^n}} \sin\sqrt{\lambda_j^n}\,t + \frac{1}{\sqrt{\lambda_{j0}^n}} \int_0^t f_j^n(s) \sin\sqrt{\lambda_j^n}(t-s)\,ds$$

where

$$y_{0j}^n = \frac{1}{\lambda_j^n} \langle A^{1/2} v_j^n, A^{1/2} y_0 \rangle_H \quad \text{and} \quad \dot{y}_{0j}^n = \langle v_j^n, \dot{y}_0 \rangle_H$$

and its first derivative reads

$$\dot{y}_j^n(t) = -y_{0j}^n \sqrt{\lambda_j^n} \sin\sqrt{\lambda_j^n}t + \dot{y}_{0j}^n \cos\sqrt{\lambda_j^n}t + \int_0^t f_j^n(s) \cos\sqrt{\lambda_j^n}(t-s)ds$$

The problem of time-minimal null-controllability as being formulated in Section 1 is formulated here in the same way with (1.1) and (1.2) being replaced by $(1.1)_n$ and $(1.2)_n$, respectively, (1.4) by

$$\|f^n\|_{2,T} = \left(\int_0^T \langle f^n(t), f^n(t)\rangle_{\mathbb{R}^n} dt\right)^{1/2} \leq M \tag{1.4$_n$}$$

and (1.5) by

$$y^n(T) = \dot{y}^n(T) = \Theta_n \tag{1.5$_n$}$$

with Θ_n being the zero vector of $\mathbb{R}^n$.

At first we consider, for $T > 0$ being fixed, the problem of minimum norm control. In this case the conditions $(1.5)_n$ are equivalent to

$$\frac{1}{\sqrt{\lambda_j^n}} \int_0^T f_j^n(t) \sin\sqrt{\lambda_j^n}(T-t)dt = -y_{0j}^n \cos\sqrt{\lambda_j^n}T - \frac{\dot{y}_{0j}^n}{\sqrt{\lambda_j^n}} \sin\sqrt{\lambda_j^n}T,$$

$$\int_0^T f_j^n(t) \cos\sqrt{\lambda_j^n}(T-t)dt = y_{0j}^n \sqrt{\lambda_j^n} \sin\sqrt{\lambda_j^n}T + \dot{y}_{0j}^n \cos\sqrt{\lambda_j^n}T \tag{2.4}$$

for $j = 1,\dots,n$.

We define, for every $j = 1,\dots,n$,

$$f_j^n(t) = b_j^1 \sin\sqrt{\lambda_j^n}(T-t) + b_j^2 \cos\sqrt{\lambda_j^n}(T-t). \tag{2.5}$$

Then insertion into (2.4) yields the linear 2×2-system

$$G_j^n \begin{pmatrix} b_j^1 \\ b_j^2 \end{pmatrix} = \begin{pmatrix} -y_{0j}^n \cos\sqrt{\lambda_j^n}T - \dfrac{\dot{y}_{0j}^n}{\sqrt{\lambda_j^n}} \sin\sqrt{\lambda_j^n}T \\ y_{0j}^n \sqrt{\lambda_j^n} \sin\sqrt{\lambda_j^n}T + \dot{y}_{0j}^n \cos\sqrt{\lambda_j^n}T \end{pmatrix} \tag{2.6}$$

with

$$G_j^n = \begin{pmatrix} \frac{1}{\lambda_j^n}\int_0^T \sin^2\sqrt{\lambda_j^n}(T-t)dt & \frac{1}{2\sqrt{\lambda_j^n}}\int_0^T \sin 2\sqrt{\lambda_j^n}(T-t)dt \\ \frac{1}{2\sqrt{\lambda_{j0}^n}}\int_0^T \sin 2\sqrt{\lambda_j^n}(T-t)dt & \int_0^T \cos^2\sqrt{\lambda_j^n}(T-t)dt \end{pmatrix}$$

for $j = 1,\dots,n$. Because of $\det(G_j^n) > 0$ each system (2.5) has a unique solution $(b_j^1, b_j^2)^T$ which can be written down explicitly by using Cramer's rule.

In analogy to the results in [3] it follows that $f_T^n(t) = (f_1^n(t),\dots, f_n^n(t))^T$ with $f_j^n(t)$ being defined by (2.5) where $b_j^1,b_j^2)^T$ is the unique solution of (2.6) for $j = 1,\dots,n$ yields the unique norm solution of (2.4).

Moreover, one can prove, in analogy to (1.6),

$$\|f_T^n\|_{2,T}^2 \le \frac{8\sqrt{\lambda_1^n}}{T\sqrt{\lambda_1^n} - 1} \sum_{j=1}^{n} [\lambda_j^n (y_{0j}^n)^2 + (\dot{y}_{0j}^n)^2] \qquad (1.6)_n$$

if $T \ge \frac{1}{\sqrt{\lambda_1^n}}$. This implies $\lim_{T\to\infty} \|f_T^n\|_{2,T} = 0$ and again we have the equivalence

$$\|f_T^n\|_{2,T} = M \iff T = T_n(M) \qquad (1.7)_n$$

where $T_n(M)$ denotes the minimum time for which (2.4) ($\iff$ $(1.5)_n$) can be solved under the condition $(1.4)_n$.

Since $\|f_T^n\|_{2,T}$ can be easily computed as shown above, the unique time-minimal control function $f_{T_n(M)}^n$ can be determined by solving the left-hand equation of $(1.7)_n$. Convergence of Galerkin's method can be proved, if the basis $(v_j)_{j=1,\dots,N}$ of V is chosen to be the sequence $(\varphi_j)_{j=1,\dots,N}$ of orthonormal eigenelements φ_j of A. In this case we obtain $v_j^n = \varphi_j$ and $\lambda_j^n = \lambda_j$ for all $j = 1,\dots,n$ and all $n \in \{1,\dots,N\}$ where the λ_j's are the eigenvalues of A and it follows that, for every $T > 0$,

$$\lim_{n\to\infty} \|f_{nT} - f_T\|_{2,T} = 0 \qquad (2.7)$$

where

$$f_{nT}(t) = \sum_{j=1}^{n} f_{jT}^{n}(t)\varphi_j, \quad t \in [0,T],$$

and $f_T^n(t) = (f_{1T}^n(t),\dots,f_{nT}^n(t))^T$ is the minimum norm solution of (2.4). Furthermore, we have

$$\lim_{n\to\infty} T_n(M) = T(M) \tag{2.8}$$

and

$$\lim_{n\to\infty} \| f_{nT_n(M)} - f_{T(M)} \|_X = 0 \tag{2.9}$$

where $X = L^2([0,\infty),H)$ equipped with

$$\|f\|_X = \left(\int_0^\infty \|f(t)\|_n^2 \, dt\right)^{1/2}, \quad f \in X,$$

and

$$f_{nT_n(M)}(t) = \sum_{j=1}^{n} f_{jT_n(M)}^{n}(t)\varphi_j, \quad t \in [0,T],$$

where $f_{T_n(M)}^n(t) = (f_{1T_n(M)}^n(t),\dots,f_{nT_n(M)}^n(t))^T$ is the time-minimal solution of (2.5) subject to $(1.4)_n$.

In general it is difficult and often impossible to determine the eigenelements φ_j and eigenvalues of A. In concrete applications of Galerkin's method, for instance to vibrating plates, the basis $(v_j)_{j=1,\dots,N}$ of V is chosen from suitable finite element spaces. But then it turns out that in order to ensure convergence the method has to be modified and the convergence statements also differ from the ones given above. Because of space limitations the details cannot be presented here.

2. A NUMERICAL EXAMPLE

In [4] we have applied Galerkin's method to a rectangular plate where the eigenfunctions φ_j and eigenvalues λ_j are easily available so that $v_j = \varphi_j$, $j \in \mathbb{N}$, can be chosen. In this case convergence can be proved as being mentioned above.

For the purpose of comparison we repeat the results. Let $\Omega = (0,1) \times (0,1)$, $A = \Delta^2$, Δ = Laplace operator, $D(A) = \{v \in H_0^1(\Omega) \mid z \in H_0^1(\Omega),\ \Delta^2 z \in L^2(\Omega)\}$, and

$$y_0(x_1,x_2) = x_1 x_2 (1-x_1)(1-x_2),$$

$$\dot{y}_0(x_1,x_2) = 0 \quad \text{for } x_1,\ x_2 \in (0,1).$$

The following table gives the values of $T_n(M)$ for n = 1, 4, 9, 16 and M = 0.1, 1, 10, 100:

M \ n	1	4	9	16
0.1	86.33869	86.33869	92.25277	92.25277
1	0.87079	0.87079	0.90327	0.90327
10	0.04823	0.04823	0.04841	0.04841
100	0.01096	0.01096	0.01097	0.01097

For reasons of comparison we have also chosen subspaces V_n of V which are constructed by finite elements based on a triangulation of Ω. The details of this method will be published elsewhere. In the case of the above example of a rectangular plate we obtain the following values of $T_n(M)$:

M	2	5	8	15
0.1	85.97344	88.15024	90.39041	90.48887
1	0.87005	0.88091	0.89279	0.89316
10	0.04825	0.04832	0.04839	0.04839
100	0.01096	0.01097	0.01097	0.01097

REFERENCES

[1] Collatz, L.: Eigenwertaufgaben und technische Anwendungen. Akademische Verlagsgesellschaft Geest und Portig K.-G.: Leipzig 1949.

[2] Fattorini, H.O.: The Time Optimal Problem for Distributed Control of Systems Described by the Wave Equation. In: Aziz, A.K., Wingate, J.W., Balas, M.J. (eds.): Control Theory of Systems Governed by Partial Differential Equations. Academic Press, New York, San Francisco, London, 1977.

[3] Krabs, W.: On Time-Minimal Distributed Control of Vibrating Systems Governed by an Abstract Wave Equation. Appl. Math. and Optim. 13 (1985), 137-149.

[4] Krabs, W.: On the Numerical Solution of Certain Time-Minimal Control Problems of Second Order. To appear in the Proceedings of the 12th IFIP Conference of Systems Modelling and Optimization, held at Budapest in August 1985.

[5] Lions, L.J.: Optimal Control of Systems Governed by Partial Differential Equations. Springer Verlag: Berlin, Heidelberg, New York, 1971.

SOME PROBLEMS RELATED TO BOUNDARY STABILIZATION OF PLATES

John Lagnese

Georgetown University, Washington, D.C., USA

In this note we shall consider some issues arising in the problem of *uniform* stabilization of solutions of equations describing the motion of thin plates by means of feedback controls acting on the boundary of the region occupied by the plate. Two models will be considered. The first is the one occurring in the Kirchoff theory, namely

$$u''+\beta^2\Delta^2 u = 0, \tag{1a}$$

and the other is

$$u''-\alpha^2\Delta u''+\beta^2\Delta^2 u = 0 . \tag{1b}$$

In (1a) and (1b), ' = $\partial/\partial t$ and Δ is the ordinary Laplacian in the variables $X = (x,y) \in \mathbb{R}^2$. α and β are nonzero constants, and (1a), (1b) are assumed to hold for $t>0$ and for (x,y) in a bounded, open, connected region Ω of $\mathbb{R}^2$ having a smooth boundary Γ.

We shall assume throughout that the plate is clamped along a nonempty portion Γ_0 of its boundary, so that one has boundary conditions

$$u = \frac{\partial u}{\partial n} = 0 \text{ on } \Sigma_0, \tag{2}$$

where $\Sigma_0 = \Gamma_0 \times (0,\infty)$. (n denotes the unit exterior normal to Γ.) Along the remainder $\Gamma_1 = \Gamma/\Gamma_0$ of the boundary (Γ_1 is assumed to be nonempty and relatively open in Γ), shear forces v_1 and/or bending mements v_2 (about the axis formed by the tangent to Γ_1)

are prescribed, whose function is to control the motion of the plate and, in particular, to uniformly stabilize the plate in a sense to be made precise below. One of the questions to be addressed is how to determine v_1, v_2 in feedback form in order to accomplish this goal.

For a 1-dimensional problem analogous to (1a), (2), i.e., a cantilevered beam with dynamics described by

$$u''+\beta^2 u_x^{(4)} = 0,\ 0<x<L,\ t>0,$$

and with end conditions

$$u(0,t) = u_x(0,t) = 0,\ t>0,$$

the question of uniform boundary stabilization by means of forces v_1 and moments v_2 applied at the free end has been considered by Chen *et al* [3]. The free end boundary conditions are taken to be

$$\beta^2 u_x^{(3)}(L,t) = v_1(t),\ -\beta^2 u_x^{(2)}(L,t) = v_2(t), \tag{3}$$

$$v_1(t) = k_1 u'(L,t),\ v_2(t) = k_2 u_x'(L,t) \tag{4}$$

with $k_1 \geq 0$, $k_2 \geq 0$, $k_1+k_2>0$. The choice of boundary conditions (3), (4) is motivated by consideration of the total (kinetic plus strain) energy of the beam, i.e.,

$$E(u,t) = \frac{1}{2}\int_0^L (|u'|^2+\beta^2|u_x^{(2)}|^2)dx$$

A simple calculation yields

$$E'(u,t) = -v_2(t)u_x'(L,t)-v_1(t)u'(L,t),$$

hence for the feedback law (4) one has $E'(u,t)\leq 0$. This *suggests* that (3), (4) may lead to a uniform decay of energy, that is, to an estimate

$$E(u,t) \leq Ce^{-\alpha t}E(u,0) \tag{5}$$

for all solutions of (1a), (2), (3), (4), with fixed positive constants C,α. That (5) is indeed true is proved in [3], provided $k_1>0$ is assumed. A similar result has recently been obtained by J.U. Kim [5] for solutions of the Timoshenko beam model.

For plate equations, no analogous results are known. To see what might be possible, it is worthwhile to look at related work due to Chen [1,2], Lagnese [6,7], and Lasiecka-Triggiani [8] on uniform boundary stabilization of solutions to *wave* equations. In each of these papers, the geometry of Γ_0 and Γ_1 plays a central role. These geometric conditions take various forms, depending on the feedback scheme considered. For example, for the problem (dimension is unimportant)

$$u''-\Delta u = 0 \text{ in } \Omega,\ t>0,$$

$$u = 0 \quad \text{on } \Sigma_0,$$

$$\frac{\partial u}{\partial n} = -ku' \text{ on } \Sigma_1,\ k>0,$$

the estimate (5) was established in [1] for the energy functional

$$E(u,t) = \frac{1}{2}\int_\Omega (|u'|^2+|\nabla u|^2)\,dx,$$

provided Γ_0, Γ_1 satisfy the conditions

$$(X-X_0)\cdot n\leq 0 \quad \text{on } \Gamma_0, \tag{6}$$

$$(X-X_0)\cdot n\geq \gamma>0 \text{ on } \Gamma_1 \tag{7}$$

for some X_0 in the exterior of $\overline{\Omega}$. For the same problem, it was shown in [6] that the vector field $X-X_0$ in (6), (7) can be replaced by any other sufficiently smooth vector field $\ell(X)$ having a positive definite Jacobian $(\partial \ell_i / \partial x_j)$ in $\bar{\Omega}$. A geometric condition of a different sort appears in [8]; however, in that paper the stabilizing feedback acts through a Dirichlet boundary condition rather than a Neumann condition. The techniques used in all of these papers are (nontrivial) adaptations of methods from the theory of local energy decay of solutions to the wave equation in the exterior of a reflecting obstacle. It is worth noting that because of the required smoothness of Γ, conditions (6), (7) imply that $\bar{\Gamma}_0 \cap \bar{\Gamma}_1 = \phi$. Thus the above mentioned results cannot hold for simply connected regions, unless $\Gamma_0 = \phi$.

Turning now to the Kirchoff plate model (1a), we try to find candidates for stabilizing boundary conditions by multiplying (1a) by u' and integrating over Ω. The result may be written

$$\frac{d}{dt}\int_{\Omega}(|u'|^2+|\Delta u|^2)\,dxdy = -2\int_{\Gamma_1}[u'\frac{\partial(\Delta u)}{\partial n} - \Delta u\,\frac{\partial u'}{\partial n}]\,d\Gamma$$

In view of this, it is tempting to define an "energy functional" by

$$E(u,t) = \frac{1}{2}\int_{\Omega}(|u'|^2+|\Delta u|^2)\,dxdy \tag{8a}$$

and to introduce the dissipative (with respect to $E(u,t)$) boundary conditions

$$\beta^2\frac{\partial(\Delta u)}{\partial n} = k_1 u',\ \beta^2\Delta u = -k_2\frac{\partial u'}{\partial n} \quad \text{on } \textstyle\sum_1 \tag{9a}$$

with $k_1 \geq 0$, $k_2 \geq 0$, $k_1+k_2 > 0$. The problem (1a), (2), (9a) with $E(u,t)$ defined by (8a) is formally analogous to the beam problem described above. A similar procedure applied to the equation (1b) leads to an energy functional

$$E(u,t) = \frac{1}{2}\int_{\Omega}(|u'|^2+\alpha^2|\nabla u'|^2+\beta^2|\Delta u|^2)\,dxdy \tag{8b}$$

and to boundary conditions on $\sum_1$ given by

$$\beta^2 \frac{\partial(\Delta u)}{\partial n} = k_1 u' + \alpha^2 \frac{\partial u''}{\partial n}, \quad \beta^2 \Delta u = -k_2 \frac{\partial u'}{\partial n}. \qquad (9b)$$

The hope, of course, is that the problem (1a), (2), (9a) (respectively (1b), (2), (9b)) will, under suitable restrictions on the geometry of Γ_0 and Γ_1, lead to an exponential decay rate for the corresponding energy functional (8a) (respectively, (8b)).

Unfortunately, this approach is not likely to succeed because of the form of the energy functional. To see why this is so, one has to examine the "finite energy" space in which the solution $(u(t),u'(t))$ to (for example) (1a), (2), (9a) lives. Because of the form of the energy functional (8a), this space $V \times L^2(\Omega)$, where V is the completion of the space

$$H^2_{\Gamma_0}(\Omega) = \{v \in H^2(\Omega) \mid v = \frac{\partial v}{\partial n} = 0 \text{ on } \Gamma_0\}$$

in the norm

$$\left(\int_\Omega |\Delta v|^2 dxdy\right)^{1/2}.$$

Except for *very* special geometries of Ω, Γ_0, Γ_1, V will be a *very bad* space and may contain elements which are not in *any* (standard) Sobolov space. Similarly, the finite energy space for the problem (1b), (2), (9b) with energy functional (8b) is $V \times H^1_{\Gamma_0}(\Omega)$.

A more physical reason why this approach will probably fail is found in the fact that neither (8a) nor (8b) represents (nor is equivalent to) the true total (kinetic plus potential) energy of the plate (except for special configurations). Thus, a more promising approach is to start from the true energy functional and to apply the energy principle to obtain appropriate dissipative boundary conditions. For an *isotropic*, homogeneous material whose stress-strain relation obeys Hooke's law, the true energy functional (up to a constant factor) corresponding to equation (1a)(c.f. Duvaut-Lions [4,IV, 2.33]) is given by

$$E(u,t) = \frac{1}{2}\int_{\Omega}\{|u'|^2+\beta^2[|\Delta u|^2+2(1-\mu)(u_{xy}^2-u_{xx}u_{yy})]\}dxdy \qquad (10)$$

where μ is Poisson's ration $(0<\mu<\frac{1}{2})$. The calculation of $E'(u,t)$ now yields

$$E'(u,t) = \beta^2\int_{\Gamma_1}[(\Delta u+(1-\mu)B_2u)\frac{\partial u'}{\partial n} - (\frac{\partial(\Delta u)}{\partial n} +(1-\mu)B_1u)u']d\Gamma$$

where

$$B_1u = \frac{\partial}{\partial\tau}[n_1n_2(u_{xx}-u_{yy})+(n_1^2-n_2^2)u_{xy}],$$

$$B_2u = 2n_1n_2u_{xy}-n_1^2u_{yy}-n_2^2u_{xx},$$

$n = (n_1,n_2)$, $\tau = (-n_2,n_1)$. One is therefore led to dissipative conditions

$$\beta^2(\frac{\partial(\Delta u)}{\partial n} +(1-\mu)B_2u) = k_1u' \text{ on } \Sigma_1 \qquad (11a)$$

$$\beta^2(\ u+(1-\mu)B_2u) = -k_2\frac{\partial u'}{\partial n} \text{ on } \Sigma_1 . \qquad (12a)$$

The principle *mathematical* advantage of using (10) rather than (8a) is that

$$\{\int_{\Omega}[|\Delta u|^2+2(1-\mu)(u_{xy}^2-u_{xx}u_{yy})]dxdy\}^{1/2}$$

defines a norm on $H_{\Gamma_0}^2(\Omega)$ *equivalent* to the usual Sobolov norm.

It can be proved that the problem (1a), (2), (11a), (12a) is well-posed on the space $H_{\Gamma_0}^2(\Omega)\times L^2(\Omega)$, i.e., solutions are governed by a (C_0)-semigroup of contractions on that space. Furthermore, one has the following uniform stability result.

THEOREM. *For the problem* (1a), (2), (11a), (12a), *and with* $E(u,t)$ *defined by* (10), *the estimate* (5) *is valid provided the following are true:*

(i) $k_1>0$, $k_2 \geq 0$;

(ii) *there exists* $X_0 \in \mathbb{R}^2$ *such that*

$$(X-X_0)\cdot n \leq 0 \text{ on } \Gamma_0, \tag{13}$$

$$(X-X_0)\cdot n \geq \gamma \text{ on } \Gamma_1, \tag{14}$$

where $\gamma = 0$ *if* $k_2 = 0$, $\gamma>0$ *if* $k_2>0$.

Note that if $k_2 = 0$, conditions (13) and (14) can be satisfied even if $\overline{\Gamma}_0 \cap \overline{\Gamma}_1 \neq \phi$, unlike the corresponding results for wave equations described above.

For the plate equation (1b), a similar argument leads to the energy functional

$$E(u,t) = \frac{1}{2}\int_\Omega \{|u'|^2+\alpha^2|\nabla u'|^2+\beta^2[|\Delta u|^2+(1-\mu)(u_{xy}^2-u_{xx}u_{yy})]\}dxdy.$$

The corresponding dissipative boundary conditions on $\sum_1$ are (12a) and

$$\beta^2\left(\frac{\partial(\Delta u)}{\partial n}+(1-\mu)B_1u\right) = k_1u' +\alpha^2\frac{\partial u''}{\partial n}. \tag{11b}$$

The problem (1b), (2), (12a), (11b) is well-posed in the space $H^2_{\Gamma_0}(\Omega)\times H^1_{\Gamma_0}(\Omega)$. The issue of uniform stability is, however, still open.

Acknowledgement. The author wishes to thank Professor J.L. Lions for very useful suggestions and insights concerning the models discussed in this paper.

REFERENCES

[1] Chen, G., "Energy decay estimates and exact boundary value controllability for the wave equation in a bounded domain." *J. Math. Pures Appl.* 58 (1979): 249-274.

[2] Chen, G., "A note on boundary stabilization of the wave equation." *SIAM J. Control Opt.* 19 (1981): 106-113.

[3] Chen, G., M.C. Delfour, A.M. Krall and G. Payre, "Modeling, stabilization and control of serially connected beams." *SIAM J. Control Opt.* To appear.

[4] Duvaut, G., and J.L. Lions. *Inequalities in Mechanics and Physics.* Berlin: Springer-Verlag.

[5] Kim, J.U., "Boundary control of the Timoshenko beam." To appear.
[6] Lagnese, J., "Decay of solutions of wave equations in a bounded region with boundary dissipation." *J. Differential Eqs.* 50 (1983): 163-182.
[7] Lagnese, J., "Boundary stabilization of linear elasto-dynamic systems." *SIAM J. Control Opt.* 21 (1983): 968-984.
[8] Lasiecka, K., and R. Triggiani. "Exponential uniform stabilization of the wave equation with $L^2(0,\infty;L^2(\Gamma)$ boundary feedback acting in the Dirichlet boundary conditions." *J. Differential Eqs.* To appear.

SENSITIVITY AND OPTIMAL SYNTHESIS FOR A CLASS OF LINEAR TIME-DELAY SYSTEMS[1]

E. Bruce Lee and N. Eva Wu

University of Minnesota, Minneapolis, USA

ABSTRACT

Some further results on optimal synthesis for linear time delay systems are presented. This is an extension of techniques for finite dimensional systems to infinite dimensional systems.

1. INTRODUCTION

Rather than attempting to reconstruct by means of an "observer" state information for an infinite dimensional system from a limited number of sensors and then using the reconstructed state information to implement feedback control actions [1], one can postulate the form of the combined observer and subsequent control action to seek directly a feedback controller that uses only the sensor information. We shall consider the synthesis of such a feedback controller (compensator) for linear time delay type systems. In this setting the linear infinite dimensional systems we will consider are represented by an input/output relationship; which is assumed to be a transfer function. In the case of linear delay type systems, the transfer function is the ratio of exponential polynomials in the complex Laplace transform variable s. The compensator will also be assumed to be of the linear delay type (transfer function ratio of exponential polynomials) and will be selected to provide stability and weighted sensitivity minimization. Both of these are well known

---

[1]Research supported in part by the National Science Foundation under Grant No. DMS 8413129

tasks in the finite dimensional case when the transfer function is rational (ratio of polynomials in complex Laplace transform variable s). Previously we have looked at the task of stabilization of linear delay type systems by linear compensators using the Rosenbrock approach. [2]

It has been pointed out recently by Foias, Tannenbaum and Zames [3] that weighted sensitivity H^∞-minimization is difficult to achieve even for the simplest delay type systems and demands infinite dimensional techniques.

The weighted sensitivity function is defined by

$$S_W(s) = W(s)\ (1 + P(s)K(s))^{-1} \tag{1}$$

where $P(s) \equiv P(s, e^{-h_1 s}, e^{-h_2 s}, \ldots, e^{-h_r s})$ is the transfer function model of a linear time invariant continuous-time system with r noncommensurate delay durations $h_1, h_2, \ldots, h_r$. Such a ratio of "proper" exponential functions can be represented by differential-difference equations (DDE)

$$\frac{dx(t)}{dt} = (A(d_1, \ldots, d_r)x)(t) + (B(d_1, \ldots, d_r)u)(t)$$

with output relationship (2)

$$y(t) = (C(d_1, \ldots, d_r)x)(t) + (D(d_1, \ldots, d_r)u)(t).$$

Here u(t) is the scalar input function at time t, x(t) is an n-vector at time t and y(t) is the scalar output at time t. A, B, C and D are real matrix polynomials in delay (right shift) operators $d_1, \ldots, d_r$ (i.e $d_i x(t) = x(t-h_i)$). K(s) is transfer function of the internally stabilizing feedback controller and can be associated with a quadruple (F, G, H, J) (of similar form to (2) with the quadruple (A, B, C, D)) if and only if

$$\text{rank}[sI - A(e^{-sh_1}, \ldots, e^{-sh_r}) \quad B(e^{-sh_1}, \ldots, e^{-sh_r})] = n \tag{3}$$

and

$$\text{rank} \begin{bmatrix} sI - A(e^{-sh_1}, \ldots, e^{-sh_r}) \\ C(e^{-sh_1}, \ldots, e^{-sh_r}) \end{bmatrix} = n \tag{4}$$

for all s in the closed right half complex plane (see Lee, Olbrot paper [4] for notation). (3) is called the stabilizability condition, analogous to the one for ordinary differential equation (ODE) systems and W(s) is the weighting function which is real and stable (having all poles with negative real part). Weighted sensitivity H^{∞}-minimization involves finding a feedback controller transfer function K(s) which achieves (or approximately achieves)

$$\inf\{\|S_W(s)\|_{\infty} : \text{ K stabilizing}\} \equiv \hat{a} \tag{5}$$

where $\| \ \|_{\infty}$ is the infinity norm of the space H^{∞}. The $L^2(H^2)$ theory of optimal control using "state" feedback is covered by Alekal, Brunovsky, Chyung and Lee [5].

In the paper by Foias, Tannenbaum and Zames a way of finding K when W(s) is any <u>real</u>, <u>strictly proper</u> and <u>rational</u> function and P(s) any stable minimum phase rational function multiplied by a transmission delay element e^{-sh} is given. Some results from Sarason [6] and a special case of the commutant lifting theorem from Sz Nagy-Foias [7] were used to obtain the feedback controller transfer function K(s) in the above special situation.

Since stabilizing a given unstable system while optimizing some other performance measure (in this case, the sensitivity measure) is a common task in optimal control theory and the transfer function of a delay system usually takes the form of ratio of two exponential polynomials [4,8] (see also equation (2)) rather than simply a stable, rational and minimum phase function multiplied by e^{-sh}, the method proposed by Foias et al [3] does not seem extendable to include any system beyond e^{-sh} times a stable minimum phase rational function. For this reason, we shall consider the class of real systems admitting stable Bezout coprime factorizations (i.e., for which there exist n(s), d(s), x(s) and y(s) εH^{∞} with $y(\infty) \neq 0$, $d(\infty) \neq 0$, such that $p(s) = n(s)d^{-1}(s)$ and $xn+yd = 1$).

For details about coprime factorizations, see papers by Callier and Desoer [9], Nett, Jacobs and Balas [10], and Kamen, Khargonekar and Tannenbaum [11,12]. We are particularly interested in the delay systems given by (2) with commensurate delays (r=1) and require that the associated transfer function P(s) contains no $j\omega$-axis poles and zeros ($j = \sqrt{-1}$).

In section 2 weighted sensitivity H^∞-minimization is discussed and a general approach to minimization is proposed. Section 3 gives some ideas on getting finite dimensional approximates, while section 4 contains examples and comments.

2. MAIN RESULTS

By using a Youla, Bongiono and Jabr [13] type parameterization, we obtain the following equivalent formulation to (5):

$$\hat{a} = \inf_{\hat{v}\varepsilon H^\infty} \|\hat{u} - \hat{v}\|_\infty \tag{6}$$

where $\hat{u}(s)$ and $\hat{v}(s)$ are given by

$$\begin{aligned} \hat{u}(s) &= w(s)\ y(s)\ n_1^{-1}(s)d_2(s) \\ \hat{v}(s) &= w(s)\ q(s)\ n_2(s)\ d_2(s) \end{aligned} \tag{7}$$

In (7) q(s), the free parameter from the Youla parameterization of stabilizing controllers for P(s), is now allowed to be such that $\hat{v}(s)\varepsilon H^\infty$. The possible nonproperness of the resulting optimal transfer function (compensator) can be resolved by approximating it by a proper one as suggested in Zames [14]. w(s) is a proper weighting function which is not necessarily a rational function although we shall assume so for simplicity. $n_1(s)$ is composed of the possibly infinite Blaschke product $\prod_{k=1}^{\infty} \frac{s-z_k}{s+\bar{z}_k}$ convergent and analytic in the closed right half plane, and a singular inner function [15] of the form e^{-sh} for some $h>0$. $n_2(s)$ is a proper stable function with no zeros on the open right half plane and on the jw-axis.

$d_2(s) = \frac{d(s)}{d_1(s)}$ and $d_1(s)$ is a Blaschke product in terms of all the open right half plane poles of P(s) (zeros of d(s)) which is of finite order because of the nature of time-delay systems. u(s) in (7) will be bounded on the $j\omega$-axis ($u(j\omega)\varepsilon L^\infty(-j\omega,j\omega)$ as long as P(s) does not have any pole or zero on the imaginary axis as we already assumed). Hence (6) falls within the Nehari formulation [16] and the infimum is equal to the norm of the bounded operator $L(\hat{u}) : H_2 \to H_2^\perp$

given by $(L(\hat{u})x)(s) = P_{H_2^\perp}(\hat{u}(s)\hat{x}(s))$, $\forall \hat{x} \in H^2$, where $P_{H_2^\perp}$ is a projection operator.

Suppose $\|L(u)\| = \sigma_{max}$ and σ^2_{max} is an eigenvalue of $L^*(\hat{u})L(\hat{u})$. Then by Adamjan, Arov and Krein [17] the infimum in (6) is simply associated with eigenvalues of

$$(L^*L\hat{x})(s) = \lambda\hat{x}(s). \tag{8}$$

After solving (8) for $\lambda = \sigma^2_{max}$, put $\hat{v}(s)$ as follows

$$\hat{v}(s) \equiv \hat{u}(s) - \frac{\sigma_{max}\hat{x}(-s)}{\hat{x}(s)} \tag{9}$$

If $u(t) = {}^{-1}[\hat{u}(jw)] \in L'(-\infty,\infty)$, then obviously $\hat{u}(jw) \in L^\infty(-j\infty,j\infty)$ C. Therefore by Hartman [18] it follows that $L(\hat{u})$ is compact and hence σ^2_{max} is an eigenvalue. In this case (8) and (9) will lead to the unique solution for minimization [17]. But if $\hat{u}(jw)$ is not continuous, $L(\hat{u})$ then may not be compact. The solution to (6) may not be unique and elegantly characterized by (9) unless $\|L(\hat{u})\|$ is an eigenvalue of $(L^*(\hat{u})L(\hat{u}))^{1/2}$.

For further clarification note that the function $\hat{u}(s)$ can be written as the product of the following factors $u_0(s)$ (includes rational function a ratio of exponential polynomials which is continuous on jw-axis and an infinite Blaschke product) and e^{sh} for some $h>0$. Since $\hat{u}(s)$ is bounded on the $j\omega$-axis it will be continuous on $j\omega$-axis unless $u_0(j\infty) \neq 0$ and $h \neq 0$. In this case, we can write $\hat{u}(s)$ as $\hat{u}_0(\infty)e^{sh} + (\hat{u}_0(s)-\hat{u}_0(\infty))e^{sh}$. Then $L(\hat{u}(s)) = \hat{u}_0(\infty)L(e^{sh}) + L((\hat{u}_0(s) - \hat{u}_0(\infty))e^{sh}) = L_1+L_2$ where the first operator is not compact but the second is. For example, $\hat{u}_0(s) = \frac{as+1}{bs+1}$, $a, b > 0$. Then

$$L(\hat{u}(s)) = L(\frac{a}{b} e^{sh}) + L(\frac{1 - \frac{a}{b}}{bs+1} e^{sh}) = L_1 + L_2$$

It is easy to see that any $f(t)$ on $[0,h]$ even (odd) symmetric about $\frac{h}{2}$ will have image $\frac{a}{b} f(-t)$ $(-\frac{a}{b} f(-t))$ under $L(\frac{a}{b} e^{sh})$. Thus $L(\frac{a}{b} e^{sh})$ has eigenvalues $\frac{a}{b}$ and $-\frac{a}{b}$ with infinite multiplicity. By a compact perturbation theorem [19] L and L_1 have the same essential spectrum $-\frac{a}{b}$ and $\frac{a}{b}$. Observe that

$$\frac{a}{b} \leq \inf_{\hat{v}\in H^\infty} \left\| \frac{as+1}{bs+1} e^{sh} - \hat{v} \right\|_\infty \leq \left\| \frac{as+1}{bs+1} \right\|_\infty \tag{11}$$

So $\|L\| > \|L_1\|$ if $\left\|\frac{as+1}{bs+1}\right\|_\infty > \frac{a}{b} = |\hat{u}_0(\infty)|$(i.e. $a > b$). Therefore (8) and (9) will give the unique solution to (6). However, if $a < b$ or $\left\|\frac{as+1}{bs+1}\right\|_\infty = |\hat{u}_0(\infty)|$, $\hat{v}(s) = 0$ is a solution to (6) which cannot be characterized by (9). The conclusion for this example can easily be extended to the general case. Thus we have the following result.

<u>Theorem</u> 1 (9) is <u>the</u> unique solution of (6) if $|\hat{u}_0(\infty)| < \|\hat{u}_0(s)\|_\infty$, while $\hat{v}(s) = 0$ is <u>a</u> solution if $|\hat{u}_0(\infty)| \geq \|\hat{u}_0(s)\|_\infty$.

In the case that $\|L(\hat{u})\|$ is an eigenvalue of $(L^*L)^{1/2}$ then it is the first thing we want to find. The eigenvalue will at least tell us how small the sensitivity measure can be made by the choice of the stabilizing controllers.

Equation (8) written in integral equation form is

$$\int_{-\infty}^{0} u(\tau-t) \int_{0}^{\infty} u(\tau-\ )x(\Theta)d\Theta)d\tau = \lambda x(t) \qquad t > 0 \tag{12}$$

or

$$\int_0^\infty u(t-\Theta)x(\Theta)d\Theta = \sqrt{\lambda}\, y(t) \quad t<0$$

$$\int_{-\infty}^0 u(\Theta-t)y(\Theta)d\Theta = \sqrt{\lambda}\, x(t) \quad t>0 \tag{13}$$

From (13) and using the boundedness of $\hat{u}(j\omega)$, the operator L can be represented by an infinite matrix $B = (b_{ij}) = [L\phi_i, \psi_j)_{i,j}]$ and $B : \ell^2 \to \ell^2$ with respect to orthonormal basis $\{\phi_i\}^\infty$ in $L^2(0,\infty)$ and $\{\psi_i\}^\infty$

in $L^2(-\infty,0)$, where $(\cdot,\cdot)$ is the usual inner product defined on L^2.[2] If we choose $\psi_i(t)$ so that $\psi_i(t) = \phi_i(-t)$, we have $b_{ij} = (L\phi_i, \psi_j)$

$$= \int_{-\infty}^{0} \left(\int_{0}^{\infty} u(t-\tau)\phi_i(\tau)\right)\overline{\psi}_j(t)\,dt = \int_{0}^{\infty}\int_{-\infty}^{0} u(t-\tau)\;\varphi_i(-t)\overline{\psi}_j(\tau)\,dt\,d\tau$$

$$= \int_{0}^{\infty}\int_{-\infty}^{0} u(t-\tau)\psi_i(t)\overline{\varphi}_j(\tau)\,dt\,d\tau = \overline{\int_{-\infty}^{0} \left(\int_{0}^{\infty} u(t-\tau)\;\varphi_j(\tau)\,d\tau\right)\overline{\psi}_j(t)\,dt}$$

$$= \overline{(L\varphi_j, \psi_i)} = \overline{b}_{ji}$$ Hence $B^* = B$. It is easy to show that

$\|B\| = \|L(u)\|$. Let $P_N = \text{diag}\,(\underbrace{1, 1, \ldots, 1}_{N}, 0, \ldots, 0)$ and $B_N = P_N B P_N$.

Then $\|B_N\| \nearrow \|B\|$ [3]

---

[2]Suppose $x(t) \in L^2(0,\infty)$ and $y(t) = (L(u)x)(t)$, then

$$y(t) = \sum_{1}^{\infty} (y, \psi_j)\psi_j = \sum_{1}^{\infty} (Lx, \psi_j)\psi_j$$

$$= \sum_{j=1}^{\infty} (L\sum\alpha_i\varphi_i, \psi_j) = \sum_{j=1}^{\infty}\sum_{i=1}^{\infty} \alpha_i(L\phi_i, \psi_j)\psi_i$$

since L is bounded. Let $\gamma_j = \sum_{i=1}^{\infty} \alpha_i b_{ij}$ where $b_{ij} = (L\varphi_i, \psi_j)$.

Then $\gamma = B\alpha$ with

$\gamma = (\gamma_1, \gamma_2, \ldots)^T$, $a = (\alpha_1, \alpha_2, \ldots)^T$, $B = (b_{ij})$.

[3]$\|B\alpha\|_{\ell_2} = \|\gamma\|_{\ell_2} = \|y(t)\|_{L^2(-\infty,0)}$ $\|L(u)\|\;\|x\|_{L^2(0,\infty)}$

$= \|L(u)\|\|\alpha\|_{\ell_2}$. So we have $\|B\| \leq \|L(u)\|$. Similarly, we can show $\|B\| \geq \|L(u)\|$.

$B_N = P_N B P_N$, $\|B_N\| = \|P_N B P_N\| \leq \|P_N\|\|B\|\|P_N\|$

$= \|B_N\|$ since $\|P_N\| = 1$. Moreover,

$\|B_1\| \leq \|B_2\| \leq \ldots \leq \sum\|B_N\| \leq \quad \leq \|B\|$

and $\{\|B_N\|\}$ is a sequence of increasing real numbers bounded above by $\|B\|$, hence it converges. Since $\|B\| - \|B_N\| \leq \|P_N B P_n - B\| \to 0$, $\|B_N\| \to \|B\|$.

Example: For the class of systems considered in [3] $\hat{u}(s)$ is a strictly proper stable minimum phase function $w(s)$ (note, not necessarily rational here) times e^{sh}. Since $L(u)$ only depends on the part of $u(t) = \mathcal{L}^{-1}(\hat{u}(s))$ that is defined on the negative time axis, (denote it by $u_-(t)$)

$$u_-(t) = \begin{cases} w(t+h) & t \in [-h, 0) \\ 0 & t \geq 0,\ t < -h \end{cases}$$

it is easy to see that $\|L(u)\|$ is attained by the functions defined in $L^2(0,h)$, a subspace of $L^2(0,\infty)$. For this reason, we can choose

$$\varphi_1 = \sqrt{\frac{1}{h}}\ ,\quad \varphi_2 = \sqrt{\frac{2}{h}}\ \cos\frac{\pi t}{h}\ ,\ \ldots\ ,\ \varphi_i = \sqrt{\frac{2}{h}}\cos\frac{(i-1)\pi t}{h},\ \ldots$$

for $t \in [0,h]$ and $\psi_i(t) = \varphi_i(-t)$. Thus

$$b_{ij} = (L(u)\varphi_i, \psi_j) = \int_{-h}^{0} \psi_j(t)\ \left(\int_{-h}^{0} w(\tau+h)\varphi_i(t-\tau)d\tau\right)dt$$

$$= \int_{-h}^{0} w(\tau+h)\ \left(\int_{\tau}^{0} \varphi_i(t-\tau)\psi_j(t)dt\right)d\tau$$

$$= \begin{cases} \sqrt{2h}\ \dfrac{(j-1)b_{j-1} - (i-1)b_{i-1}}{((i-1)^2 - (j-1)^2)\pi} & i \neq j \quad i,j \neq 1 \\[2ex] -\dfrac{1}{\sqrt{2h}}\ \left(C_{i-1} + \dfrac{b_{i-1}h}{(i-1)\pi}\right) & i = j \neq 1 \\[2ex] -\sqrt{h}\ \dfrac{b_{j-1}}{(j-1)\pi} & i = 1 \quad j \neq 1 \\[2ex] -\sqrt{h}\ \dfrac{b_{i-1}}{(i-1)\pi} & i \neq 1 \quad j = 1 \\[2ex] -\dfrac{C_0}{\sqrt{h}} & i = j = 1 \end{cases} \qquad (14)$$

where $b_k = \sqrt{\frac{2}{h}} \int_{-h}^{0} w(\tau+h)\sin\frac{k\pi\tau}{h}\ d\tau$

$$C_0 = \sqrt{\frac{1}{h}} \int_{-h}^{0} \tau w(\tau+h)d\tau \quad \text{and} \tag{15}$$

$$C_k = \sqrt{\frac{2}{h}} \int_{-h}^{0} \tau w(\tau+h) \cos \frac{k\pi\tau}{h} d\tau$$

For $w(s) = \frac{1}{s+1}$ (same example as given in [3]), we obtain $\hat{a} \cong \|B_5\| = 0.4419$.

If $W(s)$ is not strictly proper, we can write it as $W(s) = W(\infty) + W_{sp}(s)$. Hence $B_W = B_{W(\infty)} + B_{W_{sp}}$, $B_{W(\infty)} = W(\infty)\ \mathrm{diag}\{1, -1, 1, -1, 1, -1, \ldots\}$ and $B_{W_{sp}}$ is computed by (15).

For more general systems so that $\hat{u}(s)$ contains unstable poles and term e^{sh}, Laguerre functions can be used as basis, i.e.,

$$\hat{\phi}_i(s) = \frac{(s-p)^{i-1}}{(s+p)^i}\sqrt{2p} \qquad p>0 \tag{16}$$

$$\hat{\psi}_i(s) = \hat{\varphi}_i(-s) = \frac{-(s+p)^{i-1}}{(s-p)^i}\sqrt{2p}$$

and b_{ij} is given by†

$$b_{ij} = (\hat{u}\hat{\phi}_i, \hat{\psi}_j) \tag{17}$$

---

† $b_{ij} = (L(\hat{u})\hat{\phi}_i, \hat{\phi}_j)$. By definition, $L(\hat{u})\hat{\phi}_i$

$$= P_{H_2^\perp}\{(\hat{u}_+ + \hat{u}_-)\hat{\phi}_i\} = P_{H_2^\perp}\{\hat{u}_-\hat{\phi}_i\} = \sum_{k=1}^{\infty} (\hat{u}_-\hat{\phi}_i, \hat{\psi}_k)\hat{\psi}_k(s)$$

So $(L(\hat{u})\hat{\varphi}_i, \hat{\psi}_j) = (\sum_{k=1}^{\infty} (\hat{u}_-\hat{\varphi}_i, \hat{\psi}_k)\hat{\psi}_k, \hat{\psi}_j) =$

$$= \sum_{k=1}^{\infty} (\hat{u}_-\hat{\varphi}_i, \hat{\psi}_k)(\hat{\psi}_k, \hat{\psi}_j) = (\hat{u}_-\hat{\varphi}_i, \hat{\psi}_j)$$

$$= (\hat{u}_-\hat{\varphi}_i, \hat{\psi}_j) + (\hat{u}_+\hat{\varphi}_i, \hat{\psi}_j) = (\hat{u}\hat{\phi}_i, \hat{\psi}_j)$$

In general, the computation involves an integral

$$b_{ij} = \frac{1}{2\pi} \int_{-\infty}^{\infty} e^{j\omega h}\hat{u}_0(j\omega)\hat{\varphi}_i(j\omega)\hat{\psi}_j(-j\omega)d\omega$$

which is the inverse Fourier transform of $\hat{u}_0\varphi_i(j\hat{\omega})\psi_j(-j\omega)$ evaluated at t=h. However, when $\hat{u}(s)\varphi_i(s)\psi_j(-s)$ contains either a finite number of RHP poles or a finite number of LHP poles, the above integral is better written in the form of contour integral and hence reduces to the computation of residuals at these poles

$$b_{ij} = \pm \frac{1}{2\pi j} \int_{\substack{\text{RHP} \\ \text{or LHP}}} \hat{u}(s)\hat{\varphi}_i(s)\hat{\varphi}_j(-s)ds$$

where the integrand

$$\hat{u}(s)\hat{\varphi}_i(s)\hat{\psi}_j(-s) = \hat{u}(s)\frac{(s-p)^{i+j-2}}{(s+p)^{i+j}}(2p) \tag{18}$$

We can see that the value of b_{ij} depends only on the index i+j, and b_{ij}'s are real numbers if $\hat{u}(s)$ is "real." Therefore B is a real Hankel matrix with respect to basis given in (16).

3. FINITE DIMENSIONAL APPROXIMATIONS

Because of difficulty in solving equation (8), it is worth investigating the possibility to solve (8) constructively using finite-dimensional approximations so that the real sensitivity measure can be made as close to the minimal one as we wish.

Let $z = \frac{s-p}{s+p}$, $p>0$. Thus $s = p\,\frac{1+z}{1-z}$. Let $\hat{g}(z) = \hat{u}(p\,\frac{1+z}{1-z})$.

Then $\hat{g}(z)$ can be expanded into Laurent series $\sum_{i=-\infty}^{\infty} C_i z^{-i}$ in some annular region centered at zero or at least formally so. In any case, the anti-analytic part of $\hat{u}(s)$ can be uniquely expressed in the open LHP as

$$\hat{u}_-(s) = \sum_{i=1}^{\infty} C_i\left(\frac{s+p}{s-p}\right)^{i-1} \tag{19}$$

Lemma 1 $C_{i+j-1} = b_{ij} = (\hat{u}\hat{\phi}_i, \hat{\psi}_j)$ where $\hat{\phi}_i(s)$, $\hat{\psi}_j(s)$ are given by (16).

The proof follows from the integral formula for the Laurent coefficients of

$\hat{u}_-(s)$ with $s = p\,\frac{1+z}{1-z}$. []

To approximate $\hat{u}_-(s)$ uniformly on the closed left half plane we use a lemma from [12] (see also the references therein) with some modifications to suit our purpose.

Lemma 2 Any function f(s) which is holomorphic in the open left half plane and continuous on the closed left half plane including infinity may be uniformly approximated by polynomials of the form $a_{N-1}\Theta^{N-1}(s) + a_{N-2}\Theta^{N-2}(s)+\ldots+a_1\Theta(s) + a_o$, where $a_i \varepsilon R$ and Θ is any conformal equivalence mapping left half plane onto unit disk. In other words, for any given $\varepsilon>0$ there exists a polynomial $\sum_0^{N-1} a_i\Theta^i$ such that

$$\sup_{\mathrm{Re}\, s \leq 0} \left| f(s) - \sum_{i=0}^{N-1} a_i \Theta^i(s) \right| < \varepsilon \tag{20}$$

Assume now that $u_-(t) \in L_1(-\infty,0)$, then $\hat{u}_-(s)$ satisfies the condition in Lemma 2. As pointed out in [12], the Cesaro sum

$$\frac{c_1 + \sum_1^2 c_i\Theta^{i-1} + \sum_1^3 c_i\Theta^{i-1} + \ldots + \sum_1^N c_i\Theta^{i-1}}{N}$$

$$= c_1 + c_2\left(1 - \frac{1}{N}\right)\Theta + c_3\left(1 - \frac{2}{N}\right)\Theta^2 + \ldots + c_N\left(1 - \frac{N-1}{N}\right)\Theta^{N-1}$$

$$= \sum_{i=1}^{N} c_i\left(1 - \frac{i-1}{N}\right)\Theta^{i-1}$$ converges to $\hat{u}_-$ uniformly as $N\to\infty$.

We choose $\Theta(s) = \dfrac{s+p}{s-p}$, $p > 0$ in the following development.

Let N be such that $\sup_{\mathrm{Re}\, s \leq 0} \left| \hat{u}_-(s) - \sum_1^N c_i\left(1 - \frac{i-1}{N}\right)\left(\frac{s+p}{s-p}\right)^{i-1} \right| < \frac{\varepsilon}{2}$

for some given ε. Then the coefficients of the polynomial we choose to approximate $\hat{u}_-(s)$ are $a_i = c_{i+1}(1 - \frac{i}{N})$, $i = 0,1,\ldots N-1$. The matrix B^N induced by

$\sum_1^N c_i(1 - \frac{i-1}{N})(\frac{s+p}{s-p})^{i-1}$ is then given by

$$B^N = \begin{bmatrix} c_1 & c_2(\frac{N-1}{N}) & c_3(\frac{N-2}{N}) & \cdots & c_N(\frac{1}{N}) \\ c_2(\frac{N-1}{N}) & c_3(\frac{N-2}{N}) & \cdots & c_N(\frac{1}{N}) & 0 \\ \vdots & & \ddots & & \vdots \\ c_N(\frac{1}{N}) & 0 & \cdots & & 0 \end{bmatrix}$$

Hence

$$\|B - B^N\| = \|L(\hat{u}_-(s)) - L(\sum_1^N c_i(1 - \frac{i-1}{N})(\frac{s+p}{s-p})^{i-1})\|$$

$$= \inf_{\hat{v}\in H^\infty} \|\hat{u}_-(s) - \{\sum_1^N c_i(1 - \frac{i-1}{N})(\frac{s+p}{s-p})^{i-1}\} - \hat{v}(s)\|_\infty$$

$$\leq \|\hat{u}_-(s) - \sum_1^N c_i(1 - \frac{i-1}{N})(\frac{s+p}{s-p})^{i-1}\|_\infty < \frac{\varepsilon}{2}$$

and

$$\big|\|B\| - \|B^N\|\big| < \frac{\varepsilon}{2} \tag{22}$$

by the triangle inequality.

The problem

$$\hat{a}_a = \inf_{\hat{v}\in H^\infty} \|\sum_1^N C_k(1 - \frac{k-1}{N})(\frac{s+p}{s-p})^{k-1} - \hat{v}\|_\infty \tag{23}$$

is uniquely solvable following (8) and (9), i.e., solving the eigenvalue problem

$$B^N x = \lambda x \tag{24}$$

for λ with largest absolute value, where $x = (x_1 \ldots x_N)^T$.

Then $|\lambda| = \hat{a}_a$. Put $\hat{v}_a^*$ as

$$\hat{v}_a^* = \sum_1^N C_k\left(1 - \frac{k-1}{N}\right)\left(\frac{s+p}{s-p}\right)^{k-1} - \lambda \frac{\sum_1^N x_i\hat{\phi}_i(-s)}{\sum_1^N x_i\hat{\phi}_i(s)} \tag{25}$$

where $\hat{\phi}_i$'s are given by (16). (24) and (25) solve (23).

The difference between the real sensitivity measure and the original minimum sensitivity measure must be estimated so as to determine if the approximation really works.

After the approximation, the real sensitivity measure becomes

$$\|\hat{u}_-(s) - \hat{v}_a^*\|_\infty .$$

Note that $\left|\|\hat{u}_-(s) - \hat{v}_a^*\|_\infty - \|\sum_1^N c_i\left(\frac{s+p}{s-p}\right)^{i-1}\left(1 - \frac{i-1}{N}\right) - \hat{v}_a^*\|\right| \leq \frac{\varepsilon}{2}$

On the other hand, from (22) and (23)

$$|a - \hat{a}_a| < \frac{\varepsilon}{2} .$$

Combining the above two inequalities we obtain

$$\left|\|\hat{u}_-(s) - \hat{v}_a^*\|_\infty - \hat{a}\right| < \varepsilon .$$

To summarize, we have

Theorem 2 Assume that $u_-(t) \varepsilon L^1(-\infty, 0)$. The infinum problem (6) can be solved by approximation through (23), (24) and (25). Moreover the resulting sensitivity function $S_w(s)$ can be made to satisfy

$$\|S_w(s)\|_\infty < \hat{a} + \varepsilon$$

for any $\varepsilon > 0$.

4. EXAMPLES AND REMARKS ON FURTHER EXTENSIONS

To illustrate the above results we consider an example (from [24]).

Example 1 (Application to finite dimensional systems) Consider a system with transfer function $p(s) = \frac{(s-1)^2}{(s+1)^2(2-s)}$ and the weighting function $W(s) = \frac{(s+1)}{(10s+1)}$. Denote Blaschke products by $\frac{2-s}{2+s} = B_p(s)$, and $\frac{(1-s)^2}{(1+s)^2} = B_z(s)$. Then $\hat{u}(s) = B_z^{-1} \; W \, Q_2(s)$ where

$Q_2(s) = -\frac{s^2-30s+17}{(s+1)^2}$ is computed by the authors of reference [24]. So

$$\hat{u}(s) = \frac{-(s+1)(s^2+30s+17)}{(1-s)^2(10s+1)}$$

Use $\phi_1(s) = \frac{\sqrt{2}}{s+1}$, $\phi_2(s) = \frac{\sqrt{2}}{s+1}\left(\frac{s-1}{s+1}\right)$

$$B = [(\hat{u}\phi_i, \psi_j)] = \begin{bmatrix} -1.01 & -0.05 \\ -0.05 & 0 \end{bmatrix} \text{ with } \lambda_{max} = -1.247.$$

Solve $\bar{f}B = \lambda_{max}\bar{f}$ to obtain $\bar{f} = (1.0 \quad 0.437)$. Hence

$$f(s) = \frac{\sqrt{2}}{s+1} + 0.437 \frac{\sqrt{2}}{(s+1)} \frac{(s-1)}{s+1}.$$

$$(\hat{u} - \hat{v}^*)(s) = \lambda_{max} \frac{f(-s)}{f(s)} = 1.247 \frac{(s+1)^2(s-0.391)}{(s-1)^2(s+0.391)}$$

$$\tilde{X}(s) = B_p(s)B_z(s)(\hat{u}(s)-\hat{v}^*(s)) = -1.247 \frac{(s-2)(s-0.391)}{(s+2)(s+0.391)}.$$

Note that $|b_{11}|$ is in any case a lower band on $\|\tilde{X}\|_\infty$. Here $|b_{11}| = 1.01 > \left|\frac{W(1)}{B_p(1)}\right| = 0.54$. The bound $\max_i \left|\frac{W(z_i)}{B_p(z_i)}\right|$ is given in [24].

Example 2 Consider the same example as given in section 2. $\hat{u}(s) = \frac{e^s}{s+1} = \hat{u}_-(s) + \hat{u}_+(s)$, where $\hat{u}_-(s) = \frac{e^s-e^{-1}}{s+1}$ and $\hat{u}_+(s) = \frac{e^{-1}}{s+1}$.

By lemma 2, $\hat{u}_-(s)$ can be approximated, for a given ε, uniformly on the closed left half plane by

$$\hat{u}_-(s)_{app} \equiv \sum_{i=1}^{N(\varepsilon)} C_i\left(1 - \frac{i-1}{N(\varepsilon)}\right)\left(\frac{s+p}{s-p}\right)^{i-1} \quad , \quad p > 0$$

where $C_i = (\hat{u}\hat{\phi}_{i-j+1}, \hat{\psi}_j)$, $\hat{\psi}_i$ and $\hat{\psi}_i$ are given by (16).

Numerically, the computation of C_i can go as follows. Set $W = \frac{s+p}{s-p}$. Then

$$\hat{u}_-(s) = \hat{u}_-\left(p\frac{W+1}{W-1}\right) = e^{-1}\,\frac{e^{\frac{(p+1)W + p-1}{W-1}} - 1}{(p+1)W + p-1}(W-1)$$

The above function of W is holomorphic in the unit disk and is continuous on the unit circle in the sense that for each W_o on the unit circle $\lim_{\substack{W\to W_o\\ |W|<1}} \hat{u}_-\left(p\frac{W+1}{W-1}\right)$ exists. C_i, $i=1,2,\ldots,N$ are then identified with the first N Fourier coefficients of the function

$$g(\alpha) \equiv \hat{u}_-\left(p\frac{e^{j\alpha}+1}{e^{j\alpha}-1}\right)$$

and can be evaluated through DFT of the sequence $g\left(\frac{2\pi m}{M}\right)$, $m = 0,1,2,\ldots,M-1$, for sufficiently large M.

Let N be, say, 50. Then $\hat{u}_-(s)_{app} = \sum_{i=1}^{50} C_i\left(1 - \frac{i-1}{50}\right)\left(\frac{s+1}{s-1}\right)^{i-1}$ for p=1. $\hat{v}^*{}_a(s)$ can be uniquely determined by (24) (where $B^N = B^{50}$) and (25) (where p=1). The estimation deviation from the optimal sensitivity measure $\left(\left|\|\hat{u}_-(s) - \hat{v}^*{}_a(s)\|_\infty - \inf_{\hat{v}\in H^\infty} \|\hat{u}_-(s) - \hat{v}(s)\|_\infty\right|\right)$ is less than 0.186. If N=10, this deviation is less than 0.42.

Remarks The Laplace transform approach for delay systems is covered in a book by Bellman and Cooke [20], which also contains much of the known information on the location of zeros of exponential polynomials.

We are also using the transfer function approach in the study of other infinite dimensional systems, in particular two-dimensional systems where the transfer function is the ratio of polynomials in two complex variables. Questions of stability [21], stabilization [22], and model reduction [23] have been considered.

It has been noticed that in the compensator design for a delay system the order of $\hat{u}_{-}(s)_{app}$ often has to be high in order to achieve the small preset deviation from the optimal sensitivity measure.

REFERENCES

[1] W.-S. Lu and E. B. Lee, "Output feedback controller design for linear retarded systems", Proc. V. Polish-English Seminar on Real time process control, Warsaw, 1986.

[2] E. B. Lee and A. Manitius, "Synthesis techniques for multivariable feedback systems with time delays", Tech. Report 74-19, Dept. of Comp. Sci, U. of MN. (Jan 1974).

[3] C. Foias, A. Tannenbaum and G. Zames, Weighted Sensitivity Minimization for Delay Systems, Proc. of 24th IEEE CDC, pp. 244-249, 1985.

[4] E. B. Lee and A. Olbrot, "Observability and related structural results for linear hereditary systems", Int. J. Contr., 34, pp. 1061-1078, 1981.

[5] Y. Alekal, P. Brunovsky, D. Chyung and E. B. Lee, "The quadratic problem for systems with time delays", IEEE Trans. on Automat. Contr., AC-16, pp. 673-687, 1971.

[6] D. Sarason, Generalized Interpolation in H^{∞}, Trans. AMS, Vol. 127, pp. 179-203, 1967.

[7] B. Sz. Nagy and C. Forias, Harmonic Analysis of Operators on Hilbert Space, North-Holland Publishing Company, Amsterdam, 1970.

[8] E. B. Lee, Linear hereditary systems, in Calculus of Variations and Control Theory, Acad. Press, Inc. 47-72, 1976.

[9] F. M. Callier and C. A. Desoer, An algebra of transfer functions for distributed time-invariant systems, IEEE Trans. on Circuits and Systems, CAS-25, pp. 651-662, 1978.

[10] C. N. Nett, C. A. Jacobson and M. J. Balas, Fractional representation theory: Robustness with applications to finite dimensional of a class of linear distributed systems, Proc. 22nd IEEE CDC, pp. 268-280, 1983.

[11] E. W. Kamen, P. P. Khargonekar and A. Tannenbaum, Proper stable Bezout factorizations and feedback control of linear time-delay systems, Int. J. Contr., 43, 837-857, 1986.

[12] E. Kamen, P. P. Khargonekar and A. Tannenbaum, Stabilization of time-delay systems using finite-dimensional compensators, IEEE Trans. on Automat. Contr. AC-30, pp. 75-78, 1985.

[13] D. C. Youla, J.J. Bongiono, Jr., and H. A. Jabr, Modern Wiener-Hopf design of optimal controllers: Part I, The single-input, single-output case, IEEE Trans. Automat. Contr., AC-21, pp. 3-13, 1976.

[14] G. Zames, Feedback and optimal sensitivity model reference transformations, multiplication seminorms and appropriate inverses, IEEE Trans. on Automat. Contr., AC-26, pp. 301-320, 1981.

[15] K. Hoffman, Banach spaces of analytic functions, Prentice-Hall, Inc., 1962.

[16] Z. Nehari, On bounded bilinear forms, Annals of Mathematics, 65, pp. 153-162, 1957.

[17] V. M. Adamjan, D. Z. Anov and M. G. Krein, Analytic properties of Schmidt pairs for a Hankel operator and the generalized Schur-Takagi problem, Math. USSR Sbornik, 15, pp. 31-73, 1971.

[18] P. Hartman, On complete continuous Hankel matrices, Proc. Amer. Math. Soc., 9, pp. 862-866, 1958.

[19] T. Kato, Perturbation theory for linear operations, Springer-Verlag, 1966.

[20] Bellman and Cooke, Differential-Difference Equations, Academic Press, 1963.

[21] E. B. Lee and W.-S. Lu, Stabilization of two-dimensional systems, IEEE Trans. on Automat. Contr., AC-30, pp. 405-411, 1985.

[22] M. Pee, P. Khargonekar and E. B. Lee, Further results on possible root locations of 2-D polynomials, to appear IEEE Trans. on Circuits and Systems, 1986.

[23] W.-S. Lu, E. B. Lee, and Q.-T. Zhang, Balanced approximation of two-dimensional and delay-differential systems. Submitted for publication 1986.

[24] B. A. Francis and G. Zames, On H^{∞}-optimal sensitivity theory for SISO feedback systems, IEEE Trans. on Automat. Contr., AC-29, pp. 9-16, 1984.

A CLASS OF SINGULAR CONTROL PROBLEMS

Sung J. Lee

University of South Florida, Tampa, USA

1. INTRODUCTION

In this paper we consider the problem of minimizing a class of singular control problems. The main emphasis is given in deriving the equations describing the optimal pairs. The state is governed by a ℓth order system of ordinary linear differential equations with complex matrix- coefficients in a compact interval, and is subject to a class of approximate boundary conditions. Both the cost functional and state involve very general generalized boundary conditions. The complete proof will appear in [6]. A control problem involving an initial condition together with generalized boundary conditions was considered in [3], for example. The special case of our problem when the cost functional is regular and $\ell = 1$ has been considered in [5]. Since we deal with non-standard boundary conditions (which become exact boundary conditions in many important cases), a given control generate infinitely many responses, and so the cost functional becomes multi-valued function of control. For this reason we will employ a new method based on the theory of least-squares solutions of a linear relation developed in [8], [11].

2. DIFFERENTIAL OPERATOR

Let $[t_0, t_1]$ be a compact interval. Let $L_2^k = L_2^k([t_0,t_1],\mathbb{C}^k)$. Let $\|\cdot\|$ denote the norm of this space. Let τ be the ℓth order expression acting on n x 1 complex-valued functions x defined by

$$(\tau x)(t) = x^{(\ell)} + \sum_{0}^{\ell-1} P_i(t)x^{(i)}, \quad t_0 \leq t \leq t_1.$$

Here $P_i(t)$ $(0 \leq i \leq \ell-1)$ are n x n complex matrix-valued functions which are $(\ell-1)$ times continuously differentiable on $[t_0,t_1]$, and $x^{(i)}$ denotes the ith derivative of x. Let τ^+ be the Lagrange adjoint of defined by

$$(\tau^+x)(t) = (-1)^{\ell}x^{(\ell)} + \sum_{i=0}^{\ell-1}(-1)^i(P_i^*(t)x)^{(i-1)}.$$

Let T_1 be the (maximal) operator in L_2^n defined by

$$\text{Dom } T_1 = \left\{x : x \in L_2^n,\ x^{(\ell-1)} \in AC\,[t_0,t_1],\ x^{(\ell)} \in L_2^n\right\},$$

$$T_1\,x = \tau x, \quad x \in \text{Dom } T_1.$$

Then it is well-known in the literature (see, for example, [7]) that T_1 is a densely defined closed operator in L_2^n, and so Dom T_1 becomes a Hilbert space with T_1-norm defined by

$$\|x\|_{T_1} := (\|x\|^2 + \|Tx\|^2)^{\frac{1}{2}}.$$

Thus the following theorem is a consequence of the Riesz-Fischer representation theorem.

THEOREM 2.1. F is a T_1-continuous linear operator from Dom T_1 into $\mathbb{C}^d$ if and only if there exists n x d complex matrix-valued functions $\Omega_1(t)$, $\Omega_2(t)$ whose columns are in L_2^n such that

$$F(x) = \int_{t_0}^{t_1} [\Omega_2^*\tau x + \Omega_1^*x]dt$$

for all $x \in$ Dom T_1, where Ω^* denotes transpose conjugation of Ω.

Let $\Phi(t)$ be the n x $n\ell$ fundamental matrix solution of $\tau x = 0$ such that

$$[\tilde{\Phi}(t_0)]^* \equiv [\Phi^*(t_0),\ldots(\Phi^{(\ell-1)})^*(t_0)] = I_{n\ell}.$$

Let us partition $[\tilde{\Phi}(t)]^{-1}$ by $(\tilde{\Phi}(t))^{-1} = [R(t), S(t)]$, $(t_0 \leq t \leq t_1)$ where R(t) is $n\ell$ x $n(\ell-1)$ and S(t) is $n\ell$ x n.

The following seems new in the literature even though the case $\ell=1$ is trivial and the case when n=1 is contained implicitly in [1].

THEOREM 2.2. Assume that $P_i (0 \leq i \leq \ell-1)$ are $(\ell-1)$ times continuously differentiable in $[t_0,t_1]$. Then

(i) S^* is an $n \times n$ fundamental matrix solution of $\tau^+ x = 0$.

(ii) For $f \in L_2^n$, let

$$u(t) = S^*(t)\int_t^{t_1} \Phi^*(s)f(s)ds, \quad t_0 \leq t \leq t_1.$$

Then u is a solution of $\tau^+ x = f$ such that $u^{(i)}(t_1) = 0_{n\times 1}$ for $0 \leq i \leq \ell-1$.

3. SINGULAR CONTROL PROBLEM

Let γ be a given point in $\mathbb{C}^d$. Let $U(t)$, $W(t)$, $B(t)$ $(t \in [t_0,t_1])$ be $m \times m$, $n \times n$, $n \times m$ continuous, complex, matrix-valued functions, respectively. Let $x_0 \in L_2^n$ and let $F: \text{Dom } T_1 \longrightarrow \mathbb{C}^d$ and $G: \text{Dom } T_1 \longrightarrow \mathbb{C}^{d_1}$ be T_1-continuous linear operators. Let $\mathcal{D}$ be the dynamical system consisting of all ordered pairs $\{u,x\}$ satisfying

$$u \in L_2^m \tag{3.1}$$

$$x \in \text{Dom } T_1 \text{ and } T_1 x = Bu \tag{3.2}$$

$$|F(x) - \gamma| = \operatorname*{Min}_y \left\{ |F(y) - \gamma| : y \in \text{Dom } T_1,\ T_1 y = Bu \right\} \tag{3.3}$$

where $|f| := (f^* f)^{\frac{1}{2}}$.

Let J be the cost functional defined on $\mathcal{D}$ by

$$J(u,x) = \int_{t_0}^{t_1} (|Uu|^2 + |W(x-x_0)|^2)dt + |G(x)|^2.$$

Our main concern is to find $\{u^+, x^+\} \in \mathcal{D}$, called an optimal pair, such that

$$J(u^+, x^+) \leq J(u,x) \quad \text{for all } \{u,x\} \in \mathcal{D},$$

and then describe it by equations. The element u^+ is called an optimal controller while x^+ is called an optimal response corresponding to u^+. When $\{u,x\} \in \mathcal{D}$, u is called a controller and x a response corresponding to u.

The case when $\ell = 1$ and $U(t)$ is regular for all t was studied in [5]. This paper generalizes and at the same time simplifies the corresponding results of [5].

Let us define a linear operator H on L_2^m by

$$(H(u))(t) = \Phi(t)\int_{t_0}^{t} S(s)B(s)u(s)ds, \qquad t_0 \le t \le t_1.$$

Let $F(\Phi)$ denote the $d \times n\ell$ constant matrix whose ith column is $F(\phi_i)$ where ϕ_i is the ith column of Φ. Let $(F(\Phi))^{\#}$ denote the $n\ell \times d$ Moore-Penrose generalized inverse matrix of $F(\Phi)$.
We have the following description for $\mathcal{D}$.

THEOREM 3.1. The following are equivalent:

(i) $\{u,x\} \in \mathcal{D}$.

(ii) $u \in L_2^m$, $x \in \mathrm{Dom}\, T_1$ such that $T_1 x = Bu$ and
$F(x)-\gamma = [F(\Phi)(F(\Phi))^{\#}-I][\gamma-F(H(u))]$.

(iii) $u \in L_2^m$ and $x = \Phi\alpha + H(u)$ for some $\alpha \in \mathbb{C}^{n\ell}$ which is a least-squares solution of the matrix equation (for u fixed), $(F(\Phi))(\beta) = \gamma - F(H((u))$.

(iv) $u \in L_2^m$, $x \in \mathrm{Dom}\, T_1$, $T_1 x = Bu$, $F(x)-\gamma \in \mathrm{Null}(F(\Phi))^*$
$\equiv \{\alpha \in \mathbb{C}^d : (F(\Phi))^*\alpha = 0_{n\ell \times 1}\}$.

An immediate corollary to this theorem is the following.

THEOREM 3.2 $\mathrm{Dom}\,\mathcal{D} = L_2^m$.

We now proceed to characterize all the optimal pairs. To do this let us introduce two linear operators M_1, M_2 and a linear relation as follows:

$$M_1 : L_2^m \to L_2^m \oplus L_2^n \oplus \mathbb{C}^{d_1},$$

$$M_2 : \mathrm{Null}\, F(\Phi) \subset \mathbb{C}^{n\ell} \to L_2^m \oplus L_2^n \oplus \mathbb{C}^{d_1},$$

$$M \subset L_2^m \oplus [L_2^m \oplus L_2^n \oplus \mathbb{C}^{d_1}] \text{ by}$$

$$M_1(u) := \{Uu,\ W[H(u) - \Phi(F(\Phi))^{\#}F(H(u))],\ G(H(u)) - G(\Phi)(F(\Phi))^{\#}F(H(u))\},$$

$$M_2(k) := \{0,\ W\Phi k,\ G(\Phi)k\}$$

$$M := (\mathrm{graph}\, M_1) \dotplus [\{0\} \oplus \mathrm{Range}\, M_2].$$

Let $\zeta \in L_2^m \oplus L_2^n \oplus \mathbb{C}^{d_1}$ be defined by

$$\zeta = \{0,\ W[x_0 - \Phi(F(\Phi))^{\#}\gamma],\ G(\Phi)(F(\Phi))^{\#}\gamma\}.$$

Finally, let us denote the norm of $L_2^m \oplus L_2^n \oplus \mathbb{C}^{d_1}$ also by $\|\cdot,\cdot,\cdot\|$. We have the following.

LEMMA 3.3. (i) For $\{u,x\} \in$, write

$$x = \Phi k + H(u) + \Phi(F(\Phi))^{\#}[\gamma - F(H(u))]$$
for some $k \in \text{Null } F(\Phi)$.

Then $J(u,x) = \|\mathcal{M}_1(u) + \mathcal{M}_2(k) - \zeta\|^2$.

(ii) $\{u^+,x^+\}$ is optimal if and only if $u \in L_2^m$,
$x^+ = \Phi k^+ + H(u^+) + \Phi(F(\Phi))^{\#}[\gamma - F(H(u^+))]$
for some $k^+ \in \text{Null } F(\Phi)$ such that
$\|\mathcal{M}_1(u^+) + \mathcal{M}_2(k^+) - \zeta\| \le \|\mathcal{M}_1(u) + \mathcal{M}_2(k) - \zeta\|$
for all $u \in L_2^m$, $k \in \text{Null } F(\Phi)$.

Proof This follows from the theorem 3.1 together with the definitions of $\mathcal{M}_1$ and $\mathcal{M}_2$.

DEFINITION An element $u^+ \in L_2^m$ is called a least-squares solution of the inclusion $\zeta \in \mathcal{M}(u)$ if there exists $y \in \mathcal{M}(u^+)$ such that

$\|y - \zeta\| \le \|z - \zeta\|$ $\le$ for all $z \in \text{Range } \mathcal{M}$.

Using the lemma 3.3, we have the following.

LEMMA 3.4. u^+ is an optimal controller if and only if $u^+ \in L_2^m$ and u^+ is a least-squares solution of $\zeta \in \mathcal{M}(u)$.

DEFINITION. The adjoint, $\mathcal{M}^*$, of $\mathcal{M}$ is defined by the linear relation $\{\{u,v\} : \{v,-u\} \in \mathcal{M}^{\perp}\}$ where $\mathcal{M}^{\perp}$ denotes the orthogonal complement of $\mathcal{M}$.

DEFINITION. The generalized inverse, $\mathcal{M}^{\#}$, of $\mathcal{M}$ is the linear relation $\{\{x;(I-P)(u)\} : u \in L_2^m,\ x \in L_2^m \oplus L_2^n \oplus \mathbb{C}^{d_1}, \{u,(I-P^+)(x)\} \in \mathcal{M}\}$ where P is the orthogonal projector from L_2^m onto Null and P^+ is the one from $L_2^m \oplus L_2^n \oplus \mathbb{C}^{d_1}$ onto Null $\mathcal{M}^*$. It is immediate that $\mathcal{M}^{\#}$ is the graph of a linear operator. For the introduction and development, see [9], [10], [11].

THEOREM 3.5 .

(i) u^+ is an optimal controller if and only if $u^+ \in L_2^m$ and $\zeta \in \mathcal{M}(u^+) \dotplus \text{Null } \mathcal{M}^*$.

(ii) An optimal controller exists if and only if $\zeta \in M(u^+) \dotplus \text{Null } M^*$. When an optimal controller exists, the set of all the optimal controllers is given by the algebraic coset $M^{\#}(\zeta) \dotplus \text{Null } M$, and moreover $M^{\#}(\zeta)$ is the unique optimal controller with smallest L_2^m - norm.

(iii) If U is invertible on $[t_0,t_1]$, then the optimal controller always exists and is unique.

Proof The parts (i), (ii) follow from the lemma 3.4 together with Proposition 3.1 of [8]. The part (iii) follows easily from (ii) as $\text{Null } M = \{0\}$ and Range M is closed.
Once an optimal pair is known, one can find all the optimal responses corresponding to the optimal controller.

PROPOSITION 3.6. Let $\{u,x\}$ be an optimal pair. Then z is an optimal response corresponding to u if and only if $z = x+\Phi\alpha$ for some $\alpha \in \mathbb{C}^{n\ell}$ such that

$$F(\Phi)\alpha \in \text{Null}(F(\Phi))^*$$

and

$$(G(\Phi))^*G(\Phi) + \int_{t_0}^{t_1} \overline{\Phi^*W^*W\Phi}dt \in (\text{Null } F(\Phi))^{\perp}.$$

In order to characterize optimal pairs explicitly it is necessary to have explicit forms for F and G. As we have seen in the theorem 2.1, let $\Omega_i(t)$ $(i = 1,2)$ be the n x d matrix-valued functions and let $\Lambda_i(t)$ $(i = 1,2)$ be the n x d_1 matrix-valued functions on $[t_0,t_1]$ such that

$$F(x) = \int_{t_0}^{t_1} [\Omega_2^*\tau x + \Omega_1^* x]dt,$$

$$G(x) = \int_{t_0}^{t_1} [\Lambda_2^*\tau x + \Lambda_1^* x]dt$$

for all $x \in \text{Dom } T_1$. Of course the columns of Ω_i and Λ_i belong to L_2^n. For $z \in L_2^n$, define $\eta(z) \in L_2^n$ by

$$\eta(z) = \Lambda_2(t)G(z) - \Omega_2(t)[G(\Phi))^{\#}]^{*}A(z)$$
$$+ S^{*}(t)\int_{t}^{t_1} \Phi^{*}(s)\left\{W^{*}W(z-x_0)+\Lambda_1 G(z)-\Omega_1[(F(\Phi))^{\#}]^{*}A(z)\right\}(s)ds$$

where

$$A(z) = (G(\Phi))^{*}G(z) + \int_{t_0}^{t_1} \Phi^{*}W^{*}W(z-x_0)dt.$$

We have the following.

THEOREM 3.7. u is an optimal controller if and only if $u \in L_2^m$ and there exist $x \in \mathrm{Dom}\, T_1$ such that

(i) $\{u,x\} \in \mathcal{D}$

(ii) $A(x) \in (\mathrm{Null}\; F(\Phi))^{\perp}$

(iii) $U^{*}Uu + B^{*}\eta(x) = 0$ in L_2^m.

Moreover, when (i), (ii), (iii) hold, $\{u,x\}$ is an optimal pair.

Proof One can prove this by characterizing (i) of the theorem 3.5.

In the following we will characterize the optimal pairs in terms of adjoint responses.

THEOREM 3.8. Assume further that P_0, $P_1, \ldots, P_{\ell-1}$ are $(\ell-1)$ times continuously differentiable on $[t_0, t_1]$. Then u is optimal if and only if $u \in L_2^m$ and there exist $x \in \mathrm{Dom}\, T_1$ and $\eta \in L_2^n$ such that

(i) $\{u,x\} \in \mathcal{D}$

(ii) $A(x) \in (\mathrm{Null}\; F(\Phi))$,

(iii) $U^{*}Uu + B^{*}\eta = 0$ in L_2^m,

(iv) $\eta - \Psi(x) \in \mathrm{Dom}\, T_1$,

(v) $\tau^{+}(\eta-\Psi(x)) = W^{*}W(x-x_0) + \Lambda_1 G(x)-\Omega_1[(F(\Phi))^{\#}]^{*}A(x)$,

(vi) $\dfrac{d^i}{dt^i}(\eta-\Psi(x))(t_1) = 0_{n\times 1}$, $0 \le i \le \ell-1$.

Here, $\Psi(x) = \Lambda_2 G(x)-\Omega_2[(G(\Phi))^{\#}]^{*}A(x)$.

Proof In the above theorem, let $\eta = \eta(x)-\Psi(x)$ and then apply the theorem 2.2 to η.

Remark. In the above theorem none of η and $\Psi(x)$ can be sufficiently smooth even though their difference is smooth. The equations (iv) to (vi) are similar to the classical Euler-Lagrange equations in $\eta - \Psi(x)$. The term in (ii) appears because of the multi-responseness.

REFERENCES

[1] Coddington, E.A. and Levinson, N. Theory of Ordinary Differential Equations, McGraw-Hill (1955).

[2] Casti, J.L. "The Linear-quadratic Control Problem." Some Recent and Outstanding Problems, SIAM Rev. 22 no. 4 (1980): 459-485.

[3] Chan, W.L. and S.K. NG, Minimum Principle for Systems with Boundary Conditions and Cost having Stieltjes-type Integrals, J. Opt. Theory Appl., 33 no. 4 (1981): 557-573.

[4] Lee, E.B. and Markus, L., Foundation of Optimal Control Theory, Wiley (1967), New York.

[5] Lee, S.J., Multi-response quadratic Control Problems, SIAM J. Control Opt., 24, no.4 (1986).

[6] Lee, S.J., A Class of Singular Quadratic Control Problem with Nonstandard Boundary Conditions (to appear).

[7] Lee, S.J., Nonhomogeneous Boundary Value Problems for Linear Manifolds II, Ordinary Differential Subspaces in Lp-spaces. Kyungpook Math. J., 23 no. 2 (1983): 115-128.

[8] Lee, S.J. and Nashed, M.Z., Least-squares solutions of Multi-valued Linear Equations in Hilbert spaces, J. Approx. Theory, 38 no. 4 (1983): 380-391.

[9] Lee, S.J. and Nashed, M.Z., Generalized Inverses for Linear Manifolds and Applications to Boundary Value Problems in Banach spaces, C.R. Math. Rep. Acad. Sci. Canada, 4 no. 6 (1982): 347-352.

[10] Lee, S.J. and Nashed, M.Z., Operator Parts and Generalized Inverses of Multi-valued Operators with Applications to Ordinary Differential Subspaces (preprint).

[11] Nashed, M.Z., Operator parts and generalized inverses of linear manifolds with applications, Trends in Theory and Practice of Nonlinear Differential Equations (V. Lakshmikantham ed.), Marcel Dekker (1984): 395-412.

NEAR OPTIMAL TIME BOUNDARY CONTROLLABILITY FOR A CLASS OF HYPERBOLIC EQUATIONS

Walter Littman

University of Minnesota, Minneapolis, USA

ABSTRACT: We use propagation of singularities results to obtain essentially optimal time boundary controllability for strictly hyperbolic equations with real analytic coefficients.

1. INTRODUCTION

The original proof by D. Russell [R] of the boundary controllability for the wave equation in three space dimensions, in addition to its simplicity, has the advantage of producing an explicit estimate of the time necessary for controllability. However, as soon as the equation is modified by a change of dimension, or by the addition of other terms, the method must be modified (controllability via stabilizability) to give control times which cannot be estimated explicitly. The same may be said of a number of controllability results which are corollaries of decay rate results, such as those obtained by Chen, Lagnese, Seidman and the recent results of Lasiecka and Triggiani (to appear). A number of papers do give explicit estimates of the control times. In addition to a number of results described in Russell's review article [Ru 2], we mention the results of Lagnese [La] and Littman [Li].

Here we use the theory of propagation of singularities to obtain what are essentially time-optimal results for boundary controllability for strictly hyperbolic equations with real analytic coefficients.

2. BASIC ASSUMPTIONS AND CONCEPTS

Let Ω be a C^m domain in R^{n+1} and Q be the cylindrical region $\bar{\Omega} \times (t_1,t_2)$. (Bars denote closure.) L is a strictly hyperbolic linear partial differential operator of order m with real analytic coefficients in an open set containing $\overline{Q}$. Let Q_b be the open base of $Q: \Omega \times \{t = t_1\}$, and Q_ℓ be the lateral boundary of $Q: \partial\Omega \times (t_1, t_2)$. For the sake of simplicity only (this is really not necessary) we assume that L has been extended to all of R^{n+1} as a strictly hyperbolic operator with C^∞ coefficients which are constant outside of a bounded set. H_s will denote the usual Sobolev spaces, s real.

Of central importance is the following

MIXED PROBLEM: For a particular value of $s \geq -1$ find a $u \in H_{m+s}(Q)$ such that

$$Lu = f \in H_{s+1}(Q), \quad \partial_t^j u = f_j \in H_{m-j+s}(Q_b), \quad j = 0, \ldots, m-1,$$

$$B_k u = g_k \in H_{m-r_k+s}(Q_\ell), \quad k = 1, \ldots m_+ .$$

Here the B_k are boundary operators of orders r_k. In the usual situation m is even and $m_+ = m/2$. Sakamato has shown [S] that under appropriate conditions the mixed problem is well posed. However, strictly speaking, all we shall require is the uniqueness part, for some $s \geq -1$.

We shall be concerned with the following

BOUNDARY CONTROL PROBLEM: Given initial (Cauchy) data $f_j \in H_{m+s-j}(Q_b)$ (i.e. restrictions from $H_{m+s-j}(R^n)$) find g_k that steer the solution u of the resulting mixed problem to zero Cauchy data by time $t = t_2$.

Our main result is the following

THEOREM: In addition to the "basic assumptions" and the uniqueness of the "mixed problem", we assume that every bicharacteristic curve in Q either enters or leaves Q through Q_ℓ. Given initial data $f_j \in H_{m-k+s}(Q_b)$ $(s \geq -1)$, there exist boundary controllers g_k steering the solution u of the resulting mixed problem to zero in time $t_2 - t_1$.

The resulting solution to the mixed problem (whose existence will follow from the proof of the theorem) will be in $H_{m+s}(Q)$, while the functions $g_k \equiv B_k u$ will belong to whatever spaces applicable restriction theorems dictate.

Note: If there exists a single bicharacteristic curve entering and leaving Q through Q_b and the (open) top boundary, respectively, without hitting Q_ℓ, it follows from the propagation of singularities results that the conclusion of the theorem is false. See also Ralston [R] for the corresponding result where the controls are applied only to part of the lateral boundary.

3. THE BEGINNING OF THE PROOF OF THE THEOREM.

We begin by extending the given Cauchy data on Q_b as smoothly as possible to a slightly larger set so as to have compact support. Then solve the resulting pure Cauchy problem for $t > t_1$ and call the resulting solution $v(x,t) \in H_{m+s}$. Pick t_o so that $t_1 < t_o < t_2$ and let $\varphi(t)$ be a C^∞ "cut off" function that $\equiv 0$ to the left of a small neighborhood of t_o and $\equiv 1$ to the right of that neighborhood. Let $L(v\varphi) = F(x,t)$ $(\in H_{s+1})$. The function $v\varphi$ satisfies the right Cauchy data at $t = t_1$ and at $t = t_2$ but not the right equation. To compensate for that, we try to solve the problem $Lw = F(x,t)$ in Q with w satisfying Cauchy data $\equiv 0$ on the top and bottom of Q. Notice that $F(x,t)$ vanishes near $t = t_1$ and $t = t_2$. We aim for a solution w in H_{m+s}. Then we set $z = v\varphi - w$ and read of the trace of $B_k z \equiv g_k$ on Q_ℓ.

We now begin the construction of w, which will be accomplished in several steps. However we will need some preliminary constructions. By a "trapezoidal region" (t.r.) we mean a closed bounded region in R^{n+1} having a space-like lateral boundary and upper and lower boundaries contained in planes $t = \text{constant}$.

Adjacent to the top and bottom boundaries of Q we construct two trapezoidal regions T_1 and T_2. T_2 is a "top heavy" t.r. whose bottom boundary (or base) coincides with the top boundary of Q (and thus is contained in the plane $t = t_2$). The top boundary of T_2 is contained in the plane $t = t_4 > t_2$. The lateral boundary of T_2 is a C' space-like surface with outward pointing normal directed "down". Thus if we solve a pure initial value problem in $t_2 < t < t_4$ with initial data having support on the base of T_2, the support of the solution will stay in T_2 for the indicated t values. We perform the analogous construction of a "bottom heavy" t.r. whose top boundary coincides with Q_b and whose bottom boundary is contained in the plane $t = t_3 < t_1$. The lateral boundary of T_3 is such that if we solve the backward Cauchy problem for $Lu = 0$ in $t_3 < t < t_1$, with initial data having support in Q_b, the support of the solution will be in T_1 for the indicated values of t.

We define the compact $S = Q \cup T_1 \cup T_2$ and for all real r define H_r^o to be the space of distributions in $H_r(R^{n+1})$ with support in S. Let H_r^{oo} be the analogous space with support in $T = T_1 \cup T_2$. We then define the factor space $\hat{H}_r = H_r^o / H_r^{oo}$. Elements of $\hat{H}_r$ may be viewed as restrictions of elements of H_r^o to the complement of T. Roughly speaking, they may be thought of as elements of $H_r(Q)$ having zero Cauchy data on Q_ℓ.

The main ingredient in the proof of the theorem is the following

BASIC ESTIMATE: If $u \in \hat{H}_{r+m-2}$ and $L^* u \in \hat{H}_r$ then $u \in \hat{H}_{r+m-1}$ and there exists a C independent of u such that

$$|u|_{\hat{H}_{r+m-1}} \leq C |L^* u|_{\hat{H}_r} .$$

From the basic estimate the theorem now follows easily: By a Hahn-Banach type argument we obtain that L maps $(\hat{H}_r)'$ onto $(\hat{H}_{r+m-1})'$, the primes denoting dual spaces. Now since the function $F(x,t)$ introduced earlier belongs to $H_{s+1}(R^{n+1})$ and vanishes outside a slab of the type $t_1 + \epsilon < t < t_2 - \epsilon$, F may be identified as an element F_1 of $(\hat{H}_{-s-1})'$. Thus, taking $r = -s - m$, it follows from this that there exists a $u \in (\hat{H}_{-s-m})'$ such that $Lu = F_1$. This can be reinterpreted as follows: There exists a $w \in H_{s+m}(R^{n+1})$ vanishing in T and satisfying $Lw = F$ in the interior of S. This is the required w. We can now choose a value for s (possibly negative) so that all the $B_k w$ have Sobolev space traces g_k on $\partial\Omega \times (t_1, t_4)$. The g_k will automatically vanish for $t > t_2$.

4. PROOF OF THE BASIC ESTIMATE.

(a) From the theory of propagation of singularities [DH] we know that if u is a distribution in the open slab $t_1 < t < t_2$ with $\operatorname{supp} u \in Q$, $u \in H_{r+m-2}$ locally and $L^* u \in H_r$ locally, then $u \in H_{r+m-1}$ locally. Because of the assumption on the support of u, "locally" may be replaced by "locally in t" in the foregoing statement.

(b) Next we prove that if $u \in \hat{H}_{r+m-2}$, $L^* u \in \hat{H}_r$, then $u \in \hat{H}_{r+m-1}$. Under these assumptions there exists a representative $U \in H^o_{r+m-2}$ of u and a representative $G \in H^o_r$ of $L^* u$ such that $\operatorname{supp}(G - L^* U) \in T$. Again letting $t_1 < t_o < t_2$, it can be seen, by solving appropriate initial value problems, that one can find a U_1 defined for $t_3 < t < t_4$ which solves $L^* U_1 = G$, such that $U_1 \in H_{r+m-1}$ locally in t and agrees with U near $t = t_o$ (and, as a matter of fact for all $t_1 < t < t_2$). From the construction of the sets T and S it follows that $\operatorname{supp} U_1 \in S$. Next let $\varphi(t)$ be a cut-off function of t which equals one in $t_1 < t < t_2$ and zero in a complement of an interval of the type $(t_3 + \varepsilon, t_4 - \varepsilon)$, ε sufficiently small, and set $U_2 = \varphi(t) U_1$. Thus $U_2 \in H^o_{r+m-1}$. Now U_2 is a representative of u since $U = U_2$ outside of T. Hence $u \in \hat{H}_{s+m-1}$.

(c) Applying the closed graph theorem to the identity map from the Hilbert space with norm $|L^* u|_{\hat{H}_r} + |u|_{\hat{H}_{r+m-2}}$ to the Hilbert space with norm $|u|_{\hat{H}_{r+m-1}}$ we get the estimate

$$|u|_{\hat{H}_{r+m-1}} \leq C(|L^* u|_{\hat{H}_r} + |u|_{\hat{H}_{r+m-2}}) .$$

(d) From Hörmander's result on the propagation of analytic wave front sets for solutions to $L^* u = 0$ (see [H 1] p. 694) it follows that every distribution solution to $L^* u = 0$ in $t_1 < t < t_2$ with support in Q must vanish. This, together with the analogue of the Rellich compactness theorem for the spaces $\hat{H}_r$ (cf. [H 2] p. 479) enables us to eliminate the second term of the right hand side of the estimate of (c), yielding the basic estimate.

Bibliography

[DH] Duistermatt, J.J. and L. Hörmander, Fourier integral Operators II, Acta Math. 128 (1972) 183-269.

[H 1] Hörmander, L., Uniqueness theorems and wave front sets, C.P.A.M. 24 (1971) 79-183.

[H 2] ____________ The analysis of linear partial differential operators III, Berlin, Springer Verlag 1985.

[La] Lagnese, J., Boundary value control of a class of hyperbolic equations in a general domain, SIAM J. Control and Optimization 15, (1977) 973-983.

[Li] Littman, W., Boundary control theory for hyperbolic and parabolic equations with constant coefficients, Annali Sc. N. Sup. Pisa Ser. IV 5, (1978) 567-580.

[Ra] Ralston, J., Gaussian beams and the propagation of singularities, MAA Studies in Mathematics 23, W. Littman (Ed.) (1982) 206-248.

[Ru 1] Russell, D.L., A unified boundary controllability theory, Studies in Applied Math., 52, (1973) 189-211.

[Ru 2] ____________ Controllability and stabilizability theory, SIAM Rev. 20 (1978) 639-739.

[S] Sakamoto, R., Hyperbolic boundary value problems, Cambridge Univ. Press 1978.

ACKNOWLEDGEMENT: This research was partially supported by NSF grant 84-13129.

STABILITY ENHANCEMENT BY STATE FEEDBACK

Nhan Levan

University of California, Los Angeles, USA

1. INTRODUCTION

This paper will study the problem of "enhancing" stability of a distributed parameter system using a linear state feedback. In other words, suppose that a system is already stable --in a suitable sense, what we ask is that is it possible to find a feedback so that the closed loop system will be stable in a "stronger" sense? We begin in section 2 by investigating the effect of a feedback on the modes of a system. Particular attention will be given to the class of dissipative and compact feedbacks. We then turn to a sufficient condition for weak and strong stabilities of uniformly bounded Hilbert space semigroups of bounded linear operators. This then leads to a simple--perhaps, the simplest, proof of weak and strong stabilizabilities of contraction semigroups. Approximate stabilizability via the Steady State Riccati Equation (SSRE) is then simply demonstrated. A "negative" result regarding strong stabilizability of contraction semigroups is then obtained before the enhancement problem is studied in details.

2. THE MAIN RESULTS

Let a distributed parameter system be described by the equation

$$\dot{x} = Ax + Bu, \tag{2-1}$$

where A is the generator of a strongly continuous semigroup of bounded linear operators $T(t)$, $t \geq 0$, over a Hilbert space H (the state space) --with inner product

$[.,.]$ and norm $||.||$, and B is a bounded linear operator from a second Hilbert space U (the control space) to H. Let $F: U \to H$ be a bounded linear feedback then the open loop system becomes

$$\dot{x} = (A + BF)x. \qquad (2-2)$$

The closed loop semigroup, i.e., the one generated by $A + BF$, is denoted by $S(t)$, $t \geq 0$.

We begin by exploring the effect of a feedback on the modes of a system. Thus let

$$A\phi = \lambda\phi.$$

Then

$$(A + BF)\phi = (\lambda I + BF)\phi.$$

Therefore

$$S(t)\phi = e^{\lambda t} e^{BFt} \phi, \quad t \geq 0.$$

Now, since the semigroup e^{BFt}, $t \geq 0$, is uniformly continuous, there exists a constant $\omega \geq 0$ such that

$$||e^{BFt}|| \leq e^{\omega t}, \quad t \geq 0.$$

Therefore

$$||S(t)\phi|| \leq e^{(Re.\lambda + \omega)t} ||\phi||. \qquad (2-3)$$

It then follows from this inequality that, if

$$0 > Re.\lambda \geq -\omega$$

then the mode ϕ will be "destabilized" by the feedback F, while if it is already unstable, $Re.\lambda \geq 0$, then it will remain unstable under F --unless further constaints are imposed. Also if

$$Re.\lambda \leq -\omega$$

then, of course, the corresponding mode remains stable under F.

A class of feedbacks which is of great importance is the class of dissipative ones, i.e., those F's for which

$$2Re.[BFx,x] = [(BF + F^*B)x,x] \leq 0, \qquad (2-4)$$

equivalently,

$$||e^{BFt}|| \leq 1, \quad t \geq 0. \qquad (2-5)$$

In this case (2-3) becomes

$$||S(t)\phi|| \leq e^{(Re.\lambda)t}||\phi||, \quad t \geq 0, \qquad (2-6)$$

i.e., the feedback F has no effect on the modes of the open loop system.

PROPOSITION 1

Let A be dissipative and generate a contraction semigroup $T(t)$, $t \geq 0$, over H. If $F = -B^*$ then any mode of the closed loop system which does not belong to the null space $N(B^*)$ of B^* is stable. Moreover, if μ is the corresponding eigenvalue then

$$\text{Re}.\mu \leq -||B^*||^2. \tag{2-7}$$

Similarly, if $F = -Q^2B^*$, where $Q = Q^*: U \to U$, then

$$\text{Re}.\mu \leq -||Q||^2\ ||B^*||^2. \tag{2-8}$$

Proof

The semigroup e^{-BB^*t}, $t \geq 0$, is clearly contractive and $A - BB^*$ is dissipative. Let

$$(A - BB^*)\psi = \mu\psi.$$

Then, since A is dissipative,

$$\text{Re}.[A\psi,\psi] = \text{Re}.\mu\ ||\psi||^2 + ||B^*\psi||^2 \leq 0.$$

Therefore (2-7) follows. Exactly the same argument holds for (2-8). This finishes the proof of the Proposition.

We note that if the operator B is compact then

$$\text{Re}.\mu \leq -\gamma_{Max},$$

where γ_{Max} is the largest eigenvalue of BB^*. Also, if B is not compact then a compact feedback can still be "constructed" by requiring Q to be compact. Note also that if any mode of the closed loop system lies in $N(B^*)$ then, of course, it is a mode of the open loop system. Thus if we require that all unstable modes of the open loop system be controllable then, clearly, the open loop system and the closed loop system cannot share any common eigenvectors.

We now turn to sufficient conditions for weak and strong stabilities for the class of uniformly bounded C_o semigroups. We recall

THEOREM 1 (Datko, [1])

Let $T(t)$, $t \geq 0$, be a strongly continuous semigroup

with generator A in a Hilbert space H. Then the following statements are equivalent:

(i) $T(t)$, $t \geq 0$, is exponentially stable:
$||T(t)|| \leq Me^{-\omega t}$, $t \geq 0$,
for some $M \geq 1$ and some $\omega > 0$.

(ii) There exists a non-negative operator P, $P \geq 0$, such that, for x in the domain $\mathcal{D}(A)$ of A:

$$2\mathrm{Re}.[PAx,x] = -[Wx,x], \tag{2-9}$$

where $W \geq cI$ for some constant $c > 0$.

(iii) For every x in H:
$\int_0^\infty ||T(t)x||^2 \, dt < \infty$.

As far as we know, no such results exist for weak stability and strong stability. A semigroup $T(t)$, $t \geq 0$, is said to be w(weakly)-stable if

for x and y in H: $[T(t)x,y] \to 0$, $t \to \infty$,

and s(strongly)-stable if:

for x in H: $||T(t)x|| \to 0$, $t \to \infty$.

We now prove.

THEOREM 2

Let $T(t)$, $t \geq 0$, be a uniformly bounded strongly continuous semigroup, $||T(t)|| \leq M$ for $t \geq 0$, where $M \geq 1$. Let (A,B) denote the system

$\dot{x} = Ax + Bu$.

If the pair (A,B) is controllable and

$$\text{for } x \text{ in } H: \int_0^\infty ||B^*T(t)^*x||^2 \, dt < \infty, \tag{2-10}$$

then the semigroup $T(t)$, $t \geq 0$, is w-stable.

Proof

Let P be the operator defined by

$[Px,x] = \int_0^\infty ||B^*T(t)^*x||^2 \, dt$ for x in H.

Then, of course, P is self-adjoint and positive --by the fact that (A,B) is controllable. Thus it follows that

$$[PT(t)^*x,T(t)^*x] = \int_t^\infty ||B^*T(\sigma)^*x||^2 \, d\sigma. \tag{2-11}$$

Therefore

$$\lim_{t \to \infty} [PT(t)^*x,T(t)^*x] = 0 \text{ for } x \text{ in } H.$$

Let Q be the positive square root of P, $Q^2 = P$, then

for x in H: $||QT(t)^*x|| \to 0$, $t \to \infty$.

We have, for $t \to \infty$:

$$[QT(t)^*x,z] = [T(t)^*,Qz] \le ||QT(t)^*x|| \; ||z|| \to 0.$$

Therefore, since $T(t)^*$, $t \ge 0$, is also uniformly bounded and the range of Q is dense in H,

$[T(t)^*x,y] \to 0$, $t \to \infty$, for x and y in H.

Finally, if $T(t)^*$, $t \ge 0$, is w-stable then so is $T(t)$, $t \ge 0$. This finishes the proof.

COROLLARY 1

If the generator A has compact resolvent then the conditions of Theorem 2 are sufficient for the uniformly bounded semigroup $T(t)$, $t \ge 0$, to be s-stable.

Proof

We only have to note that if A has compact resolvent then w-stability implies s-stability.

Returning to (2-11) we obtain, by differentiating both sides with respect to t, for x in $\mathcal{D}(A^*)$:

$$2\text{Re}.[PA^*T(t)^*x,T(t)^*x] = -\,||B^*T(t)^*x||^2.$$

Therefore, for x in $\mathcal{D}(A^*)$:

$$2\text{Re}.[PA^*x,x] = -\,||B^*x||^2 = -\,[BB^*x,x], \tag{2-12}$$

which, of course, is the same as (2-9) of Theorem 1, except that the operator BB^* is, in general, not strictly positive-- as in the case of W.

Suppose now that there exists $P \ge 0$ satisfying (2-12) then it is clear that

$$[PT(t)^*x,T(t)^*x] - [Px,x] = -\int_0^t ||B^*T(\sigma)^*x||^2 \, d\sigma.$$

Therefore

for x in H: $\int_0^t ||B^*T(\sigma)^*x||^2 \, d\sigma \le [Px,x]$.

It then follows that

for x in H: $\int_0^\infty ||B^*T(\sigma)^*x||^2 \, d\sigma < \infty$.

Thus we have shown that

THEOREM 3

Let $T(t)$, $t \ge 0$, be a uniformly bounded C_o semigroup over H, with generator A. A necessary and sufficient condition for the integral

$${}_0\int^\infty ||B^*T(t)^*x||^2 \, dt, \text{ for every } x \text{ in } H,$$

to converge is the existence of a non-negative operator P satisfying the equation:

$$2\text{Re.}[PA^*x,x] = -[BB^*x,x], \quad \text{for } x \text{ in } \mathcal{D}(A^*).$$

We have obtained sufficient conditions for weak stability of uniformly bounded semigroups. These conditions are analogs of those of Theorem 1-- for e-stability. It is not clear whether these conditions are also necessary in general. However we can state the following special results.

THEOREM 4

Let $T(t)$, $t \geq 0$, be a contraction semigroup with generator A in H and suppose that A has compact resolvent. Let $B: U \to H$ be compact and suppose that (A,B) is controllable. Then the following conditions are equivalent:

(i) $T(t)$, $t \geq 0$, is s-stable.

(ii) For x in H: ${}_0\int^\infty ||B^*T(t)^*x||^2 \, dt < \infty$.

(iii) There exists a positive operator P such that for x in $\mathcal{D}(A^*)$: $2\text{Re.}[PA^*x,x] = -[BB^*x,x]$.

Proof

We only need to show that (i) implies (ii). First, recall that a contraction operator is completely nonunitary (cnu) if $\{0\}$ is the only subspace which reduces the operator to a unitary one. It is clear that if the semigroup $T(t)$, $t \geq 0$, is s-stable then it is cnu, hence so is the adjoint semigroup $T(t)^*$, $t \geq 0$. Finally, in [2] we have shown that if B is compact and if the semigroup $T(t)^*$, $t \geq 0$, is cnu then the integral of (ii) certainly converges. This finishes the proof of the Theorem.

We now apply the above results to stabilizability of Hilbert space contraction semigroups.

THEOREM 5

Let A be the generator of a contraction semigroup over H. If the pair (A,B) is controllable then it is weakly stabilizable by the feedback $-B^*$.

Proof

Let $S(t)$, $t \geq 0$, denote the semigroup generated by $A - BB^*$ then, of course, it is contractive. Hence so is the adjoint semigroup $S(t)^*$, $t \geq 0$. Thus we have, for x in $\mathcal{D}(A^*)$:

$$\frac{d}{dt}||S(t)^*x||^2 = 2\text{Re.}[(A^* - BB^*)S(t)^*x, S(t)^*x],$$
$$= 2\text{Re.}[A^*S(t)^*x, S(t)^*x] - 2||B^*S(t)^*x||^2.$$

Therefore, for x in $\mathcal{D}(A^*)$:

$$||S(t)^*x||^2 - ||x||^2 = {}_0\int^t 2\text{Re.}[A^*S(\sigma)^*x, S(\sigma)^*x]\, d\sigma$$
$$- 2\,{}_0\int^t ||B^*S(\sigma)^*x||^2\, d\sigma, \quad t \geq 0.$$

Then, since $S(\sigma)^*x$ lies in $\mathcal{D}(A^*)$ for each x in $\mathcal{D}(A^*)$ and since A^* is also dissipative, the intergrand of the first integral on the right hand side is nonpositive. Thus it follows that

for x in $\mathcal{D}(A^*)$: $2\,{}_0\int^t ||B^*S(\sigma)^*x||^2\, dt \leq ||x||^2$.

Therefore ${}_0\int^\infty ||B^*S(t)^*x||^2\, dt$ converges for every x in H. This, by Theorem 2, shows that the semigroup $S(t)$, $t \geq 0$, is w-stable, i.e., the pair (A,B) is weakly stabilizable by the feedback $-B^*$.

We must note that this proof is, perhaps, the simplest and most straightforward proof of weak stabilizability of Hilbert space contraction semigroups. Earlier proofs used the LaSalle Invariance Principle,[3], or the Nagy-Foias Canonical Decomposition, [4], [5].

We now consider another situation where (2-12) comes "free". Let $P \geq 0$ be a non-negative solution of the Steady State Riccati Equation (SSRE), for x in $\mathcal{D}(A)$:

$$[Ax,Px] + [Px,Ax] - [PBB^*Px,x] + [Rx,x] = 0, \qquad (2\text{-}13)$$

where R is non-negative. It follows easily from this equation that

$$2\text{Re.}[P(A - BB^*P)x,x] = -[(R + PBB^*P)x,x], \quad x \text{ in } \mathcal{D}(A).$$

Then, since both R and PBB^*P are non-negative, the operator $W = R + PBB^*P$ is non-negative. Next, as in the above, it is easy to see that

$$\text{for } x \text{ in } H: \quad {}_0\int^{\infty} ||\sqrt{W}\, S(t)x||^2 \, dt < \infty,$$

where $S(t)$, $t \geq 0$, is the semigroup generated by $A - BB^*P$. Let V be defined by

$$[Vx,x] = {}_0\int^{\infty} ||\sqrt{W}\, S(t)x||^2 \, dt, \quad x \text{ in } H. \tag{2-14}$$

Then, cleraly, V is positive as soon as either R is positive, or the pair (A^*,R) is controllable. In these cases we can conclude, as in the above, that the semigroup $S(t)$, $t \geq 0$, is w-stable on the range of V which is dense in H. It is not at all clear whether the semigroup is uniformly bounded or not. Thus, in general, stabilizability by the feedback $-B^*P$ is only weakly on a dense subspace --thus it can be called approximately weak stabilizability. Note that the semigroup $S(t)$, $t \geq 0$, satisfies the property:

$$\text{for } x \text{ in } H: \quad [PS(t)x, S(t)x] \leq [Px,x], \quad t \geq 0.$$

Suppose now that P is positive and let $Q^2 = P$ then, for x in H:

$$||QS(t)x|| \leq ||Qx||, \quad \text{for } t \geq 0.$$

Now define, for each x in H:

$$QS(t)x = Z(t)Qx, \quad t \geq 0.$$

Then $Z(t)$, $t \geq 0$, is a well-defined semigroup of contractions on the range $\mathcal{R}(Q)$ of Q. Hence on all of H by the fact that $\mathcal{R}(Q)$ is dense in H. Thus the semigroup $S(t)$, $t \geq 0$, is a quasi-affine transform of a contraction semigroup. Using this fact one can also show that $S(t)$, $t \geq 0$, is w-stable on the dense range of P. For details we refer to [6].

To proceed further, we now show a "negative" result regarding strong stabilizability of Hilbert space contraction semigroups.

THEOREM 6

Let $T(t)$, $t \geq 0$, be a contraction semigroup with generator A in H. If the resolvent of A is not compact then it is not possible to strongly stabilize the semigroup by the feedback $-B^*$, or the feedback $-Q^2B^*$.

Proof

Let $Z(t)$, $t \geq 0$, be a contraction semigroup with generator A_Z then, for x in $\mathcal{D}(A_Z)$:

$$||Z(t)x||^2 - ||x||^2 = {}_0\int^\infty 2\text{Re}.[A_Z Z(t)x, Z(t)x]\ dt.$$

Therefore if $Z(t)$, $t \geq 0$, is s-stable then

$$||x||^2 = -{}_0\int^\infty 2\text{Re}.[A_Z Z(t)x, Z(t)x]\ dt, \quad x \text{ in } \mathcal{D}(A_Z).$$

Consider the bilinear form:

$$[x,y]_n = -[A_Z x, y] - [x, A_Z y], \text{ for } x \text{ and } y \text{ in } \mathcal{D}(A_Z).$$

Then $\mathcal{D}(A_Z)$ is a pre-Hilbert space. Let H be the completion of $\mathcal{D}(A_Z)$, in the norm $||.||_n$, modulo the zero vectors. We have

$$||x||^2 = {}_0\int^\infty ||Z(t)x||_n^2\ dt, \quad \text{for } x \text{ in } \mathcal{D}(A_Z).$$

This shows that the transformation V (say) defined by

$$V: \mathcal{D}(A_Z) \to L^2([0,\infty); H),$$

$$(Vx)(t) = y(t), \quad y(t) = Z(t)x, \quad t \geq 0,$$

is an isometry which can be extended to all of H --by the fact that $\mathcal{D}(A_Z)$ is dense in H. More is true, in fact it follows easily that $Z(t)$, $t \geq 0$, is unitarily equivalent to the restriction to an invariant subspace of the left shift semigroup $L(t)$, $t \geq 0$, over the space $L^2([0,\infty); H)$, [7],

$$L(t)f = g, \quad g(s) = f(s+t), \quad \text{for } t \text{ and } s \geq 0.$$

The above is also sufficient for a contraction semigroup to be s-stable.

Now returning to the Theorem. Let $S(t)$, $t \geq 0$, be the contraction semigroup with generator $A - BB^*$. Then, from the above, if the semigroup is s-stable we can have

$$S(t) = U^* P\, L(t)\, P\, U, \quad t \geq 0,$$

where P is the orthogonal projection onto an invariant subspace of $L(t)$, $t \geq 0$, and $U: H \to L^2([0,\infty); H)$ is unitary. Next, it is clear that

$$S(t)^* = U^* P\, R(t)\, P\, U, \quad t \geq 0,$$

where $R(t) = L(t)^*$, $t \geq 0$, is the right shift semigroup

--which, of course, is isometric. We have, for x in $\mathcal{D}(A^*)$:

$$\text{Re.}[(A^* - BB^*)x,x] = \text{Re.}[U^*PA_R PUx,x],$$
$$= \text{Re.}[A_R PUx,PUx],$$

where A_R is the generator of the right shift semigroup. Then, since the right shift semigroup is isometric, it is plain that

$$\text{Re.}[(A^* - BB^*)x,x] = 0, \quad \text{for } x \text{ in } \mathcal{D}(A^*),$$

or

$$\text{Re.}[A^*x,x] = ||B^*x||^2.$$

But, since A^* is also dissipative, it must follow that

$$B^*x = 0 \quad \text{for all } x \text{ in } \mathcal{D}(A^*).$$

Therefore $B^* = 0$ which is not possible. This finishes the proof of the Theorem.

Note that this Theorem does not apply to the case of a self-adjoint contraction semigroup, since w-stability now implies s-stability.

We now turn to the stability enhancement problem of Hilbert space contraction semigroups. Suppose now that a contraction semigroup is already w-stable. Then, by Theorem 6, it is not possible to strongly enhance the semigroup by the feedback $-B^*$. It remains to study the case in which A has compact resolvent, and the semigroup is w-stable --hence it is s-stable also. We begin with the Lemma.

LEMMA 1

If A generates a contraction semigroup $T(t)$, $t \geq 0$, and has compact resolvent, and if the semigroup is w-stable. Then $A - BB^*$ generates a s-stable contraction semigroup $S(t)$, $t \geq 0$.

Proof

Since the semigroup $T(t)$, $t \geq 0$, is also s-stable, it is completely nonunitary, i.e., its maximal unitary subspace

$$H_u(T) = \{x \text{ in } H: ||T(t)x|| = ||x|| = ||T(t)^*x||, \ t \geq 0\}$$

is trivial. Let $H_u(S)$ be the maximal unitary subspace of $S(t)$, $t \geq 0$, then for x in $\mathcal{D}(A) \cap H_u(S)$:

$$\text{Re.}[(A - BB^*)x,x] = 0 \implies \text{Re.}[Ax,x] = ||B^*x||^2 = 0.$$

Therefore, since $\mathcal{D}(A)$ is dense in H,

$$H_u(S) \subseteq \mathcal{N}(B^*).$$

It then follows that $T(t)x = S(t)x$ for every x in $H_u(S)$. This, by the maximality of $H_u(T)$, implies that

$$H_u(S) \subseteq H_u(T).$$

Hence $H_u(S)$ is trivial, as soon as $H_u(T)$ is trivial, i.e., the semigroup $S(t)$, $t \geq 0$, is also completely nonunitary. Finally, we know that a completely nonunitary contraction semigroup is w-stable, [7], and since the resolvent of $A - BB^*$ is also compact, the semigroup $S(t)$, $t \geq 0$, is s-stable as expected. This completes the proof.

It follows from the above that, for the class of contraction semigroups, if the feedback $-B^*$ is used to enhance stability then we only need to consider the case in which the semigroup is w-stable, and its generator has compact resolvent, hence it is also s-stable.

Suppose now that the contraction semigroup $S(t)$, $t \geq 0$, with generator $A - BB^*$ is exponentially stable then, by Datko's results, there exists a positive operator P such that

$$2\text{Re.}[P(A - BB^*)x,x] = - ||x||^2, \quad \text{for } x \text{ in } \mathcal{D}(A),$$

or

$$2\text{Re.}[PAx,x] = - ||x||^2 + [Wx,x],$$

where

$$W = PBB^* + BB^*P.$$

It then follows that

$$[PT(t)x,T(t)x] - [Px,x] = - \int_0^t ||T(\sigma)x||^2 \, d\sigma + \int_0^t [WT(\sigma)x,T(\sigma)x] \, d\sigma.$$

Then, since the semigroup $T(t)$, $t \geq 0$, is s-stable, we obtain by letting $t \to \infty$:

$$[Px,x] = \int_0^\infty [(I - W)T(t)x,T(t)x] \, dt. \tag{2-15}$$

Now, since $T(t)$, $t \geq 0$, is only s-stable, the operator $I - W$ cannot be strictly positive, otherwise (2-15) would imply that $T(t)$, $t \geq 0$, is exponentially stable. However, we can have, again from (2-15),

$$I < PBB^* + BB^*P,$$

by the fact that P is positive. Suppose now that

$$PBB^* + BB^*P = cI$$

for some constant $c > 0$. Then, of course, 0 cannot be in the spectrum of BB^*. In other words, the existence of P implies that we must have $BB^* > 0$. We therefore conclude that if the feedback $-B^*$ exponentially enhances a strongly stable contraction semigroup $T(t)$, $t \geq 0$, then it is necessary that $N(B^*) = \{0\}$. This, of course, is very restrictive!

The enhancement problem for the general class of C_o semigroups is being studied. This will be reported elsewhere.

REFERENCES

[1] Datko, R. "Extending a Theorem of A.M. Liapunov to Hilbert Space." *J. Math. Anal. Appl.*, 32(1970): 610-616.

[2] Levan, N. "Stability of an Exponentially Stabilizable System." *IEEE Trans. Auto. Contr.*, AC-29(1984): 939-941.

[3] Slemrod, M. "A Note On Complete Controllability and Stabilizability for Linear Control Systems In Hilbert Space." *SIAM Contr. & Optim.*, 12(1974): 500-508.

[4] Benchimol, C.D. "A Note On Weak Stabilizability of Contraction Semigroups." *SIAM Contr. & Optim.*, 16(1978): 373-379.

[5] Levan, N., and Rigby, L. "Strong Stabilizability of Contractive Systems On Hilbert Space." *SIAM Contr. & Optim.*, 17(1979): 23-35.

[6] Levan, N. "Approximate Stabilizability Via The Algebraic Riccati Equation." *SIAM Contr. & Optim.*, 23(1985): 153-160.

[7] Fillmore, P.A. *Notes On Operator Theory*. New York: Van Nostrand, 1970.

NUMERICAL SOLUTION OF SOME PARABOLIC BOUNDARY CONTROL PROBLEMS BY FINITE ELEMENTS

U. Mackenroth

Mathematisches Institut der Universität Bayreuth
The Federal Republic of Germany

1. INTRODUCTION

In this paper we are concerned with error estimates and convergence properties of Ritz-type approximations for a certain class of optimal parabolic boundary control problems. The main feature of these problems is the fact that the final state may be constrained.

Let Ω be a bounded open domain in $\mathbb{R}^N$ with smooth boundary Γ, and suppose that A is a second order uniformly elliptic operator:

$$A := -\sum_{i,j=1}^{N} \frac{\partial}{\partial x_i}\left(a_{ij}\frac{\partial}{\partial x_j}\right) + a_o .$$

The coefficients a_{ij}, a_o are assumed to be smooth and symmetric functions on $\bar{\Omega}$. Let $\mu \geqslant 0$, $\nu \geqslant 0$, $\alpha > 0$, $m > 0$, $y_T \in L^2(\Omega)$, $Q =]0,T[\times \Omega$, $\Sigma =]0,T[\times \Gamma$ and suppose that C is a closed convex set of $L^2(\Omega)$.
By $\|\cdot\|$ we always denote the L^2-norm, regardless on which space it is considered. With these data we consider the following optimal control problem.

(P) Minimize

$$\mu\|y(T) - y_T\|^2 + \nu\|u\|^2$$

subject to $u \in L^\infty(\Sigma)$, $y \in C([0,T];L^2(\Omega))$ such that

$$\frac{\partial y}{\partial t} + Ay = 0, \tag{1.1}$$

$$\alpha y_{|\Sigma} + \frac{\partial y}{\partial n_A} = u, \tag{1.2}$$

$$y(0) = 0, \tag{1.3}$$

$$|u| \leqslant m, \tag{1.4}$$

$$y(T) \in C.$$

The typical examples we have in mind are the following:

(1) $\mu = 1,\ \nu = 0,\ C = L^2(\Omega)$;

(2) $\mu = 0,\ \nu = 1,\ C \neq L^2(\Omega)$,

(2a) $C = \{v \in L^2(\Omega) \mid \|v - z_T\| \leq \rho\}$,

(2b) $C = \{v \in C(\bar{\Omega}) \mid \|v - z_T\|_\infty \leq \rho\}$,

(2c) $C = \{v \in C(\bar{\Omega}) \mid v \geq z_T\}$.

Herein, we assume $\rho > 0$, $z_T \in L^2(\Omega)$ in case (2a) resp. $z_T \in C(\bar{\Omega})$ in the cases (2b), (2c). We mention that for $u \in L^\infty(\Sigma)$ we always have $y(T) \in C(\bar{\Omega})$.

Of course the case $\mu = 1,\ \nu > 0,\ C = L^2(\Omega)$ is also of interest, but is analyzed in great detail in Malanowski [6]. The case (2a) has some similarity to the time optimal control problems analyzed numerically in Lasiecka [4] and Knowles [3]. Problems with a full state constraint are considered in Alt/Mackenroth [1].

2. DISCRETIZATION AND ERROR ESTIMATES

Our first aim is to give a discretization of (P). We consider here only a semidiscretization: the time variable remains continuous. The main step is to replace $U = L^\infty(\Sigma)$ by

$$U_h := L^\infty(0,T;U_h^\Gamma),$$

where U_h^Γ denotes a suitable finite dimensional subspace of $L^\infty(\Gamma)$. For the approximation of the partial differential equation we use finite element subspaces $V_h \subset H^1(\Omega)$ which fulfill the following assumptions (c always denotes a generic constant):

(2.1) $$\inf_{w \in V_h} \{\|v - w\| + h\|v - w\|_{H^1(\Omega)}\} \leq c\, h^s \|v\|_{H^s(\Omega)} \qquad \forall\, v \in H^s(\Omega),\ \forall\, s \in [1,2];$$

(2.2) $$\|v\|_{H^1(\Omega)} \leq c\, h^{-1} \|v\| \qquad \forall\, v \in V_h;$$

(2.3) the family $\{V_h\}_{h>0}$ is dense in $L^2(\Omega)$.

An example of such spaces is given by an appropriate family of piecewise lineare finite elements.

Let now a be the well-known bilinear form associated with the partial differential equation (1.1), (1.2), denote by $(\cdot,\cdot)$ the inner product of an L^2-space and define $y_h(u) = y_h$ as the solution of

$$\frac{d}{dt}(v,y_h(t)) + a(v,y_h(t)) = (v_{|\Gamma},u(t)) \quad \forall\, v \in V_h, \ \forall'\, t \in [0,T]$$

and (1.3) (where of course in addition $y_h(t) \in V_h$ is required). Then we have for each $u \in L^\infty(0,T;L^2(\Gamma))$ (cf. Knowles [3], Schatz, Thomée, Wahlbin [7])

$$(2.4) \qquad \|y_h(u)(t) - y(u)(t)\| \leqslant c\, h^{3/2-\delta}\|u\|_{L^\infty(0,T;L^2(\Gamma))} \qquad \forall\, t \in [0,T].$$

Herein, $y(u)$ denotes the solution of (1.1) - (1.3) and $\delta > 0$ may be chosen arbitrarily small.

Finally, we replace C by a discrete version C_h. In the case (2a) we simply set $C_h = C$. In the cases (2b) and (2c) we choose for every $h > 0$ suitable points $x_1,\dots,x_{l_h} \in \bar\Omega$ (this will be made more precisely later on) and define in case (2b)

$$(2.5) \qquad C_h := \{v \in C(\bar\Omega) \mid |v(x_i) - z_T(x_i)| \leqslant \rho \quad i = 1,\dots,l_h\}$$

and analogously in case (2c).

The discrete problem is now given as follows:

(P_h) Minimize

$$\mu\|y_h(u)(T) - y_T\|^2 + \nu\|u\|^2$$

subject to $u \in L^\infty(0,T;U_h^\Gamma)$, (1.4), and $y_h(u)(T) \in C_h$.

Next we shall analyze convergence properties for the discrete problems. In doing so, the following Slater condition wil be fundamental.

(SL) There is a function $\bar u \in U_{ad}$ such that $y(\bar u)(T) \in \operatorname{int} C$.

Suppose that U is a Banach space such that $L^\infty(\Sigma) \subset U \subset L^2(\Sigma)$ with continuous inclusions. Let $Su := y(u)(T)$ for every $u \in U$ and assume that U is chosen in such a way that we have $S \in \mathcal{L}(U,E)$, where E is one of the spaces $L^2(\Omega)$, $C(\bar\Omega)$, and moreover that $S^*(E^*) \subset \tilde U$ holds where $\tilde U$ is a separable Banach space such that $\tilde U^* = U$. Let $U_{ad} :=$ $\{u \in U \mid |u| \leqslant m\}$ be closed and bounded with respect to U and define

$$U_{ad}^h := U_h \cap U_{ad}.$$

The following lemma and the following proposition are now immediate consequences of the corresponding results in Alt/Mackenroth [1].

LEMMA 1. Suppose that the following assumptions hold.

(i) There is a sequence $\{\bar{u}_h\}_{h>0}$ with $\bar{u}_h \in U^h_{ad}$ for every $h > 0$ such that $\lim_{h\to 0} \|\bar{u}_h - \bar{u}\|_U = 0$ and $\lim_{h\to 0} \|y_h(\bar{u}_h)(T) - y(\bar{u})(T)\|_E = 0$ (where $\bar{u}$ is as in (SL)).

(ii) $C \subset C_h \quad \forall\, h > 0.$

Then there is $h_o > 0$ such that for every $h \in \,]0,h_o]$ the problem (P_h) has an optimal solution u_h.

For every $u \in U$ let

$$f(u) \;.\!= \mu\|y(u)(T) - y_T\|^2 + \nu\|u\|^2,$$

$$f_h(u) \;.\!= \mu\|y_h(u)(T) - y_T\|^2 + \nu\|u\|^2.$$

PROPOSITION 1. In addition to the assumptions of Lemma 1 suppose that $\{y_h(u) \mid u \in M\}$ is bounded if $M \subset U_{ad}$ is bounded.

Let u_o be an optimal solution of (P) and let $\{w_h\}_{h>0}$ be a sequence with $w_h \in U^h_{ad}$ for every $h > 0$ such that $\lim_{h\to 0} \|w_h - u_o\|_U = 0$. Then there are $h_o > 0$, a sequence $\{v_h\}_{h>0}$ which is bounded with respect to U such that $v_h \in U^h_{ad}$, $y_h(u_h)(T) \in C_h$ for every $h > 0$, and a number $c > 0$ such that for every $h \leq h_o$ the following estimates hold:

(a) $f_h(u_h) - f(u_o) \leq c\|y_h(v_h)(T) - y(v_h)(T)\|_E + c\|y_h(w_h)(T) - y(w_h)(T)\|_E + c\|w_h - u_o\|_U,$

(b) $f(u_o) - f_h(u_h) \leq c\|y_h(u_h)(T) - y(u_h)(T)\|_E + c\, d(y_h(u_h)(T), C),$

(c) $2\nu\|u_h - u_o\|^2 \leq f_h(u_h) - f(u_o) + c\|y_h(u_h)(T) - y(u_h)(T)\|_E + c\, d(y_h(u_h)(T), C).$

We apply now this result to (P) in various situations.

THEOREM 1. Suppose that μ, ν, C are as in (1). Then each discrete problem (P_h) has an optimal solution u_h. Let $\{U^\Gamma_h\}_{h>0}$ a family of spaces of piecewise constant functions on Γ such that $U^\Gamma_{h_1} \subset U^\Gamma_{h_2}$ for $h_1 > h_2$ and such that $\bigcup_{h>0} U^\Gamma_h$ is dense in $L^2(\Omega)$. Then we have

$$\lim_{h\to 0} |f_h(u_h) - f(u_o)| = 0.$$

Moreover, if y_T cannot be reached by a feasible control u, then we have in addition

$$\lim_{h\to 0} \|u_h - u_o\| = 0.$$

Proof. We apply Proposition 1 with $U = L^2(0,T;L^2(\Gamma))$ and $E = L^2(\Omega)$. Due to our assumptions there exists a sequence $\{\hat{w}_h\}_{h>0}$ in U with $\hat{w}_h \in L^2(0,T;U_h^\Gamma)$ for every $h > 0$ and $\lim_{h\to 0} \|\hat{w}_h - u_o\| = 0$. Let $Q : L^2(\Gamma) \to L^2(\Gamma)$ be the projection of $L^2(\Gamma)$ on $\{v \in L^2(\Gamma) \mid |u| \leqslant m\}$ and define $w_h := Q\hat{w}_h$. Since Q is Lipschitz continuous, we have $w_h \in U_{ad}^h$ and $\lim_{h\to 0} \|w_h - u_o\| = 0$. The first assertion is now immediate from Proposition 1 and (2.4).

It is well-known that under the above assumption concerning reachability u_o is bang-bang (cf. Schmidt/Weck [8]). Thus u_o is unique and from the first assertion it is clear that $\{u_h\}_{h>0}$ converges weakly to u_o. The second assertion follows now from Mackenroth [5], Lemma 2. ∎

We consider from now on the situation described in (2). In particular, we assume $\mu = 0$, $\nu = 1$. In the next theorem we shall make use of the optimality conditions. Let (SL) hold with respect to $C(\bar{\Omega})$. Then u_o is optimal if and only if u_o is feasible and if there is a regular Borel measure μ such that

$$(2.6) \qquad u(t,\xi) = \begin{cases} p(t,\xi) & \text{if } |p(t,\xi)| \leqslant m \\ m & \text{if } p(t,\xi) > m \\ -m & \text{if } p(t,\xi) < -m \end{cases} \qquad \forall\ (t,\xi) \in \Sigma$$

where p is given by

$$(2.7) \qquad p(t,x) = \sum_{k=1}^{\infty} e^{-\lambda_k (T-t)} \int_\Omega v_k d\mu \; v_k(x) \qquad \forall\ (t,x) \in [0,T[\times \bar{\Omega}$$

(λ_k and v_k denote the eigenvalues and eigenfunctions of the corresponding eigenvalue problem) and where the relation

$$(2.8) \qquad \sup_{v\in C} \int_{\bar{\Omega}} v \, d\mu = \int_{\bar{\Omega}} y(u_o)(T) \, d\mu$$

holds. The function p fulfills

$$-\frac{\partial p}{\partial t} + A p = 0, \qquad \alpha p_{|\Sigma} + \frac{\partial p}{\partial n_A} = 0.$$

Let now U_h^Γ be a space of piecewise linear finite elements on Γ which is chosen in such a way that

$$\text{(2.9)} \qquad \|v - P_h v\| \leq c\, h \|v\|_{H^1(\Gamma)} \qquad \forall\, v \in H^1(\Gamma).$$

Herein, P_h denotes the orthogonal projection from $L^2(\Gamma)$ onto U_h^Γ. Such a choice is possible (cf. Aziz [2]).

THEOREM 2. Suppose that we have $C_h = C$ for every $h > 0$ where C is given by (2.a). Let (SL) hold and $0 \notin C$. Then we have for each sufficiently small h

$$|f_h(u_h) - f(u_o)| \leq c\, h,$$

$$\|u_h - u_o\| \leq c\, h^{1/2}.$$

Proof. Since in this case C has interior points with respect to the norm of $L^2(\Omega)$, the measure μ of (2.8) must be absolutely continuous. Denote by w_o the function of $L^2(\Omega)$ which represents μ. Then it is quite elementary to see that (2.8) is equivalent to

$$(y(u_o)(T) - z_T, w_o) = \|y(u_o)(T) - z_T\| \|w_o\|, \quad \|y(u_o)(T) - z_T\| = \rho.$$

Thus, since $0 \notin C$, (2.8) is equivalent to the existence of $\lambda > 0$ such that

$$w = \lambda(y(u_o)(T) - z_T), \quad \|y(u_o)(T) - z_T\| = \rho.$$

From this we conclude $w \in H^1(\Omega)$. Hence from (2.7) we get

$$p(T) = w \in H^1(\Omega).$$

This implies $p \in H^{1,2}([0,T] \times \Omega)$ and therefore

$$p_{|\Sigma} \in L^2(0,T;H^{3/2}(\Gamma)) \subset L^2(0,T;H^1(\Gamma)).$$

Thus using (2.6) we see

$$\text{(2.10)} \qquad u_o \in L^2(0,T;H^1(\Gamma)).$$

We apply now Proposition 1 with $U = L^2(0,T;L^2(\Gamma))$ and $E = L^2(\Omega)$. Let $w_h = Q\, P_h u_o$ with Q as in the preceding proof. We get from (2.10) and (2.9)

$$\|w_h - u_o\| \leq c\, h \|u_o\|_{L^2(0,T;H^1(\Gamma))}.$$

The first inequality follows now directly from Lemma 1, Proposition 1(a), (b) and (2.4) since the sequences $\{w_h\}_{h>0}$, $\{v_h\}_{h>0}$, $\{u_h\}_{h>0}$ lie in U_{ad} and are consequently uniformly bounded in $L^{\infty}(0,T;L^2(\Gamma))$. The second inequality is then a direct consequence of Proposition 1(c). ∎

In the general case where the endpoint constraint set has an interior only with respect to $C(\bar{\Omega})$ it seems not to be possible to derive an order of convergence directly. We obtain such a rate only after having regularized (P) suitably. To this end define for small $\varepsilon > 0$ the sets $\Sigma_{\varepsilon} :=]T-\varepsilon,T[\times \Gamma$ and

$$U_{ad}^{\varepsilon} := \{u \in L^{\infty}(\Sigma) \mid |u| \leq m,\ u(t,\xi) = 0 \quad \forall' \ (t,\xi) \in \Sigma_{\varepsilon}\}.$$

The regularized problem (P_{ε}) is defined as (P) with the only exception that U_{ad} is replaced by U_{ad}^{ε}. By $(P_{\varepsilon,h})$ we denote a discretization of (P_{ε}) in the sense described above. The following strenghtend version of the Slater condition will be needed.

(SL_{ε_o}) There exists $\varepsilon_o > 0$ and $\bar{u}_{\varepsilon_o} \in U_{ad}^{\varepsilon_o}$ such that $y(\bar{u}_{\varepsilon_o})(T) \in \operatorname{int} C$.

THEOREM 3. Suppose that (SL_{ε_o}) holds. Then the following assertions are true.

(a) Let $\varepsilon \in]0,\varepsilon_o]$ and denote by u_{ε} the optimal solution of (P_{ε}). Then for small $\delta > 0$ the following error estimates hold:

$$|f(u_{\varepsilon}) - f(u_o)| \leq c\ \varepsilon^{1/(N+1+\delta)},$$

$$\|u_{\varepsilon} - u_o\| \leq c\ \varepsilon^{1/2(N+1+\delta)}.$$

(b) Suppose that $(P_{\varepsilon,h})$ is obtained from (P_{ε}) by discretizing only the controls and let U_h^{Γ} be as in Theorem 2. Then for sufficiently small $h > 0$ $(P_{\varepsilon,h})$ has an optimal solution $u_{\varepsilon,h}$ and we have

$$|f(u_{\varepsilon,h}) - f(u_{\varepsilon})| \leq c_{\varepsilon}\ h,$$

$$\|u_{\varepsilon,h} - u_{\varepsilon}\| \leq c_{\varepsilon}\ h^{1/2}.$$

Proof. (a) From Tröltzsch [9], Lemma 5.6.6, we see that $S \in \mathcal{L}(U,E)$ holds for

$$q = N+1+\delta,\ U = L^q(\Sigma),\ E = C(\bar{\Omega}).$$

Let now

$$w_{\varepsilon}(t,\xi) = \begin{cases} u_o(t,\xi) & \text{if } (t,\xi) \in \Sigma \setminus \Sigma_{\varepsilon}, \\ 0 & \text{elsewhere.} \end{cases}$$

Using Proposition 1 (where ε takes on the role of the discretization parameter h and where the formulation has to be slightly modified) we obtain

$$\|w_\varepsilon - u_o\| \leq c\|w_\varepsilon - u_o\|_{L^q(\Sigma)} \leq c\ \varepsilon^{1/q}$$

and also the second inequality of (a).

(b) For $\varepsilon > 0$ we can take $E = C(\bar{\Omega})$ and

$$U = \{u \in L^2(\Sigma) \mid u(t,\xi) = 0 \quad \forall' \ (t,\xi) \in \Sigma_\varepsilon\}.$$

Equation (2.6) holds now on $\Sigma \setminus \Sigma_\varepsilon$, but for $t \in [0,T-\varepsilon]$ we have $p_{|\Gamma}(t) \in H^1(\Gamma)$ and thus $u_\varepsilon \in L^2(0,T;H^1(\Gamma))$. The proof is now finished by an argumentation similar to that of Theorem 2. ∎

Theorem 3(b) can be generalized to the case where also the partial differential equation is discretized if one uses a suitable finite element method which converges in the maximum norm.

It is also possible to discretize C_h and to get a rate of convergence for the term $d(y_h(u_h)(T),C)$. For example, if one uses piecewise linear finite elements for V_h and chooses for the points x_i in the set (2.5) the knots of the finite elements, then it is easy to see that $d(y_h(u_h)(T),C)$ converges linearly to 0.

3. NUMERICAL RESULTS

In this section we shortly describe the numerical results for two typical examples. The numerical solution of (P_h) requires of course a discretization of the time variable t. This gives "fully discrete" basis functions for the controls. For each of them, the parabolic equation is solved approximately by using piecewise linear finite elements for x and an adequate finite difference method for t. It is obvious that in this way the discrete problem can be rewritten as a quadratic programming problem with constraints, which then can be solved directly on a computer by a suitable algorithm.

In the first example let μ,ν,C as in (1). Let $A = -\Delta$,

$$\Omega := \{(x_1,x_2) \in \mathbb{R}^2 \mid x_1,x_2 \in]0,1[\}$$

and suppose that the control is excercised on the set

$$\Gamma_o = \{(x_1,x_2) \in \bar{\Omega} \mid x_1 = 1\}.$$

The function which has to be approximated is

$$y_T(x_1,x_2) = \frac{1}{2}x_1x_2 + \frac{1}{4} \qquad \forall\ (x_1,x_2) \in \bar{\Omega}.$$

A discrete optimal control is shown in Fig. 1 and for an enlarged space of discrete controls a second one in Fig. 2. The discrete controls are chosen as piecewise constant functions. In both pictures, the right axis is Γ_o whereas the left one coinsides with $[0,T]$. They clearly show that the discrete controls approximate a bang-bang control. The optimal value for the first one is $- 0.152375 + \|y_T\|^2$ and $- 0.152345 + \|y_T\|^2$ for the second one which is only slightly better.

Fig. 1

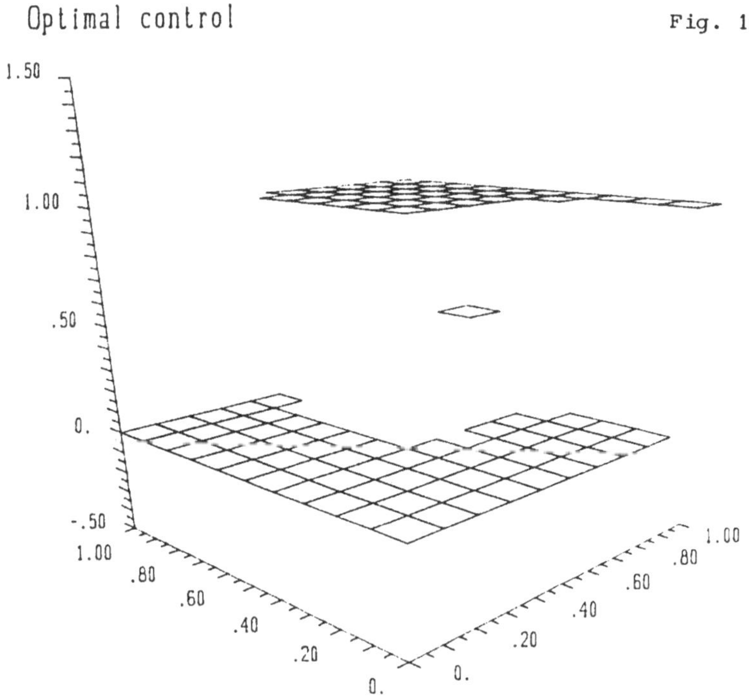

In the second example let μ,ν,C as in (2a) with a certain small ρ. Let $A = -\frac{\partial^2}{\partial x^2}$ and $\Omega =]0,1[$. The optimal control is approximated by piecewise linear functions. A discrete solution with a rather fine grid is shown in Fig. 3.

The quadratic problems were solved by Gill/Murray's program SOL/QPSOL. The numerical solution has been carried out together with W. Alt from University Bayreuth on a VAX 11/780 computer.

Optimal control

Fig. 2

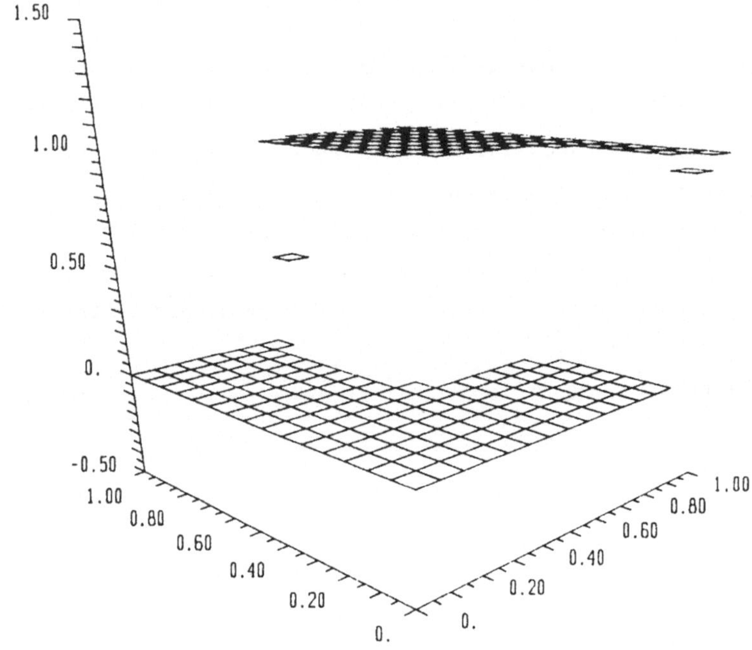

Optimale Steuerung

Fig. 3

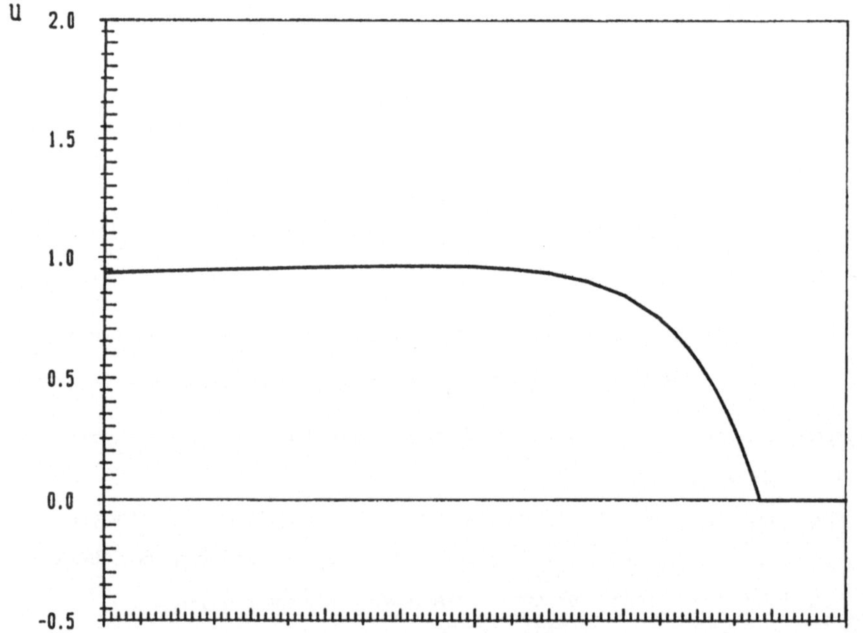

REFERENCES

[1] Alt, W. and Mackenroth, U. "Numerical solution of parabolic optimal control problems with an integral state constraint". Submitted.

[2] Aziz, A.K. (ed.) *The Mathematical Foundations of the Finite Element Method with Applications to Partial Differential Equations.* New York: Academic Press, 1972.

[3] Knowles, G. "Finite element approximation of parabolic time optimal control problems". *SIAM J. Control and Optimization* 20 (1982): 414-427.

[4] Lasiecka, I. "Ritz-Galerkin approximation of the time optimal boundary control problem for parabolic systems with Dirichlet boundary conditions". *SIAM J. Control and Optimization* 22 (1984): 477-500.

[5] Mackenroth, U. "Some remarks on the numerical solution of bang-bang type optimal control problems". *Numer. Funct. Anal. and Optimiz.* 5 (1982-83): 457-484.

[6] Malanowski, K. "Convergence of approximations vs. regularity of solutions for convex, control-constrained optimal control problems". *Appl. Math. Optim.* 8 (1981): 69-95.

[7] Schatz, A.H., Thomée, V. and Wahlbin, L.B. "Maximum norm stability and error estimates in parabolic finite element equations". *Comm. Pure Applied Math.* 33 (1980): 265-304.

[8] Schmidt, E.P.J.G. and Weck, N. " On the boundary behaviour of solutions to elliptic and parabolic equations - with applications to boundary control for parabolic equations". *SIAM J. Control and Optimization* 16 (1978): 593-598.

[9] Tröltzsch, F. *Optimality Conditions for Parabolic Control Problems and Applications.* Leipzig: Teubner, 1984.

INVARIANCE UNDER NONLINEAR PERTURBATIONS FOR REACHABLE AND ALMOST-REACHABLE SETS*

Thomas I. Seidman

Univeristy of Maryland Baltimore County,
Catonsville, Maryland, USA

0. PREPARATORY REMARKS

The material presented here is based partly on the results of [9] (stimulated by [7]) and partly on joint work with K. Naito, in progress. The latter portion, concerning invariance of almost-reachable sets, constitutes an extension of the talk actually presented in Gainesville. It should be noted that this work is very much in the line of development of the theses of Henry [4], Wong [11], and Carmichael [2]; see [3] and the references therein for more on the history of the "fixpoint theory approach to controllability." At this point it seems appropriate to acknowledge both the support of the AFOSR and of UMBC and also the successful efforts of Professor Lasiecka (and others) in organizing a most stimulating meeting.

1. INTRODUCTION

Consider a quasilinear control system written in the abstract form

$$\dot{x} + A(t)x = \varphi(t,x) + B(t)v, \qquad x(0) = \xi_0 \tag{1.1}$$

with $A(\cdot)$ linear, generating an evolution operator $T(t,s)$ so

$$x(t) = \int_0^t T(t,s)[\varphi(s,x(s)) + B(s)v(s)]\, ds \tag{1.2}$$

where, for simplicity of exposition, we have taken ξ_0 to be 0. We view (1.2) as an integral equation for x and define a (mild) solution of (1.1) as a solution of (1.2). This may be viewed as a

---

[1]This research has been partially supported by the Air Force Office of Scientific Research under grant number AFOSR-82-0271.

perturbation of the linear equation: $\dot{x} + Ax = Bv$ $(\varphi = 0)$. The *reachable set* for (1.1) or (1.2) is

$$\mathcal{K}_\varphi := \{x(\tau): (1.2) \text{ for some } v \in \mathcal{V}\} \tag{1.3}$$

where we have fixed a terminal time $\tau > 0$ and denoted the set of admissible controls by $\mathcal{V}$; the *almost-reachable set* is just the closure $\overline{\mathcal{K}}_\varphi$. All this is, of course, purely formal until we introduce spaces and topologies and impose hypotheses ensuring that (1.2) has solutions for $v \in \mathcal{V}$ and that these are continuous (at least at $t = \tau$) so the point evaluation implicit in (1.3) is feasible, e.g.,

$$L_t: f \mapsto \int_0^t T(t,s)f(s)\, ds \tag{1.4}$$

is continuous to a state space X_1 (in which one considers $\mathcal{K}_\varphi, \overline{\mathcal{K}}_\varphi$) for $f = \varphi(x) + Bv$ and $t = \tau$.

Our object, then, is to show the invariance of the reachable or the almost-reachable set under suitable hypotheses: $\mathcal{K}_\varphi = \mathcal{K}_0$ or $\overline{\mathcal{K}}_\varphi = \overline{\mathcal{K}}_0$. In comparison with the earlier work cited, our underlying controllability hypothesis takes the form of requiring that this is already known for certain *affine* perturbations: $\varphi(t,x) = f(t)$ with $f \in \mathcal{Z}$. We will, here, take these to be

$$\mathcal{Z} := L^p([0,\tau] \to Z), \qquad \mathcal{V} := L^{p'}([0,\tau] \to V) \tag{1.5}$$

$(1 < p,p' < \infty$; Z, V Banach spaces) and note that one can then easily show the equivalence of the desired controllability hypotheses ($\mathcal{K}_f = \mathcal{K}_0$ *or* $\overline{\mathcal{K}}_f = \overline{\mathcal{K}}_0$; each for all $f \in \mathcal{Z}$) with the conditions

$$(H_1) \quad L_\tau \mathcal{Z} \subset L_\tau B\mathcal{V} = \mathcal{K}_0 \quad \textit{or} \quad (H_1') \quad L_\tau \mathcal{Z} \subset \overline{\mathcal{K}}_0 \tag{1.6}$$

which are a bit more convenient for application.

Introducing state spaces X_1 and X_2 we set $\mathcal{X}_1 := C([0,\tau] \to X_1)$, $\mathcal{X}_2 := L^{\bar{p}}([0,\tau] \to X_2)$ with $1 \le \bar{p} < p$ (take $\bar{r} \le \bar{p}/p < 1$). Impose the usual measurability conditions and:

$$(H_2) \quad (i) \quad \|T(t,s)\|_{Z\to X_2} \le \rho_Z(t-s), \qquad \rho_Z \in L^{\bar{q}} \quad (1/\bar{q} + 1/p \le 1 + 1/\bar{p});$$

$$(ii) \quad \|T(t,s)B(s)\|_{V\to X_2} \le \rho_V(t-s), \qquad \rho_V \in L^{\bar{q}'} \quad (1/\bar{q}' + 1/p' \le 1 + 1/\bar{p}).$$

(H_3) (*i*) $\|\varphi(t,\xi)\|_Z \le \alpha + \beta\|\xi\|_{X_2}^{\bar{r}}$, $\alpha \in L^p$;

(*ii*) $\|T(t,s)[\varphi(s,\xi)-\varphi(s,\xi')]\|_{X_2} \le \rho_X(t-s)\|\xi-\xi'\|_{X_2}$, $\rho_X \in L^1$.

(H_4) (*i*) $\|T(t,s)\|_{Z\to X_1} \le \hat{\rho}_Z(t-s)$, $\hat{\rho}_Z \in L^q$ $(1/q + 1/p = 1)$;

(*ii*) $\|T(t-s)B(s)\|_{V\to X_1} \le \hat{\rho}_V(t-s)$, $\hat{\rho}_V \in L^{q'}$ $(1/q' + 1/p' = 1)$;

(*iii*) $\|T(t,s)-T(t',s)\|_{t\to X_1} \le \varepsilon$, $\|[T(t,s)-T(t',s)]B(s)\|_{V\to X_1} \le \varepsilon$ for $0 \le s \le t'-\varepsilon$, $t' < t \le \tau$ with $\varepsilon = \varepsilon(h) \to 0$ as $h = t-t' \to 0$.

A fairly standard set of arguments then gives existence of solutions of (1.2) -- i.e., the map

$$S_\varphi : \mathcal{V} \to \mathcal{X}_2 : v \mapsto x(\cdot) = \text{solution of (1.2)} \tag{1.7}$$

is well-defined -- with S_φ and

$$F_\varphi : \mathcal{V} \to \mathcal{Z} : v \mapsto x := S_\varphi(v) \mapsto f := \varphi(\cdot, x) \tag{1.8}$$

continuous and a growth condition

$$\|F_\varphi(v)\|_{\mathcal{Z}} \le a + b\|v\|_{\mathcal{V}}^{\bar{r}} \tag{1.9}$$

as well as the desired continuity of L_t to X. We state this formally.

THEOREM 1: Assume (H_2), (H_3). Then S_φ, F_φ (as in (1.7), (1.8)) are well-defined and continuous with (1.9). If one also assumes (H_4), then $L_t: \mathcal{Z} \to X_1$ and $L_tB: \mathcal{V} \to X_1$ are continuous and the solution $x = S_\varphi(v)$ is actually in $\mathcal{X}_1$ for $v \in \mathcal{V}$ with a (uniform) modulus of continuity to X_1 depending only on $\|v\|_{\mathcal{V}}$ (and the specifics of (H_2)-(H_4)).

Proof: See [9] for details. We note for later reference that the argument involves using an exponential weighting $e^{-\mu t}$ (with μ large enough) for the $\mathcal{X}_2$-norm so that the composition $L\Phi$ is contractive where Φ is the Nemytsky operator induced by the nonlinear perturbation φ and L is the linear map: $[Lv](t) := L_tv$. □

2. INVARIANCE OF THE REACHABLE SET

We wish to show that (H_1) implies the invariance $\mathcal{K}_\varphi = \mathcal{K}_0$ under the hypotheses (H_2)-(H_4) together with any of a number of supplementary conditions which ensure that:

(C) F_φ is a (well-defined, continuous) compact map: $\mathcal{V} \to \mathcal{Z}$.

It is easy to see, given Theorem 1, that (H_1) implies $\mathcal{K}_\varphi \subset \mathcal{K}_0$. To show that also $\mathcal{K}_0 \subset \mathcal{K}_\varphi$ we will use the Schauder Fixpoint Theorem and will need (C).

LEMMA 1: Let $L_\tau: \mathcal{Z} \to X_1$ and, $L_\tau B: \mathcal{V} \to X_1$ be linear and continuous. Then (H_1) implies existence of $C: \mathcal{Z} \to \mathcal{V}$ such that

$$\|C(f)\|_{\mathcal{V}} \le \alpha \|f\|_{\mathcal{Z}}, \qquad L_\tau f + L_\tau BC(f) = 0. \tag{2.1}$$

Proof: Let $\mathcal{V}_0 := \mathcal{V}/\mathcal{N}(L_\tau B)$ and $A: [f,w] \mapsto L_\tau f + L_\tau Bw: \mathcal{Z} \times \mathcal{V}_0 \to X_1$; A is well-defined and continuous so $\mathcal{N}(A)$ is closed. By (H_1) and the definition of $\mathcal{V}_0$, $\mathcal{N}(A)$ is the graph of an everywhere defined linear map $C_0: \mathcal{Z} \to \mathcal{V}_0$, continuous by the Closed Graph Theorem. The Michael Selection Theorem [6] gives $C_1: \mathcal{V}_0 \to \mathcal{V}$ continuous and of linear growth for a right inverse of the canonical projection: $\mathcal{V} \to \mathcal{V}_0$. Take $C = C_1 \circ C_0$. □

THEOREM 2: Assume (H_2)-(H_4) together with some supplementary condition giving (C). Then (H_1) implies the invariance of the reachable set: $\mathcal{K}_\varphi = \mathcal{K}_0$.

Proof: We need only prove that $\xi_* \in \mathcal{K}_0$ (so $\xi_* = L_\tau Bv_*$ for some $v_* \in \mathcal{V}$) implies $\xi_* \in \mathcal{K}_\varphi$ under the hypotheses. Consider

$$\Psi = \Psi_\varphi: \hat{v} \mapsto C(F_\varphi(\hat{v}+v_*)): \mathcal{V} \to \mathcal{V} \tag{2.2}$$

which is defined, continuous, and compact by Theorem 1, Lemma 1, and the condition (C). Using (1.9) and (2.1) one has $\|\Psi(\hat{v})\| = \mathcal{O}(\|\hat{v}\|^{\bar{r}})$ so there must be an invariant ball in $\mathcal{V}$. Application of the Schauder Theorem gives existence of a fixpoint $\hat{v}$ of Ψ and we set $v := \hat{v} + v_*$, $x := S_\varphi(v)$, $f := F_\varphi(v)$. Then x satisfies (1.1) with this v so $x(\tau) \in \mathcal{K}_\varphi$. Also, x satisfies: $\dot{x} + Ax = f + Bv$ so

$$x(\tau) = L_\tau f + L_\tau Bv = (L_\tau f + L_\tau B\hat{v}) + L_\tau Bv_*.$$

Since $L_\tau Bv_* = \xi_*$ and (2.2) gives $\hat{v} = C(f)$, the definition of C gives $x(\tau) = \xi_*$ so $\xi_* \in \mathcal{K}_\varphi$. □

Our major effort, here, is to provide suitable supplementary hypotheses giving the compactness of F_φ. We provide four such hypotheses. The first two impose compactness conditions on the evolution operator $T(\cdot)$ or on its composition with B; the third is a compactifying condition on φ. Each of these works in the context of $\mathfrak{X}_2$ (to which (H_4) is now irrelevant) and uses the Aubin Compactness Theorem [1] for which we need also the mild condition

$$\text{A, B in (1.1) are (uniformly in t) bounded operators from } X_2, V, \text{ respectively, to some (compatible) space } \tilde{X} \tag{2.3}$$

since, given bounds on v, $x = S_\varphi(v)$, $f = F_\varphi(v)$, this bounds $\dot{x}$ in $\tilde{\mathfrak{X}} := L^{\tilde{p}}(\to\tilde{X})$. The fourth condition is also a compactifying condition on φ, but now in the context of $\mathfrak{X}_1$, depending on a form of the Arzela-Ascoli Theorem. For the first two conditions we also use a compactness theorem from [8]:

LEMMA 2: Let $\Gamma: Y \times \mathfrak{B} \to Y$ with Y complete metric. Suppose

(*i*) uniform contractivity of Γ with respect to Y,

(*ii*) $\Gamma(\mathfrak{A},\mathfrak{B})$ precompact in Y for $\mathfrak{A}$ compact in Y.

Then the set of fixpoints $\{y \in Y: y = \Gamma(y,b),\ b \in \mathfrak{B}\}$ is precompact in Y. □

This will be employed with $Y := \mathfrak{X}_2$, $\mathfrak{B} :=$ (ball in $\mathfrak{V}$), and $\Gamma: [x,v] \mapsto L\varphi(x) + S_0 v$; since, as noted in the 'proof' of Theorem 1, we have already verified and used the contractivity of $L\Phi$ (for a suitable $\mathfrak{X}_2$-norm), it will only be necessary to verify the hypothesis (*ii*) of Lemma 2.

THEOREM 3: Assume (H_2), (H_3) and supplement or modify these as in any of the following (assuming also (H_4) in connection with (C_4)):

(C_1) For some X_3 with a compact embedding $X_2 \hookrightarrow X_3$: for $\delta > 0$ there is M_δ such that

$$\|T(t,t-\delta)\|_{X_3 \to X_2} \leq M_\delta \qquad \text{for } \delta \leq t \leq \tau. \tag{2.4}$$

(C_2) Strengthen (H_2-(*ii*)), replacing X_2 with some X_4 such that $X_4 \hookrightarrow X_2$ is a compact embedding.

(C_3) Strengthen the growth condition (H_3-(*i*)) by replacing X_2 with some space X_3 for which $X_2 \hookrightarrow X_3$ is a compact embedding.

(C_4) With $X_1 = X_2$ reflexive and an embedding $X_2 \hookrightarrow X_3$ continuous from weak to norm topologies, assume

$$\|\varphi(s,\xi)\|_Z \leq \alpha_M(s) \qquad \text{for } \|\xi\|_{X_3} \leq M \tag{2.5}$$

with each $\alpha_M \in L^p$.

Then (C) holds: $F_\varphi \colon \mathcal{V} \to \mathcal{Z}$ is well-defined, continuous, and compact.

Proof: We proceed case by case. Fixing M, we set $\mathfrak{B} := \{v \in \mathcal{V}: \|v\|_{\mathcal{V}} \leq M\}$.

Case 1: Employ the Aubin Theorem to get precompactness of $\mathfrak{S}_0 := S_0(\mathfrak{B})$ in $\mathcal{X}_3 := L^{\bar{p}}(\to X_3)$, noting that $\mathfrak{S}_0$ is bounded in $\mathcal{X}_2$ and (3.3) then bounds $\{\dot{x}: x \in \mathfrak{S}_0\}$. Consider the map: $y \mapsto y_\delta$ with $y_\delta(t) := T(t,t-\delta)y(t-\delta)$ ($:= 0$ for $t < \delta$) for $y \in \mathfrak{S}_0$. Then (C_1) ensures continuity of this map (any $\delta > 0$) so the image $\mathfrak{I}_\delta$ is precompact in $\mathcal{X}_2$, hence totally bounded. Noting that, for $y \in \mathfrak{S}_0$,

$$y(t) - y_\delta(t) = \int_{(t-\delta)_+}^{t} T(t,s)B(s)v(s)\, ds \qquad (y := S_0 v),$$

(H_2-(*ii*)) ensures uniform approximation of $\mathfrak{I}_\delta$ to $\mathfrak{S}_0$: one has $\|y-y_\delta\| \leq \varepsilon(\delta)$ with $\varepsilon(\delta) = (\text{const.})\cdot(L^{q'}$-norm of restriction of ρ_V to $[0,\delta])$. Thus $\mathfrak{S}_0$ is totally bounded, hence precompact. As $L\Phi$ is continuous and $\Gamma(\mathfrak{A},\mathfrak{B}) = L\Phi(\mathfrak{A}) + \mathfrak{S}_0$, this gives (*ii*) of Lemma 2 and so the precompactness of $S_\varphi(\mathfrak{B})$ in $\mathcal{X}_2$. As $F_\varphi(\mathfrak{B}) = \Phi(S_\varphi(\mathfrak{B}))$ and Φ is continuous, this gives the result.

Case 2: The same estimates used for Theorem 1 to prove boundedness of $\mathfrak{S}_0$ in $\mathcal{X}_2$ from (H_2-(*ii*)) now -- using (C_2) -- bound $\mathfrak{S}_0$ in

$\mathfrak{X}_4 := L^{\bar{p}}([0,\tau] \to X_4)$. Noting (2.3), the Aubin Theorem then gives precompactness of $\mathfrak{S}_0$ in $\mathfrak{X}_2$ so Lemma 2 applies as in Case 1.

Case 3: The estimates used in Theorem 1 give boundedness of $\mathfrak{S}_\varphi := S_\varphi(\mathfrak{B})$ in $\mathfrak{X}_2$ and so of $F_\varphi(\mathfrak{B})$ in $\mathfrak{Z}$. Using (2.3), the Aubin Theorem gives precompactness of $\mathfrak{S}_\varphi$ in $\mathfrak{X}_3$ and -- using (C_3) -- Krasnoselski's Theorem now gives continuity of $\Phi: \mathfrak{X}_3 \to \mathfrak{Z}$ whence $F_\varphi(\mathfrak{B}) = \Phi(\mathfrak{S}_\varphi)$ is precompact in $\mathfrak{Z}$.

Case 4: Assuming (H_4) as well, now, we recall from Theorem 1 that, as $F_\varphi(\mathfrak{B})$ is bounded in $\mathfrak{Z}$, we have $\mathfrak{S}_\varphi$ bounded in $\mathfrak{X}_1 := C([0,T] \to X_1)$ with a uniform modulus of continuity. In particular, $y(t) \in \hat{M}\mathfrak{B}_1$ ($\mathfrak{B}_1$:= unit ball of $X_1 = X_2$) for all $y \in \mathfrak{S}_\varphi$ and $t \in [0,T]$. As $\mathfrak{B}_1$ is weakly compact in X_1 it is compact in X_3 so the Arzela-Ascoli Theorem gives precompactness of $\mathfrak{S}_\varphi$ in $C([0,T] \to X_3)$. Using (C_4), Krasnoselski's Theorem gives continuity for $\Phi: C(\to X_3) \to \mathfrak{Z}$ and so precompactness of $F_\varphi(\mathfrak{B}) = \Phi(\mathfrak{S}_\varphi)$. □

COROLLARY: (H_1)-(H_4) with any of (C_1)-(C_4) give $\mathfrak{K}_\varphi = \mathfrak{K}_0$. □

Remark 1: If $\mathfrak{K}$ were closed then an argument like that for Lemma 1 gives existence of $\bar{C}: \mathfrak{K} = \bar{\mathfrak{K}} \to \mathcal{V}$, continuous with $\xi + L_\tau B\bar{C}(\xi) = 0$ for $\xi \in \bar{\mathfrak{K}}$. One can then replace (2.2) by

$$\bar{\Psi} = \bar{\Psi}_\varphi: \hat{v} \mapsto \bar{C}(L_\tau F_\varphi(v+v_*)): \mathcal{V} \to \mathcal{V} \tag{2.6}$$

and proceed as in Theorem 2 if $L_\tau F_\varphi$ were compact, e.g., if

(C_5) $\qquad L_\tau$ is compact from $\mathfrak{Z}$ to X_2

since we already know F_φ is bounded. □

3. INVARIANCE OF THE ALMOST-REACHABLE SET

Approximate controllability is somewhat more delicate since the hypothesis (H_1') is much weaker than (H_1). Indeed, if one had (H_1) then Theorem 3 would apply under (H_2)-(H_4) with any of (C_1)-(C_4) and, clearly, $\mathfrak{K}_\varphi = \mathfrak{K}_0$ gives $\bar{\mathfrak{K}}_\varphi = \bar{\mathfrak{K}}_0$. While we will comment later on a direction for possible further extension (cf., Remark 2), we present an invariance result under the rather strong condition that φ be bounded in its dependence on ξ (i.e., $\bar{r} = 0$ in $(H_3$-$(i))$).

LEMMA 3: Let $L_\tau B: \mathcal{V} \to X_1$ be continuous and $\mathcal{F}$ precompact in X_1 with $\mathcal{F} \subset \bar{\mathcal{K}}_0$:= (closure of $L_\tau B\mathcal{V}$). Then, for any $\varepsilon > 0$ there is a (compact, convex) polyhedron $\mathcal{V}^\varepsilon \subset \mathcal{V}$ such that

$$\varepsilon \geq \|\mathcal{F}-L_\tau B\mathcal{V}^\varepsilon\| := \sup_{f\in\mathcal{F}} \inf_{v\in\mathcal{V}^\varepsilon} \|f-L_\tau Bv\|_{X_1} \qquad (3.1)$$

Proof: For $\xi \in \bar{\mathcal{F}} \subset \bar{\mathcal{K}}_0$ and $\varepsilon > 0$ there exists $v = v^\varepsilon(\xi)$ such that $\|\xi-L_\tau Bv\| \leq \varepsilon/2$. Taking a finite cover of $\bar{\mathcal{F}}$ by $\varepsilon/2$-balls with centers $\{\xi_k: k=1,\ldots,K(\varepsilon)\}$, one has (3.1) with $\mathcal{V}^\varepsilon = \text{hull}\{v^\varepsilon(\xi_k)\}$. □

THEOREM 4: Assume (H_2), (H_4), $(H_3\text{-}(ii))$, and

(*i*) $\mathcal{Z}_0 := \Phi(\mathcal{X}_2)$ is bounded in $\mathcal{Z}$ (e.g., $(H_3\text{-}(i))$ with $\bar{r} = 0$);

(*ii*) $\mathcal{F} := L_\tau \mathcal{Z}_0$ is precompact in X_1 (e.g., (C_5)).

Then (H_1') (i.e., $L_\tau \mathcal{Z} \subset \bar{\mathcal{K}}_0$ -- although $\mathcal{F} \subset \bar{\mathcal{K}}_0$ is sufficient) implies invariance of the almost-reachable set: $\bar{\mathcal{K}}_\varphi = \bar{\mathcal{K}}_0$.

Proof: As before, it is easy to see that $\mathcal{K} \subset \bar{\mathcal{K}}_0$ so $\bar{\mathcal{K}} \subset \bar{\mathcal{K}}_0$ and we need only prove that, given $\xi_* \in \mathcal{K}_0$ (so $\xi_* = L_\tau Bv_*$) and $\varepsilon > 0$, there is a $\xi \in \mathcal{K}_\varphi$ with $\|\xi-\xi_*\| \leq \varepsilon$. If we set

$$\Psi(\hat{v}) = \Psi^\varepsilon_\varphi(v) := \{\tilde{v} \in \mathcal{V}^\varepsilon: \|L_\tau F_\varphi(\hat{v}+v_*)-L_\tau B\tilde{v}\| \leq \varepsilon\}, \qquad (3.2)$$

then a fixpoint $\hat{v} \in \Psi(\hat{v})$ would give $\|\xi-\xi_*\| \leq \varepsilon$, as desired, for $\xi := [S_\varphi(\hat{v}+v_*)](\tau) \in \mathcal{K}_\varphi$. In (3.2) we are of course, taking $\mathcal{V}^\varepsilon$ as in Lemma 3, using (*ii*). Thus, for each $\hat{v} \in \mathcal{V}^\varepsilon$ we know that $\Psi(\hat{v}) \neq \emptyset$ and, as $L_\tau B$ is linear, that $\Phi(\hat{v})$ is convex; the continuity of $L_\tau F_\varphi$ and $L_\tau B$ ensure that each $\Phi(\hat{v})$ is closed and $\Phi(\cdot)$ is (set-valued) upper semicontinuous. Hence, the Kakutani Fixpoint Theorem [5] applies to ensure existence of a fixpoint $\hat{v} \in \Psi(\hat{v})$. □

Remark 2: Suppose, rather than imposing (*i*), (*ii*) for Theorem 4, one wished to impose (H_3) with (C_5). The difficulty now is finding a suitable invariant set in $\mathcal{V}$. For any ball $\mathcal{B}_M := \{v \in \mathcal{V}: |v|_\mathcal{V} \leq M\}$ one would have $LF_\varphi(\mathcal{B}_M) \subset (a + bM^{\bar{r}})\mathfrak{C}$ with $\mathfrak{C}$ compact in X_1. For any $\tilde{\varepsilon} > 0$ one could proceed as in Lemma 3 to obtain a polyhedron

$\mathcal{V}^{\tilde{\varepsilon}}$ with $\|\mathfrak{C}-L_\tau B\mathcal{V}^{\tilde{\varepsilon}}\| \leq \tilde{\varepsilon}$ whence (4.2) gives $\Psi(\hat{v}) \neq \emptyset$ for $\varepsilon := (a+bM^{\bar{r}})\tilde{\varepsilon}$ and $(\hat{v}+v_*) \in \mathfrak{N}_M$; now set $\tilde{M}(\tilde{\varepsilon}) := \max\{\|v\|: v \in \mathcal{V}^{\tilde{\varepsilon}}\}$. We can proceed as in the proof above of Theorem 4 provided $\|v_*\| + \tilde{M}(\tilde{\varepsilon}) \leq M$. This gives the desired result ($\mathfrak{K}_0 \subset \bar{\mathfrak{K}}_\varphi$) provided

$$\tilde{\varepsilon}\tilde{M}(\tilde{\varepsilon})^{\bar{r}} \to 0 \qquad \text{as} \quad \tilde{\varepsilon} \to 0 \tag{3.3}$$

so one can let $\varepsilon \to 0$. It is not easy to estimate $\tilde{M}(\cdot)$, making it difficult to verify (3.3) unless $\bar{r} = 0$. The condition (3.3) seems related to n-width considerations and it is hoped to be able to explore this approach further. □

Remark 3: From the form of (1.1) it is evident that v itself is irrelevant in affecting x but rather $w := Bv$ and the map $\tilde{S}_0$ defined by $[\tilde{S}_0 w](t) := L_t w$. Suppose, rather than admitting all $v \in \mathcal{V}$ one were to restrict attention to $v \in \mathcal{V}_0$; set $\mathfrak{M}_0 := \{Bv: v \in \mathcal{V}_0\}$. From this point, the nature of $\mathcal{V}_0 \subset \mathcal{V}$ is irrelevant, as is the linearity of B; we only require the "segmentation property":

(SP) $w,w' \in \mathfrak{M}_0$ implies $w_* \in \bar{\mathfrak{M}}_0$ where, for arbitrary $s \in (0,\tau)$, we set $w_* := \{w \text{ on } [0,s];\ w' \text{ on } (s,\tau]\}$.

Here we topologize $\mathfrak{M}_0$ in $\mathfrak{M} := L^{\hat{p}}((0,\tau] \to W)$ for some reflexive W and $1 < \hat{p} < \infty$ and take $\bar{\mathfrak{M}}_0$ to be the *weak* closure of $\mathfrak{M}_0$ in $\mathfrak{M}$. By [10], we note that $\bar{\mathfrak{M}}_0$ is convex, given (SP). Now, if L_τ is continuous to X_1 from $\mathfrak{M}_w$ ($\mathfrak{M}$ with its weak topology) then $\mathfrak{K}_0 := L_\tau B(\mathcal{V}_0) = L_\tau\mathfrak{M}_0$ and $\mathfrak{K}_* := L_\tau\bar{\mathfrak{M}}_0$ have the same closure $\bar{\mathfrak{K}}_0$. If, in addition, $\tilde{S}_0$ is continuous from $\mathfrak{M}_w$, then one might as well take $w \in \bar{\mathfrak{M}}_0$ as the control when investigating the almost-reachable set. In particular, if $\bar{\mathfrak{M}}_0$ is a subspace of $\mathfrak{M}$ then most of the previous analysis would apply. □

REFERENCES

[1] Aubin, J.P. 1963. Un théorème de compacité (A compactness theorem). *CRAS de Paris* 265:5042-5043.

[2] Carmichael, N. 1982. Functional analysis and aspects of nonlinear control theory. Ph.D. diss., Univ. Warwick, Coventry, England.

[3] Carmichael, N., and M.D. Quinn. 1985. Fixed point methods in nonlinear control. In *Distributed Parameter Systems*, ed. F. Kappel, K. Kunisch, and W. Schappacher, 24-51. Lecture Notes in Control and Inf. Sci. 75. Berlin Heidelberg New York Tokyo: Springer-Verlag.

[4] Henry, J. 1978. Quelques problèmes de contrôlabilité de systèmes paraboliques (Some controllability problems for parabolic systems). Thèse de Doctorat d'Etat, Univ. Paris VI.

[5] Kakutani, S. 1941. A generalization of Brouwer's fixed point theorem. *Duke Math. J.* 8:457-459.

[6] Michael, E. 1963. Continuous selections, I. *Annals of Math.* 63:361-382.

[7] Naito, K. Controllability of semilinear control systems, I. *SIAM J. Control Optim.* Forthcoming.

[8] Seidman, T.I. Two compactness lemmas. In *Proc. of the First Howard Univ. Sympos. on Nonlinear Semigroups, Evolution Oper.*, ed. W. Gill and W. Zachary. Forthcoming.

[9] Seidman, T.I. Invariance of the reachable set under nonlinear perturbations. Forthcoming.

[10] Seidman, T.I. A theorem on convexity. Forthcoming.

[11] Wong, H.D. 1979. Controllability for nonlinear differential equations in infinite dimensional space. Ph.D. diss., Univ. Minnesota, Minneapolis, Minnesota.

SENSITIVITY ANALYSIS OF OPTIMAL CONTROL PROBLEMS FOR PARABOLIC SYSTEMS

Jan Sokolowski

Systems Research Institute, Warsaw, Poland

Abstract

This paper is concerned with the sensitivity analysis of solutions of optimal control problems for parabolic systems. The method of sensitivity analysis proposed in [24] is exploited throughout. The right-derivative of an optimal control with respect to the parameter is given as the unique solution of an auxiliary optimal control problem. The material derivative method is used for the shape sensitivity analysis of a boundary control problem for a parabolic system.

Key words. differential stability, shape sensitivity analysis, boundary control, optimality system, metric projection

1. INTRODUCTION

This paper is concerned with the differential sensitivity analysis of control constrained optimal control problems for systems described by parabolic partial differential equations. Throughout this paper the method of sensitivity analysis proposed in [22-26] is used. This method is combined with the material derivative method [30,33,34] in order to handle the sensitivity analysis with respect to the perturbations of the domain of integration of partial differential equations under consideration. We refer the reader to [14,15] for related results on the sensitivity analysis of control problems for systems described by ordinary differential equations. Related results on the sensitivity

*) The writing of this paper was completed while the author was visiting the Department of Mathematics, University of Florida, Gainesville, Florida 32611.

analysis of optimization problems are presented in [1-5,5,17,19-21,27]. The differential stability of optimal solutions of optimal control problems for distributed parameter systems is considered in [6,11,12,16,22-26,28]. Related results on the shape sensitivity analysis of unilateral problems are given in [29-32].
We refer the reader to [8,9,11-13,18] for results on the optimal control of distributed parameter systems.
The outline of the paper is as follows.
In section 2 a convex, control constrained, optimal control problem for one dimensional heat equation is considered. The differential stability of an optimal control with respect to the perturbations of a point source location is investigated.
Section 3 is devoted to the shape sensitivity analysis of a quadratic, control constrained, optimal control problem for a system described by the heat equation with Neumann boundary conditions.
The Euler and Lagrange derivatives of an optimal control in the direction of a vector field are derived in the form of optimal solutions of auxiliary optimal control problems for the heat equation.
Throughout the paper standard notation is used [10].

2. Sensitivity Analysis of Optimal Control Problems.

In this section, we present our method to study sensitivity analysis in the case of a simple model problem.
Let us consider a system described by a parabolic initial-boundary value problem of the form:

find an element $y = y(\underset{\sim}{u};x,t)$, $\underset{\sim}{u} \in L^2(0,T;R^n)$,
$(x,t) \in Q = (0,1) \times (0,T)$ such that

$$\left\{\begin{array}{ll} \frac{\partial y}{\partial t}(\underset{\sim}{u};x,t) - \frac{\partial^2 y}{\partial x^2}(\underset{\sim}{u};x,t) = \sum_{i=1}^{n} u_i(t)\delta(x - x_i), \text{ in } Q & (2.1) \\ y(\underset{\sim}{u};0,t) = y(\underset{\sim}{u};1,t) = 0, \text{ in } (0,T) & (2.2) \\ y(\underset{\sim}{u};x,0) = 0, \text{ in } (0,1) & (2.3) \end{array}\right.$$

where $\underset{\sim}{u} = \mathrm{col}(u_1,\ldots,u_n)$ is control, $x_1,\ldots,x_n$, $0 < x_1 < x_2 < \ldots < x_n < 1$ are given points in $(0,1)$. Since $\delta(x - x_i) \in H^{-1}(0,1)$, $i = 1,2,\ldots,n$ then $\sum_{i=1}^{n} u_i(t)\delta(x - x_i) \in L^2(0,T;H^{-1}(\Omega))$ and there exists a unique solution $y(\underset{\sim}{u}) \in W(0,T)$ of system (2.1)-(2.3), for any $\underset{\sim}{u} \in L^2(0,T;R^n)$. Here we denote

$$W(0,T) = \{\phi \in L^2(0,T;H_o^1(0,1)) \,|\, \frac{\partial\phi}{\partial t} \in L^2(0,T;H^{-1}(0,1))\} \tag{2.4}$$

In order to define an optimal control problem, we first introduce a cost functional $J(\underset{\sim}{u})$, $\underset{\sim}{u} \in L^2(0,T;R^n)$ and a set of admissible controls $\mathcal{U}_{ad} \in L^2(0,T;R^n)$ given by

$$\begin{aligned} J(\underset{\sim}{u}) &= \tfrac{1}{2}\int_0^T\int_0^1 [\max\{y(\underset{\sim}{u};x,t) - z_d(x,t),0\}]^2 dx\, dt \\ &+ \frac{\alpha}{2}\sum_{i=1}^{n}\int_0^T [u_i(t)]^2 dt, \quad \alpha > 0, \end{aligned} \tag{2.5}$$

$$\mathcal{U}_{ad} = \{\underset{\sim}{u} \in L^2(0,T;R^n) \,|\, \sum_{i=1}^{n} u_i^2(t) \leq 1, \text{ for a.e. } t \in (0,T)\}, \tag{2.6}$$

respectively.

Let us consider the following optimal control problem

Problem (P):

Find an element $\underset{\sim}{u} \in \mathcal{U}_{ad}$ such that

$$J(\underset{\sim}{u}) \leq J(\underset{\sim}{v}), \quad \forall \underset{\sim}{v} \in \mathcal{U}_{ad} \tag{2.7}$$

It follows by standard argument that an optimal solution of Problem (P) is given by a unique solution of the following system.

Optimality system for problem (P):

Find $(\underset{\sim}{u},y,p) \in \mathcal{U}_{ad} \times W(0,T) \times H^{2,1}(Q)$ such that the following system is verified:

State equation:

$$\begin{cases} \frac{\partial y}{\partial t}(x,t) - \frac{\partial^2 y}{\partial x^2}(x,t) = \sum_{i=1}^{n} u_i(t)\delta(x - x_i), \text{ in } Q & (2.8) \\ y(0,t) = y(1,t) = 0, \text{ in } (0,T) & (2.9) \\ y(x,0) = 0, \text{ in } (0,1) & (2.10) \end{cases}$$

Adjoint-state equation:

$$\begin{cases} -\frac{\partial p}{\partial t}(x,t) - \frac{\partial^2 p}{\partial x^2}(x,t) = \max\{y(x,t) - z_d(x,t),0\}, & (2.11) \\ p(0,t) = p(1,t) = 0, \text{ in } (0,T) & (2.12) \\ p(x,T) = 0, \text{ in } (0,1) & (2.13) \end{cases}$$

Optimality conditions:

$$\sum_{i=1}^{n} \int_0^T (\alpha u_i(t) - p(x_i,t))(v_i(t) - u_i(t))dt \geq 0, \forall \underset{\sim}{v} \in \mathcal{U}_{ad} \tag{2.14}$$

Remark 2.1

Let us observe that condtion (2.14) is equivalent to the following condition

$$\underset{\sim}{u} = P_{\mathcal{U}_{ad}}(\underset{\sim}{f}) \tag{2.15}$$

where $P_{\mathcal{U}_{ad}}$ denotes the metric projection in the space $L^2(0,T;R^n)$ onto the set $\mathcal{U}_{ad}$, $\underset{\sim}{f} = (f_1,\ldots,f_n)$,

$$f_i(t) = \frac{1}{\alpha} p(x_i,t), \; t \in (0,T), \; i = 1,\ldots,n \tag{2.16}$$

■

2.1 Differential Stability of Metric Projection in the Space $L^2(0,T;R^n)$ onto the Set $\mathcal{U}_{ad}$.

We briefly recall a related result [22] on the differential stability of metric projection in the space $L^2(0,T;R^n)$ onto the set

of admissible controls $\mathcal{U}_{ad}$, which we will use in the sequel.
Let there be given an element $\underset{\sim}{f}^{\varepsilon} \in L^2(0,T;R^n)$, $\varepsilon \in [0,\delta]$, such that for $\varepsilon > 0$, ε small enough

$$\underset{\sim}{f}^{\varepsilon} = \underset{\sim}{f}^{o} + \varepsilon \underset{\sim}{f}' + o(\varepsilon), \text{ in } L^2(0,T;R^n) \tag{2.17}$$

where $\|o(\varepsilon)\|_{L^2(0,T;R^n)}/\varepsilon \to 0$ with $\varepsilon \downarrow 0$, $\underset{\sim}{f}' \in L^2(0,T;R^n)$ is a given element. We denote

$$\underset{\sim}{u}^{\varepsilon} = P_{\mathcal{U}_{ad}}(\underset{\sim}{f}^{\varepsilon}) \tag{2.18}$$

<u>Lemma 2.1</u>

For $\varepsilon > 0$, ε small enough

$$\underset{\sim}{u}^{\varepsilon} = \underset{\sim}{u}^{o} + \varepsilon \underset{\sim}{q} + o(\varepsilon), \text{ in } L^2(0,T;R^n), \tag{2.19}$$

where $\|o(\varepsilon)\|_{L^2(0,T;R^n)}/\varepsilon \to 0$ with $\varepsilon \downarrow 0$. An element $\underset{\sim}{q} \in L^2(0,T;R^n)$ is given by a unique solution of the following variational inequality: $\underset{\sim}{q} = (q_1,\dots,q_n) \in S$,

$$\sum_{i=1}^{n} \int_0^T (1 + \lambda_i(t))q_i(t)(v_i(t) - q_i(t))dt \geq \tag{2.20}$$

$$\sum_{i=1}^{n} \int_0^T f_i'(t)(v_i(t) - q_i(t))dt,$$

$$\forall \underset{\sim}{v} = (v_1,\dots,v_n) \in S$$

where $S \subset L^2(0,T;R^n)$ is a cone of the form

$$S = \{\underset{\sim}{\phi} \in L^2(0,T;R^n) \,|\, \phi_i(t) \geq 0, \text{ for a.e. } t \in \Xi_i^o, \quad \phi_i(t) = 0, \text{ for a.e. } t \in \Xi_i^+\} \tag{2.21}$$

Here we denote

$$\Xi = \{t \in (0,T) \mid \sum_{i=1}^{n} [u_i^o(t)]^2 = 1\} \tag{2.22}$$

$$\lambda_i(t) = \begin{cases} 0, & t \in (0,T) \setminus \Xi \\ u_i(t) - f_i(t), & t \in \Xi \end{cases} \tag{2.23}$$

$$\Xi_i^o = \{t \in \Xi \mid \lambda_i(t) = 0\} \tag{2.24}$$

$$\Xi_i^+ = \Xi \setminus \Xi_i^o \tag{2.25}$$

The proof of Lemma 2.1 follows from the results presented in [22].

2.2. Differential Stability of Solutions of Optimal Control Problem.

Let us consider the following parabolic initial - boundary value problem:

$$\frac{\partial \eta_\varepsilon}{\partial t}(\underset{\sim}{u};x,t) - \frac{\partial^2 \eta_\varepsilon}{\partial x^2}(\underset{\sim}{u};x,t) = u_1(t)\delta(x - x_1^\varepsilon) + \sum_{i=2}^{n} u_i(t)\delta(x - x_i), \text{ in } Q, \tag{2.26}$$

where $x_1^\varepsilon = x_1 + \varepsilon$, $\varepsilon \in [0,\delta)$.

We denote

$$F_\varepsilon(x,t) = u_1(t)\delta(x - x_1^\varepsilon) + \sum_{i=2}^{n} u_i(t)\delta(x - x_i)$$

$$F_\varepsilon \in L^2(0,T;H^{-1}(0,1)), \ \varepsilon \in [0,\delta) \tag{2.27}$$

For $\varepsilon > 0$, ε small enough, it follows that

$$F_\varepsilon = F_o + \varepsilon F' + o(\varepsilon), \text{ in } L^2(0,T;H^{-2}(0,1)) \tag{2.28}$$

where $F'(x,t) = u_1(t)\delta'(x - x_1)$.

In view of (2.28), we obtain

$$\eta_\varepsilon = \eta_o + \varepsilon\eta' + o(\varepsilon), \text{ in } L^2(Q) \tag{2.29}$$

where $\|o(\varepsilon)\|_{L^2(Q)}/\varepsilon \to 0$ with $\varepsilon \downarrow 0$. An element η' in (2.29) is given by a unique solution of the following integral identity

$$\eta' \in L^2(Q) : \int_0^T \int_0^1 \eta'\left(-\frac{\partial\phi}{\partial t} - \frac{\partial^2\phi}{\partial x^2}\right) dx\, dt = \int_0^T u_1(t)\frac{\partial\phi}{\partial x}(x_1,t)dt, \tag{2.30}$$

$\forall\phi \in H^{2,1}(Q) \cap L^2(0,T;H_o^1(0,1)),\ \phi(x,T) = 0,$ in $(0,1)$

Let us consider the following optimal control problem:

Problem (P_ε):

Find an element $\underset{\sim}{u}_\varepsilon \in \mathcal{U}_{ad}$ which minimizes the cost functional

$$J_\varepsilon(\underset{\sim}{u}) = \tfrac{1}{2}\int_0^T \int_0^1 [\max\{\eta_\varepsilon(u;x,t) - z_d(x,t),0\}]^2 dx\, dt + \frac{\alpha}{2}\sum_{i=1}^{n} \int_o^T [u_i(t)]^2 dt, \tag{2.31}$$

over the set $\mathcal{U}_{ad}$. ■

An optimal solution $\underset{\sim}{u}_\varepsilon$ of Problem (P_ε) is given by a unique solution of the following optimality system.

Optimality system for problem (P_ε):

Find $(\underset{\sim}{u}^\varepsilon, y^\varepsilon, p^\varepsilon) \in \mathcal{U}_{ad} \times W(0,T) \times H^{2,1}(Q)$ which verify the following system:

State equation:

$$\frac{\partial y^\varepsilon}{\partial t}(x,t) - \frac{\partial^2 y^\varepsilon}{\partial x^2}(x,t) = u_1^\varepsilon(t)\delta(x - x_1^\varepsilon) + \sum_{i=2}^{n} u_i^\varepsilon(t)\delta(x - x_i), \text{ in } Q \tag{2.32}$$

$$y^{\varepsilon}(0,t) = y^{\varepsilon}(1,t) = 0, \text{ in } (0,T) \tag{2.33}$$

$$y^{\varepsilon}(x,0) = 0, \text{ in } (0,1) \tag{2.34}$$

Adjoint-state equation:

$$-\frac{\partial p^{\varepsilon}}{\partial t}(x,t) - \frac{\partial^2 p^{\varepsilon}}{\partial x^2}(x,t) = \max\{y^{\varepsilon}(x,t) - z_d(x,t), 0\}, \tag{2.35}$$

$$p^{\varepsilon}(0,t) = p^{\varepsilon}(1,t) = 0, \text{ in } (0,T) \text{ in } Q \tag{2.36}$$

$$p^{\varepsilon}(x,0) = 0, \text{ in } (0,1) \tag{2.37}$$

Optimality conditions:

$$\underset{\sim}{u}^{\varepsilon} \in \mathcal{U}_{ad} : \sum_{i=1}^{n} \int_0^T (\alpha u_i^{\varepsilon}(T) - p^{\varepsilon}(x_i^{\varepsilon},t))(v_i(t) - u_i^{\varepsilon}(t))dt \geqslant 0, \tag{2.38}$$

$$\forall \underset{\sim}{v} = (v_1, \dots, v_n) \in \mathcal{U}_{ad}$$

In (2.38) we denote $x_1^{\varepsilon} = x_1 + \varepsilon$, $x_i^{\varepsilon} = x_i$, $i = 2,\dots,n$. ■

In what follows we will apply Lemma 2.1 to problem (2.38) i.e., in the particular case of elements $\underset{\sim}{f}^{\varepsilon}$, $\varepsilon \in [0,\delta)$, given by

$$\underset{\sim}{f}^{\varepsilon}(t) = (\frac{1}{\alpha} p^{\varepsilon}(x_1,t), \dots, \frac{1}{\alpha} p^{\varepsilon}(x_n,t)) \tag{2.39}$$

This implies that in the definition of a cone $S \subset L^2(0,T;R^n)$ defined by (2.21)-(2.25), we will have (2.23) in the form

$$\lambda_i(t) = \begin{cases} 0, & t \in (0,T) \setminus \Xi \\ u_i^o(t) - \frac{1}{\alpha} p^o(x_i,t), & t \in \Xi \end{cases} \tag{2.40}$$

Theorem 2.1

For $\varepsilon > 0$, ε small enough

$$\underset{\sim}{u}^{\varepsilon} = \underset{\sim}{u}^{o} + \varepsilon \underset{\sim}{q} + o(\varepsilon), \text{ in } L^2(0,T;R^n) \tag{2.41}$$

where $\|o(\varepsilon)\|_{L^2(0,T;R^n)}/\varepsilon \to 0$ with $\varepsilon \downarrow 0$.

An element $\underset{\sim}{q}$ in (2.41), is given by a unique solution of the following optimality system:

<u>Optimality system for problem (P'):</u>

Find $(\underset{\sim}{q},z,w) \in S \times L^2(Q) \times H^{2,1}(Q)$ such that the following system is verified:

<u>State equation:</u>

$$\begin{cases} \dfrac{\partial z}{\partial t}(x,t) - \dfrac{\partial^2 z}{\partial x^2}(x,t) = \sum\limits_{i=1}^{n} q_i(t)\delta(x - x_i) + u_1^o(t)\delta'(x - x_1), & (2.42) \\ \qquad\qquad\qquad\qquad\qquad\qquad \text{in } \mathcal{D}'(Q) & \\ z(0,t) = z(1,t) = 0, \text{ in } (0,T) & (2.43) \\ z(x,0) = 0, \text{ in } (0,1) & (2.44) \end{cases}$$

<u>Adjoint-state equation:</u>

$$\begin{cases} -\dfrac{\partial w}{\partial t}(x,t) - \dfrac{\partial^2 w}{\partial x^2}(x,t) = z(x,t)\chi_1(x,t) + & \\ \qquad + \max\{z(x,t),0\}\chi_2(x,t), \text{ in } Q & (2.45) \\ w(0,t) = w(1,t) = 0, \text{ in } (0,T) & (2.46) \\ w(x,T) = 0, \text{ in } (0,1) & (2.47) \end{cases}$$

<u>Optimality conditions:</u>

$\underset{\sim}{q} \in S$:

$$\sum_{i=1}^{n} \int_0^T (1 + \lambda_i(t))q_i(t)(\phi_i(t) - q_i(t))dt \geq \sum_{i=1}^{n} \int_0^T w(x_i,t)(\phi_i(t) - q_i(t))dt, \qquad (2.48)$$

$$\forall \underset{\sim}{\phi} = (\phi_1,\ldots,\phi_n) \in S$$

Here we denote

$$\chi_1(x,t) = \begin{cases} 1, & y^o(x,t) > z_d(x,t) \\ 0, & y^o(x,t) \leq z_d(x,t) \end{cases} \tag{2.49}$$

$$\chi_2(x,t) = \begin{cases} 1, & y^o(x,t) = z_d(x,t) \\ 0, & y^o(x,t) \neq z_d(x,t) \end{cases} \tag{2.50}$$

Proof:

It can be verified, using an abstract result presented in [22] (Proposition 3, p. 106) that for $\varepsilon > 0$, ε small enough

$$\|\underset{\sim}{u}^\varepsilon - \underset{\sim}{u}^o\|_{L^2(0,T;R^n)} \leq C\varepsilon \tag{2.51}$$

hence there exists an element $\underset{\sim}{v} \in L^2(0,T;R^n)$, such that

$$\underset{\sim}{u}^\varepsilon = \underset{\sim}{u}^o + \varepsilon\underset{\sim}{v} + r(\varepsilon), \text{ in } L^2(0,T;R^n) \tag{2.52}$$

where $r(\varepsilon)/\varepsilon \to 0$ weakly in $L^2(0,T;R^n)$ with $\varepsilon \downarrow 0$.

From (2.52), in view of (2.29), (2.32)-(2.34), it follows that

$$y^\varepsilon = y^o + \varepsilon z + o(\varepsilon), \text{ in } L^2(Q) \tag{2.53}$$

where z verifies the following equation

$$\frac{\partial z}{\partial t}(x,t) - \frac{\partial^2 z}{\partial x^2}(x,t) = \sum_{i=1}^{n} v_i(t)\delta(x - x_i) + u_i^o(t)\delta'(x - x_i), \quad \text{in } \mathcal{D}'(Q) \tag{2.54}$$

$$z(0,t) = z(1,t) = 0, \text{ in } (0,T) \tag{2.55}$$

$$z(x,0) = 0, \text{ in } (0,1) \tag{2.56}$$

Furthermore by (2.53), in view of (2.35)-(2.37), it follows that

$$p^\varepsilon = p^o + \varepsilon w + o(\varepsilon), \text{ in } H^{2,1}(Q) \tag{2.57}$$

where $w \in H^{2,1}(Q)$ verifies the system (2.43)-(2.45). From (2.57) we have

$$p^{\varepsilon}(x_i,\cdot) = p^{o}(x_i,\cdot) + \varepsilon w(x_i,\cdot) + o(\varepsilon), \tag{2.58}$$

$$\text{in } L^2(0,T)$$

So that we can apply Lemma 2.1 to the variational inequality (2.46) and obtain

$$\underset{\sim}{u}^{\varepsilon} = \underset{\sim}{u}^{o} + \varepsilon \underset{\sim}{q} + o(\varepsilon), \text{ in } L^2(0,T;R^n), \tag{2.59}$$

where $\|o(\varepsilon)\|_{L^2(0,T;R^n)}/\varepsilon \to 0$ with $\varepsilon \downarrow 0$. An element $\underset{\sim}{q}$ is given by a unique solution of the variatonal inequality (2.20).

From (2.59), in view of (2.52), it follows that we have actually the equality

$$\underset{\sim}{q} = \underset{\sim}{v} \tag{2.60}$$

This completes the proof ■

3. Shape Sensitivity Analysis.

This section is concerned with shape sensitivity analysis of solutions of boundary optimal control problems for systems described by parabolic equations.

We first define a family of domains $\{\Omega_\varepsilon\} \subset R^n$, $\varepsilon \in [0,\delta)$, depending on a vector field $V \in C(0,\delta;C^1(R^n;R^n))$.

3.1. Family of domains $\{\Omega_\varepsilon\}$.

In order to derive the form of the so-called Euler and Lagrange derivatives of an optimal control in the direction of a vector field

$$V(\cdot,\cdot) \in C(0,\delta;C^1(R^n;R^n)) \tag{3.1}$$

we define a family of domains $\{\Omega_\varepsilon\} \subset R^n$, $\varepsilon \in [0,\delta)$ as follows [33,34].

Let $\Omega \subset R^n$ be a domain with smooth boundary $\Gamma = \partial\Omega$. Let us denote by

$$T_\varepsilon : R^n \to R^n,\ \varepsilon \in [0,\delta) \qquad (3.2)$$

a mapping of the form

$$T_\varepsilon(X) = x(\varepsilon),\ X \in R^n,\ \varepsilon \in [0,\delta) \qquad (3.3)$$

where

$$\begin{cases} \dfrac{dx}{ds}(s) = V(s,x(s)),\ s \in (0,\delta) \\ x(0) = X \end{cases} \qquad (3.4)$$

We denote

$$\Omega_\varepsilon = T_\varepsilon(V)(\Omega) = \{x \in R^n \mid \exists\ X \in \Omega \text{ such that } x(0) = X,\ x(\varepsilon) = x\} \qquad (3.5)$$

We will denote by $DT_\varepsilon(x)$ the Jacobian matrix of mapping (3.2) evaluated at a point $x \in R^n$, $DT_\varepsilon^{-1}(x)$ is the inverse of matrix $DT_\varepsilon(x)$ and $*DT_\varepsilon^{-1}(x)$ is the transpose of matrix $DT_\varepsilon^{-1}(x)$.
Furthermore, we denote

$$\gamma_\varepsilon(x) = \det(DT_\varepsilon(x)),\ x \in \overline{\Omega} \qquad (3.6)$$

$$A_\varepsilon(x) = \gamma_\varepsilon(x) DT_\varepsilon^{-1}(x) . *DT_\varepsilon^{-1}(x),\ x \in \overline{\Omega} \qquad (3.7)$$

$$\sigma_\varepsilon(x) = \| \gamma_\varepsilon(x) *DT_\varepsilon^{-1}(x).\ \underset{\sim}{n}(x) \|_{R^n},\ x \in \partial\Omega \qquad (3.8)$$

3.2. Metric projection in $L^2(\Sigma)$ onto K.

We denote $\Sigma = \partial\Omega \times (0,T)$, $T > 0$ a given constant. Let $K \subset L^2(\Sigma)$ be a set of admissible controls of the form

$$K = \{u \in L^2(\Sigma) \mid 0 < u(x,t) < M, \text{ for a.e. } (x,t) \in \Sigma\} \qquad (3.9)$$

Let us consider the following variational inequality

$$\begin{cases} u_\varepsilon \in K \\ \int_\Sigma (a_\varepsilon u_\varepsilon - f_\varepsilon)(\phi - u_\varepsilon)d\Sigma \geq 0, \ \forall\phi \in K \end{cases} \tag{3.10}$$

where $a_\varepsilon \in L^\infty(\Sigma)$, $f_\varepsilon \in L^2(\Sigma)$, $\varepsilon \in [0,\delta)$ such that for $\varepsilon > 0$, ε small enough

$$a_\varepsilon = a_o + \varepsilon a' + o(\varepsilon), \text{ in } L^\infty(\Sigma) \tag{3.11}$$

$$f_\varepsilon = f_o + \varepsilon f' + o(\varepsilon), \text{ in } L^2(\Sigma) \tag{3.12}$$

here $a' \in L^\infty(\Sigma)$, $f' \in L^2(\Sigma)$ are given elements.

Lemma 3.1

For $\varepsilon > 0$, ε small enough

$$u_\varepsilon = u_o + \varepsilon q + o(\varepsilon), \text{ in } L^2(\Sigma) \tag{3.13}$$

where $\|o(\varepsilon)\|_{L^2(\Sigma)}/\varepsilon \to 0$ with $\varepsilon \downarrow 0$. An element q of (3.13) is given by unique solution of the variational inequality

$$\begin{cases} q \in S(\Sigma) \\ \int_\Sigma (a_o q + a' u_o - f')(\phi - q)d\Sigma \geq 0, \\ \qquad\qquad \forall\phi \in S(\Sigma) \end{cases} \tag{3.14}$$

Here $S(\Sigma) \subset L^2(\Sigma)$ is a closed, convex cone of the form:

$$S(\Sigma) = \{\phi \in L^2(\Sigma) \mid \phi(x,t) \geq 0, \text{ for a.e. } (x,t) \in \Xi_1,$$

$$\phi(x,t) \leq 0, \text{ for a.e. } (x,t) \in \Xi_2, \tag{3.15}$$

$$\int_\Sigma (u_o(x,t) - f_o(x,t))\phi(x,t)d\Sigma = 0\}$$

where

$$\Xi_1 = \{(x,t) \in \Sigma | u_o(x,t) = 0\} \tag{3.16}$$

$$\Xi_2 = \{(x,t) \in \Sigma | u_o(x,t) = M\} \tag{3.17}$$

The proof of Lemma 3.1 is given e.g. in [23].

3.3 Optimal Control Problem (P_ε).

We define an optimal control problem (P_ε) in a cylinder $Q_\varepsilon = \Omega_\varepsilon \times (0,T)$, $\varepsilon \in [0,\delta)$, where the domain $\Omega_\varepsilon \subset R^n$ is defined by (3.5). We first introduce a state equation and a cost functional of the following form

State equation:

Find an element $y = y(u;x,t)$, $u \in L^2(\Sigma_\varepsilon)$, $(x,t) \in Q_\varepsilon$, such that

$$\frac{\partial y}{\partial t} - \Delta y = 0, \text{ in } Q_\varepsilon \tag{3.18}$$

$$\frac{\partial y}{\partial n_\varepsilon} = u, \text{ on } \Sigma_\varepsilon, \tag{3.19}$$

$$y(u;x,0) = 0, \text{ in } \Omega_\varepsilon \tag{3.20}$$

here $\underset{\sim}{n}_\varepsilon$, $\varepsilon \in [0,\delta)$, is a unit, outward, normal vector on $\partial\Omega_\varepsilon$.

Cost functional:

$$J_\varepsilon(u) = \tfrac{1}{2}\int_{\Omega_\varepsilon} [y(u;x,T) - z_d(x)]^2 dx + \frac{\alpha}{2}\int_{\Sigma_\varepsilon} (u(x,t))^2 d\Sigma, \ \alpha > 0 \tag{3.21}$$

here $z_d \in H^1(R^n)$ is a given element.

We assume that a set $K(\Sigma_\varepsilon)$ of admissible controls is defined by

$$K(\Sigma_\varepsilon) = \{u \in L^2(\Sigma_\varepsilon) | 0 \le u(x,t) \le M, \text{ for a.e. } (x,t) \in \Sigma_\varepsilon\} \tag{3.22}$$

Let us consider the following optimal control problem:

<u>Problem (P_ε):</u>

Find an element $u_\varepsilon \in K(\Sigma_\varepsilon)$ such that

$$J_\varepsilon(u_\varepsilon) \leq J_\varepsilon(u), \ \forall u \in K(\Sigma_\varepsilon) \tag{3.23}$$

■

It can be verified that an optimal solution of problem (P_ε) is given by the unique solution of the following optimality system.

<u>Optimality system for problem (P_ε).</u>

Find $(u_\varepsilon, y_\varepsilon, p_\varepsilon) \in K(\Sigma_\varepsilon) \times W_\varepsilon(0,T) \times W_\varepsilon(0,T)$ which verify the following system:

<u>State equation:</u>

$$\begin{cases} \dfrac{\partial y_\varepsilon}{\partial t} - \Delta y_\varepsilon = 0, \text{ in } Q_\varepsilon & (3.24) \\[2ex] \dfrac{\partial y_\varepsilon}{\partial n_\varepsilon} = u_\varepsilon, \text{ on } \Sigma_\varepsilon & (3.25) \\[2ex] y_\varepsilon(x,0) = 0, \text{ in } \Omega_\varepsilon & (3.26) \end{cases}$$

<u>Adjoint-state equation:</u>

$$\begin{cases} -\dfrac{\partial p_\varepsilon}{\partial t} - \Delta p_\varepsilon = 0, \text{ in } Q_\varepsilon & (3.27) \\[2ex] \dfrac{\partial p_\varepsilon}{\partial n_\varepsilon} = 0, \text{ on } \Sigma_\varepsilon & (3.28) \\[2ex] p_\varepsilon(x,T) = y_\varepsilon(x,T) - z_d(x), \text{ in } \Omega_\varepsilon & (3.29) \end{cases}$$

Optimality conditions:

$$u_\varepsilon \in K(\Sigma_\varepsilon) : \int_{\Sigma_\varepsilon} (\alpha u_\varepsilon - p_\varepsilon)(\phi - u_\varepsilon) d\Sigma \geq 0, \qquad \forall \phi \in K(\Sigma_\varepsilon) \tag{3.30}$$

Here we set

$$W_\varepsilon(0,T) = \{\phi \in L^2(0,T;H^1(\Omega_\varepsilon) | \frac{\partial\phi}{\partial t} \in L^2(0,T;(H^1(\Omega_\varepsilon))'))\} \tag{3.31}$$

Moreover, we denote

$$u^\varepsilon \overset{\text{def}}{=} u_\varepsilon \circ T_\varepsilon \in L^2(\Sigma), \ \forall \varepsilon \in [0,\delta) \tag{3.32}$$

Let us observe, that

$$u^\varepsilon \in K(\Sigma), \ \forall \varepsilon \in [0,\delta) \tag{3.33}$$

Furthermore, it can be verified that an element u^ε defined by (3.32) is given by the unique solution of the following optimal control problem.

Problem (P^ε):

Find an element $u^\varepsilon \in K(\Sigma)$ which minimizes the cost functional

$$J^\varepsilon(u) = 1/2 \int_\Omega [\eta^\varepsilon(u;x,T) - z_d^\varepsilon(x)]\gamma_\varepsilon(x)dx + \frac{\alpha}{2} \int_\Sigma (u(x,t))^2 \sigma_\varepsilon(x) d\Sigma, \tag{3.34}$$

over the set $K(\Sigma)$. ■

Here an element $\eta^\varepsilon = \eta^\varepsilon(u)$ is given by a unique solution of the following state equation

$$\begin{cases} \gamma_\varepsilon(x)\dfrac{\partial\eta^\varepsilon}{\partial t}(u;x,t) - \mathrm{div}(A_\varepsilon(x).\nabla\eta^\varepsilon(u;x,t)) = 0, & (3.35) \\ \qquad\qquad\qquad\qquad\qquad\qquad\qquad\qquad \text{in } Q & \\ \langle A_\varepsilon(x).\nabla\eta^\varepsilon(u;x,t), \underset{\sim}{n}(x)\rangle_{R^n} = \sigma_\varepsilon(x)u(x,t), \text{ on } \Sigma & (3.36) \\ \eta^\varepsilon(u;x,0) = 0, \text{ in } \Omega & (3.37) \end{cases}$$

Theorem 3.1 [28]

For $\varepsilon > 0$, ε small enough

$$u^\varepsilon = u^o + \varepsilon\dot{u}(\Sigma) + o(\varepsilon), \text{ in } L^2(\Sigma) \tag{3.38}$$

where $\|o(\varepsilon)\|_{L^2(\Sigma)}/\varepsilon \to 0$ with $\varepsilon \downarrow 0$.

The Euler derivative $\dot{u} = \dot{u}(\Sigma) \in L^2(\Sigma)$ of an optimal control u^o in the direction of a vector field $V(\cdot,\cdot)$ is given by a unique solution of the following optimality system.

Optimality system for problem ($\dot{P}$):

Find $(\dot{u},z,w) \in S(\Sigma) \times W(0,T) \times W(0,T)$ such that the following system is verified

State equation:

$$\int_\Omega\{\frac{\partial z}{\partial t}\phi + \nabla z.\nabla\phi\}dx = \int_{\partial\Omega} \dot{u}\phi \, d\Gamma - \int_{\partial\Omega} v_\tau \frac{\partial u_o}{\partial\tau}\phi \, d\Gamma$$
$$+ \int_{\partial\Omega} v_n\{(Hu_o - \frac{\partial y_o}{\partial t})\phi - \frac{\partial y_o}{\partial\tau}\frac{\partial\phi}{\partial\tau}\}d\Gamma, \tag{3.39}$$

$$\text{for a.e. } t \in (0,T), \ \forall\phi \in H^2(\Omega) \tag{3.40}$$

$z(x,0) = 0$, in Ω

Adjoint-state equation:

$$\int_\Omega \{-\frac{\partial w}{\partial t}\phi + \nabla w.\nabla\phi\}dx = \int_{\partial\Omega} v_n\{\frac{\partial p_o}{\partial t}\phi - \frac{\partial p_o}{\partial \tau}\frac{\partial \phi}{\partial \tau}\}d\Gamma, \tag{3.41}$$

for a.e. $t \in (0,T$, $\forall\phi \in H^2(\Omega)$

$$w(x,T) = z(x,T), \text{ in } \Omega \tag{3.42}$$

Optimality conditions:

$$\dot{u} \in S(\Sigma) : \int_\Sigma (\alpha\dot{u} - w)(\phi - \dot{u})d\Sigma \geq 0, \ \forall\phi \in S(\Sigma) \tag{3.43}$$

here we denote $\frac{\partial\phi}{\partial\tau} = \langle\nabla_\Gamma\phi,\underset{\sim}{\tau}\rangle_{R^n}$, $\forall\phi \in H^1(\partial\Omega)$, $\underset{\sim}{\tau}$ is a unit tangent vector on $\partial\Omega$, $v_n(x) = \langle V(0,x),\underset{\sim}{n}(x)\rangle_{R^n}$, $v_\tau(x) = \langle V(0,x),\underset{\sim}{\tau}(x)\rangle_{R^n}$, $x \in \partial\Omega$. Cone $S(\Sigma)$ is defined by (3.15) with $f_o(x,t) = p_o(x,t)$, $(x,t) \in \Sigma$.

The proof of Theorem 3.1 is given in [28]. Briefly it uses Lemma 3.1 in order to differentiate with respect to the parameter ε, at $\varepsilon = 0^+$, the optimality conditions in the optimality system derived for problem (P^ε). We refer the reader to [28] for the details. Let $\tilde{u}_\varepsilon$ denotes an extension of an optimal control $u_\varepsilon \in L^2(\Sigma_\varepsilon)$ to an open neighborhood of $\partial\Omega_\varepsilon$ in R^n such that $\tilde{u}_\varepsilon|_\Sigma \in L^2(\Sigma)$.

Theorem 3.2.

For $\varepsilon > 0$, ε small enough

$$\tilde{u}_\varepsilon|_\Sigma = u_o + \varepsilon u' + o(\varepsilon), \text{ in } L^2(\Sigma) \tag{3.44}$$

where $\|o(\varepsilon)\|_{L^2(\Sigma)}/\varepsilon \to 0$ with $\varepsilon \downarrow 0$.

The Lagrange derivative $u' \in L^2(\Sigma)$ of an optimal control $u_o \in L^2(\Sigma)$ in the direction of a vector field $V(\cdot,\cdot)$ is given by a unique solution of the following optimality system

Optimality system for problem (P'):

Find $(u',z,w) \in S(\Sigma) \times W(0,T) \times W(0,T)$ such that the following system is verified

State equation:

$$\int_\Omega \{\frac{\partial z}{\partial t}\phi + \nabla z.\nabla\phi\}dx = \int_{\partial\Omega} u'\phi \, d\Gamma + \\ + \int_{\partial\Omega} v_n\{(Hu_o - \frac{\partial y_o}{\partial t})\phi - \frac{\partial y_o}{\partial \tau}\frac{\partial \phi}{\partial \tau}\}d\Gamma, \tag{3.45}$$

for a.e $t \in (0,T$, $\forall\phi \in H^2(\Omega)$

$$z(x,0) = 0, \text{ in } \Omega \text{ (3.46)}$$

Adjoint-state equation:

$$\int_\Omega \{-\frac{\partial w}{\partial t}\phi + \nabla w.\nabla\phi\}dx = \int_{\partial\Omega} v_n\{\frac{\partial p_o}{\partial t}\phi - \frac{\partial p_o}{\partial \tau}\frac{\partial \phi}{\partial \tau}\}d\Gamma, \tag{3.47}$$

for a.e. $t \in (0,T)$, $\forall\phi \in H^2(\Omega)$

$$w(x,T) = z(x,T), \text{ in } \Omega \tag{3.48}$$

Optimality conditions:

$$u' \in S(\Sigma) : \int_\Sigma (\alpha u' - w)(\phi - u')d\Sigma \geq 0, \tag{3.49}$$

$$\forall\phi \in S(\Sigma)$$

Proof of Theorem 3.2 is given in [28]. ■

From (3.45)-(3.49) it follows that the Lagrange derivative $u' \in L^2(\Sigma)$ is given by a unique solution of the following optimal control problem.

Problem (P'):

Find an element $u' \in L^2(\Sigma)$ which minimizes the cost functional

$$J(u) = \tfrac{1}{2}\int_\Omega [z(u;x,T)]^2 dx + \frac{\alpha}{2}\int_\Sigma [u(x,t)]^2 d\Sigma + \int_\Sigma v_n(x)\{\frac{\partial p_o}{\partial t}(x,t)z(u;x,t) - \frac{\partial p_o}{\partial \tau}(x,t)\frac{\partial z}{\partial \tau}(u;x,t)\}d\Sigma, \tag{3.50}$$

over the set $S(\Sigma)$ of admissible controls. ■

Here we denote by $z = z(u;x,t)$, $u \in L^2(\Sigma)$, $(x,t) \in Q$, a unique solution of the following state equation

$$\begin{cases} \int_\Omega \{\frac{\partial z}{\partial t}\phi + \nabla z.\nabla\phi\}dx = \int_{\partial\Omega} u\phi\, d\Gamma + \int_{\partial\Omega} v_n\{(Hu_o - \frac{\partial y_o}{\partial t})\phi - \frac{\partial y_o}{\partial \tau}\frac{\partial \phi}{\partial \tau}\}d\Gamma, & (3.52) \\ \text{for a.e. } t \in (0,T),\ \forall\phi \in H^2(\Omega) & \\ z(x,0) = 0, \text{ in } \Omega & (3.53) \end{cases}$$

Remark 3.1

We refer the reader to [28] for the related results on shape sensitivity analysis of boundary optimal control problems for systems described by parabolic equations with Dirichlet boundary conditions.

References

[1] Bendsøe M.P., Olhoff N., Sokolowski J. (1985) Sensitivity analysis of problems of elasticity with unilateral constraints. J. Struct. Mech. 13(2), 201-222.

[2] Fiacco A.V. (1983) Introduction to sensitivity and stability analysis in nonlinear programming, Academic Press, New York.

[3] Fitzpatrick S., Phelps R.R. (1982) Differentiability of the metric projection in Hilbert space. Trans. Amer. Math. Soc. 270: 483-501.

[4] Haraux A. (1977) How to differentiate the projection on a convex set in Hilbert space. Some applications to variational inequalities. J. Math. Soc. Japan, 29(4) 615-631.

[5] Holmes R.B. (1973) Smoothness of certain metric projections on Hilbert space. Trans. Amer. Math. Soc. 184: 87-100.

[6] Holnicki P., Sokolowski J., Zochowski A. (1985) Sensitivity analysis of an optimal control problem arising from air quality control in urban area. In: System Modelling and Optimization, Proceeding of the 12th IFIP conference, Budapest, Hungary, Springer Verlag (in press).

[7] Jittorntrum K. (1984) Solution point differentiability without strict complementarity in nonlinear programming. In: A.V. Fiacco (ed.) Mathematical Programming Studies 21. North-Holland, Amsterdam 127-138.

[8] Lasiecka I., Triggiani R. (1984) Dirichlet boundary control problems for parabolic equations with quadratic cost: analyticity and Riccati's feedback synthesis. SIAM J. on Control and Optimization 21(1) 41-67.

[9] Lions J.L. (1968) Contrôle optimal de systemes gouvernes par des equations aux derivées partielles. Dunod, Paris.

[10] Lions J.L., Magenes E. (1968) Problémes aux limites non homogénes et applications. Vol. 1, Dunod, Paris.

[11] Lions J.L. (1973) Perturbations singulieres dans les problemes aux limites et en contrôle optimal. Lectures Notes in Mathematics, Vol. 323, Springer Verlag, Berlin.

[12] Lions J.L. (1981) Some methods in the mathematical analysis of systems and their control. Gordon and Breach, New York.

[13] Lions J.L. (1983) Contrôle des systems distribues singuliers. Dunod, Paris.

[14] Malanowski K. (1984) Differential stability of solutions to convex, control constrained optimal control problems. Applied Mathematics and Optimization, 12: 1-14.

[15] Malanowski K. (1984) On differentiability with respect to parameter of solutions to convex optimal control problems subject to state space constraints. Applied Mathematics and Optimization 12: 231-245.

[16] Malanowski K. Sokolowski J. (1985) Sensitivity of solutions to convex, control constrained optimal control problems for distributed parameter systems. Journal of Mathematical Analysis and Applications (to appear).

[17] Mignot F. (1976) Contrôle dans les inequations variationelles elliptiques. J. Functional Analysis, 22, 130-185.

[18] Simon J. (1983) Characterization d'un espace fonctionnel intervenant en contrôle optimal. Annales Faculté des Sciences Toulouse, 5: 149-169.

[19] Sokolowski J. (1981) Sensitivity analysis for a class of variational inequalities. In: Haug E.J., Cea J. (eds), Optimization of distributed parameter structures, vol. 2, Sijthoff & Noordhoff, Alphen aan den Rijn, The Netherlands, 1600-1609.

[20] Sokolowski J. (1983) Optimal control in coefficients of boundary value problems with unilateral constraints. Bulletin of the Polish Academy of Sciences, Technical Sciences, vol. 31 (1-12): 71-81.

[21] Sokolowski J. (1984) Sensitivity analysis of Signorini variational inequality. In: Bojarski B. (ed) Banach Center Publications, Polish Scientific Publisher, Warsaw. (to appear).

[22] Sokolowski J. (1985) Differential stability of solutions to constrained optimization problems. Appl. Math. Optim. 13: 97-115.

[23] Sokolowski J. (1985) Differential stability of control constrained optimal control problems for distributed parameter system. In: F. Kappel, K. Kunisch, W. Schappacher (eds) Distributed Parameter Systems, Proceedings of the 2nd International Conference, Vorau, Austria 1984, Lectures Notes in Control and Information Sciences, Vol. 75, Springer Verlag, 382-399.

[24] Sokolowski J. (1985) Sensitivity analysis and parametric optimization of optimal control problems for distributed parameter systems. Zeszyty Naukowe Politechniki Warszawskiej, seria Elektronika No. 73, 152 pages (in Polish).

[25] Sokolowski J. (1985) Differential stability of solutions to boundary optimal control problems for parabolic systems. In: System Modelling and Optimization, Proceedings of the 12th IFIP conference, Budapest, Hungary, Springer Verlag (in press).

[26] Sokolowski J. Sensitivity analysis of control constrained optimal control problems for distributed parameter systems (to be published).

[27] Sokolowski J. Sensitivity analysis of contact problems with friction (to appear)

[28] Sokolowski J. Sensitivity analysis of boundary optimal control problems for parabolic systems (to appear).

[29] Sokolowski J., Zolesio J.P. (1985), Dérivéé par rapport au domaine de la solution d'un problème unilatéral. C. R. Acad. Sc. Paris, t. 301, Série I, N°4: 103-106.

[30] Sokolowski J., Zolesio J.P. (1985) Shape sensitivity analysis of unilateral problems. Publication Mathematiques No. 67 Université de Nice.

[31] Sokolowski J., Zolesio J.P. (1985) Shape sensitivity analysis of an elastic-plastic torsion problem. Bulletin of the Polish Academy of Sciences, Technical Sciences.

[32] Zolesio J.P. (1984) Shape controlability for free boundaries. In: Thoft-Christensen P. (ed) System modelling and optimization, LNCIS vol. 59, Springer Verlag, 354-361.

[33] Zolesio J.P. (1979) Identification de domaines par deformations. Thèse d'Etat, Université de Nice.

[34] Zolesio J.P. (1981) The material derivative (or speed) method for shape optimization. In: Haug, E.J., Cea J., (eds), Optimization of distributed parameter structures, vol. 2, Sijthoff & Noordhoff, Alphen aan den Rijn, The Netherlands.

SOME NONLINEAR PROBLEMS IN THE CONTROL OF DISTRIBUTED SYSTEMS

P.K.C. Wang

University of California, Los Angeles, USA

ABSTRACT: In this paper, the control problems associated with a moving flexible robot arm, and with the attitude of a space station with flexible structural components are discussed. The robot arm is modelled by a partial differential equation with a time-dependent spatial domain. It is shown that the arm motion may stabilize or destabilize the system, and its feedback control gives rise naturally to a distributed system with a free boundary. The space station is modelled by a partial differential equation coupled with a nonlinear integrodifferential equation. A nonlinear control law based on a rigid-body model is derived. Its effectiveness in the presence of elastic deformations is discussed. The paper concludes with a brief discussion of certain control problems for abstract evolution equations with time-dependent spatial domains.

1. INTRODUCTION

Recent interests in the control of high-speed robots with light flexible arms and large space structures with elastic components give rise to new classes of nonlinear problems in the control of distributed systems. In this paper, we consider two specific problems motivated from realistic physical situations. Discussion is focused on their salient features and possible methods of approach rather than on mathematical technicalities. The paper concludes with a discussion of a class of abstract control problems motivated from these specific problems.

2. CONTROL OF A MOVING FLEXIBLE ROBOT ARM

Figure 1 shows a robot with a prismatic joint and a long flexible arm which can undergo vertical translation, rotation about the z-axis, and horizontal extension or contraction. For simplicity, we only consider the bending motion in the (x,z)-plane associated with the horizontal translational motion which is modelled by a moving slender prismatic beam described by

$$\rho w_{tt} + \nu w_t + (EIw_{xx})_{xx} + \rho\ddot{\ell}(t)w_x - \rho(\ell(t) - x)\ddot{\ell}(t)w_{xx} = 0, \qquad (1)$$

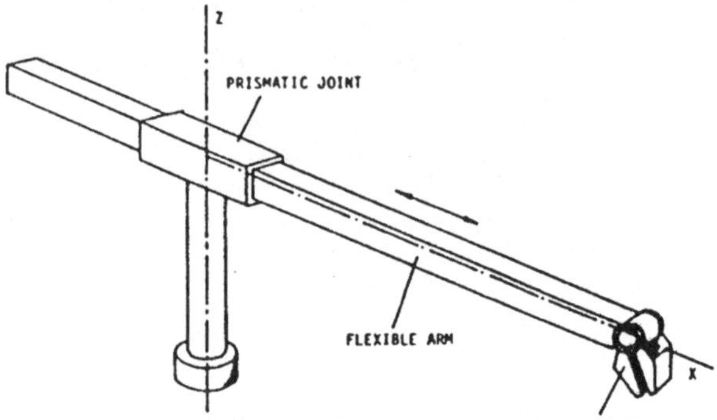

Fig.1 A robot with a long flexible arm.

defined on the time-dependent spatial domain $\Omega(t) =]0, \ell(t)[$, $t > 0$, where the instantaneous arm length $\ell = \ell(t)$ is a specified real positive C_2 function of t. The lettered subsrcipts denote partial differentiation. The coefficients $EI = EI(x)$ and $\rho = \rho(x)$ are specified real positive continuous functions of x representing the flexural rigidity and mass density respectively, and $\nu = \nu(x)$ is a given continuous function corresponding to the damping coefficient. The last two terms in the left-hand-side of (1) represent the effect of the axial force induced by the acceleration of the total mass of the beam section ahead of the point x $\Omega(t) = [0, \ell(t)]$.

The boundary conditions at the fixed and free ends are given by

$$w(t,0) = 0, \quad w_x(t,0) = 0, \quad (EIw_{xx})\big|_{x=\ell(t)} = 0, \quad (EIw_{xx})_x\big|_{x=\ell(t)} = 0, \qquad (2)$$

The initial conditions at t=0 are specified by

$$w(0,x) = w_o(x), \quad w_t(0,x) = \dot{w}_o(x), \quad x \in \Omega(0), \qquad (3)$$

where $\Omega(0) = [0, \ell(0)]$ and $\ell(0) > 0$.

A possible approach to the foregoing initial boundary-value problem is to introduce a continuous invertible time-dependent transformation $y = x/\ell(t)$ which maps the spatial domain $\Omega(t)$ onto the unit interval $I_s = [0,1]$ for every $t \geq 0$. Let $\tilde{w}(t,y) = w(t, \ell(t)y)$. Then the equation for $\tilde{w}$ is given by

$$\rho(\ell(t)y)\tilde{w}_{tt} + \nu(\ell(t)y)\tilde{w}_t = \{\rho(\ell(t)y)[\ddot{\ell}(t)\ell(t)^{-1} + y(\ddot{\ell}(t)\ell(t)^{-1}$$
$$-2\dot{\ell}(t)^2\ell(t)^{-2})] + \nu(\ell(t)y)\dot{\ell}(t)\ell(t)^{-1}y\}\tilde{w}_y$$
$$+ 2\rho(\ell(t)y)y\dot{\ell}(t)\ell(t)^{-1}\tilde{w}_{yt} - \rho(\ell(t)y)\{ y\dot{\ell}(t)\ell(t)^{-1}$$
$$+ (1-y)\ddot{\ell}(t)\ell(t)^{-1}\}\tilde{w}_{yy} - \ell(t)^{-4}[EI(\ell(t)y)\tilde{w}_{yy}]_{yy}, \tag{4}$$

defined for $t>0$ and $y \in I_s =]0,1[$, with boundary conditions

$$\tilde{w}(t,0) = 0, \qquad \tilde{w}_y(t,0) = 0, \tag{5}$$
$$[EI(\ell(t)y)\ell(t)^{-2}\tilde{w}_{yy}]|_{y=1} = 0, \quad [EI(\ell(t)y)\ell(t)^{-2}\tilde{w}_{yy}]_y|_{y=1} = 0,$$

and initial conditions at $t=0$:

$$\tilde{w}(0,y) = w_o(\ell(0)y), \quad \tilde{w}_t(0,y) = w_o(\ell(0)y) + y\dot{\ell}(0)\ell(0)^{-1}[w_o(\ell(0)y)]_y. \tag{6}$$

It is apparent from (4) that the arm motion produces a damping term $-2\rho(\ell(t)y)y\dot{\ell}(t)\ell(t)^{-1}\tilde{w}_{yt}$ whose coefficient may be positive or negative depending on the sign of $\dot{\ell}(t)$.

Although it is possible to seek solutions $\tilde{w}(t)$ to (4)-(6) such that $\tilde{w}(t) \in L^2(0,T;V(I_s))$ with $\tilde{w}_t(t) \in L^2(0,T;L^2(I_s))$, where $V(I_s)$ is the subspace of functions in the Sobolev space $H^2(I_s)$ satisfying boundary conditions (5), it is not clear how to choose a basis for $V(I_s)$ which is most suitable for the numerical computation of solutions. In the special case of a uniform arm with ρ, EI and ν being specified positive constants, it is useful to introduce a time-dependent basis $\{\phi_n(t,\cdot)$, $n=1,2,\ldots\}$ for the solution space, where the $\phi_n(t,\cdot)$ correspond to the orthonormalized eigenfunctions of the biharmonic operator $A = (EI/\rho)\cdot \partial^4/\partial x^4$ with domain $D(A)$ being the dense subspace of $L^2(\Omega(t))$ defined by

$$D(A) = \{w \in L^2(\Omega(t)) : Aw \in L^2(\Omega(t)) \text{ and } w(0), w_x(0), w_{xx}(\ell(t)),$$
$$w_{xxx}(\ell(t)) = 0\}. \tag{7}$$

Since $\Omega(t)$ is time-dependent, both the ϕ_n and their corresponding eigenvalues λ_n are also time-dependent. An explicit expression for $\phi_n(t,x)$ is given by

$$\phi_n(t,x) = A_n(t)\{g_n^{-1}(\cosh(\beta_n(t)x) - \cos(\beta_n(t)x))$$
$$- h_n^{-1}(\sinh(\beta_n(t)x) - \sin(\beta_n(t))\}, \; n=1,2,\ldots, \; 0<x<\ell(t), \tag{8}$$

where $\beta_n^4(t) = \rho\lambda_n(t)/(EI)$ and $g_n = \cosh(\kappa_n\pi) + \cos(\kappa_n\pi)$, $h_n = \sinh(\kappa_n\pi) + \sin(\kappa_n\pi)$, where $\kappa_n = \beta_n(t)\ell(t)/\pi$ is the n-th root of the equation

$$\cos(\kappa\pi) = 1/\cosh(\kappa\pi). \tag{9}$$

The normalizing coefficient $A_n(t)$ has the form $q(\kappa_n)/\sqrt{\ell(t)}$, where $q(\kappa_n)$ is a time-independent coefficient depending only on the constant κ_n.

Now, we seek a solution $w(t,x)$ of (1)-(3) in the form:

$$w(t,x) = \sum_{n=1}^{\infty}\alpha_n(t)\phi_n(t,x). \tag{10}$$

Making use of the orthogonality property of $\{\phi_n(t,\cdot), n=1,2,\ldots\}$, we obtain the following countably infinite system of linear time-varying ordinary differential equations for $\alpha_n(t)$:

$$\begin{aligned}\ddot{\alpha}_n + \nu\rho^{-1}\dot{\alpha}_n + \lambda_n(t)\alpha_n + 2\sum_{m=1}^{\infty}\dot{\alpha}_n(t)\langle\phi_{m,t}(t,\cdot),\phi_n(t,\cdot)\rangle \\ + \sum_{m=1}^{\infty}\alpha_m(t)\langle\phi_{m,tt}(t,\cdot) + \nu\rho^{-1}\phi_{m,t}(t,\cdot) - \ddot{\ell}(t)\phi_{m,x}(t,\cdot) \\ - (\ell(t)-\cdot)\ddot{\ell}(t)\phi_{m,xx}(t,\cdot),\phi_n(t,\cdot)\rangle = 0,\quad n=1,2,\ldots,\end{aligned} \tag{11}$$

with initial conditions:

$$\alpha_n(0) = \langle w_o,\phi_n(0,\cdot)\rangle, \qquad \dot{\alpha}_n(0) = \langle\dot{w}_o,\phi_n(0,\cdot)\rangle, \tag{12}$$

where $\langle\cdot,\cdot\rangle$ denotes the inner product for $L^2(\Omega(t))$.

Using the fact that

$$\|\phi_n(t,\cdot)\|^2 = \int_0^{\ell(t)}\phi_n^2(t,x)\,dx = 1, \qquad \frac{d^k}{dt^k}\|\phi_n(t,\cdot)\|^2 = 0,\ k=1,2,\ldots, \quad \text{for all } t \geq 0, \tag{13}$$

and the identities:

$$\begin{aligned}\langle(\ell(t)-\cdot)\phi_{m,xx}(t,\cdot),\phi_n(t,\cdot)\rangle &= \langle\phi_{m,x}(t,\cdot),\phi_n(t,\cdot)\rangle \\ &\quad - \langle(\ell(t)-\cdot)\phi_{m,x}(t,\cdot),\phi_{n,x}(t,\cdot)\rangle,\end{aligned} \tag{14}$$

$$\langle\phi_{n,x}(t,\cdot),\phi_n(t,\cdot)\rangle = \phi_n^2(t,\ell(t))/2,$$

(11) can be rewritten as

$$\ddot{\alpha}_n + (\nu\rho^{-1} - \dot{\ell}(t)\phi_n^2(t,\ell(t))\dot{\alpha}_n + \{\lambda_n(t) - \tfrac{1}{2}\phi_n^2(t,\ell(t))[\ddot{\ell}(t) + \dot{\ell}(t)\nu\rho^{-1}]$$

$$- \dot{\ell}(t)\phi_n(t,\ell(t))[2\phi_{n,t}(t,\ell(t)) + \dot{\ell}(t)\phi_{n,x}(t,\ell(t))] - \|\phi_{n,t}(t,\circ)\|^2$$

$$-\ddot{\ell}(t)\langle(\ell(t) - \cdot)\phi_{n,x}(t,\cdot),\phi_{n,x}(t,\circ)\rangle\}\alpha_n$$

$$= \sum_{m=1,m\neq n}^{\infty}\{\dot{\alpha}_m(t)\langle\phi_{m,t}(t,\cdot),\phi_n(t,\circ)\rangle - \alpha_m(t)\{\langle(\phi_{m,tt}(t,\circ)$$

$$+ \nu\rho^{-1}\phi_{m,t}(t,\circ),\phi_n(t,\cdot)\rangle - \langle\ddot{\ell}(t)(\ell(t) - \circ)\phi_{m,x}(t,\cdot),\phi_{n,x}(t,\circ)\rangle\},$$

$$n=1,2,\ldots \qquad (15)$$

It can be verified that for $m \neq n$,

$$\langle\phi_{m,t}(t,\cdot),\phi_n(t,\cdot)\rangle = \frac{1}{\beta_m^4(t) - \beta_n^4(t)}(\phi_{m,xxxt}\phi_n - \phi_{m,xxt}\phi_{n,x})\Big|_{x=\ell(t)},$$

$$\langle\phi_{m,tt}(t,\cdot),\phi_n(t,\cdot)\rangle = \frac{1}{\beta_n^4(t) - \beta_m^4(t)}\left(\frac{8\beta_m^3(t)\dot{\beta}_m(t)}{\beta_m^4(t) - \beta_n^4(t)}\{\phi_{m,xxxt}\phi_n\right.$$

$$\left.- \phi_{m,xxt}\phi_{n,x}\} - \phi_{m,xxxtt}\phi_n + \phi_{m,xxtt}\phi_{n,x}\right)\Big|_{x=\ell(t)}. \qquad (16)$$

Thus, some of the inner products in the right-hand-side of (16) can be reduced to function evaluations at the point $x = \ell(t)$.

To obtain numerical solutions of (1)-(3), we consider finite-dimensional Galerkin approximations of w(t,x) in the form:

$$w_N(t,x) = \sum_{n=1}^{N}\alpha_n(t)\,\phi_n(t,x), \qquad (17)$$

where $\alpha_n(t)$, $n=1,\ldots,N$ are given by the solutions of the truncated system corresponding to (15). It is possible to establish rigorously the convergence of $\{w_N\}$ to a weak solution of (1)-(3) as $N\to\infty$. Also, explicit error estimates can be obtained. Due to the fact that $\lambda_n(t) \simeq (EI/\rho)((n-\frac{1}{2})\pi/\ell(t))$ for $n > 3$, the convergence is very rapid.

To examine the effect of arm motion on stability, we consider the time rate-of-change of the instantaneous total energy $\mathcal{E}(t)$ given by

$$\dot{\mathscr{E}}(t) = \frac{d}{dt}\left\{\frac{1}{2}\int_0^{\ell(t)} (\rho w_t^2 + EI(w_{xx})^2)\, dx\right\}$$

$$= \frac{1}{2}\,\dot{\ell}(t)\,\rho(\ell(t))\,w_t^2(t,\ell(t)) - \int_0^{\ell(t)} \nu\, w_t^2\, dx$$

$$+ \ddot{\ell}(t)\int_0^{\ell(t)} \rho\, w_t\{w_x - (\ell(t)-x)w_{xx}\}\, dx. \tag{18}$$

We observe that in the case of uniform motion ($\ddot{\ell}(t) = 0$ for all t), $\dot{\mathscr{E}}(t) \leqslant 0$ for $\dot{\ell}(t) < 0$. This implies that the contraction of the arm has a stabilizing effect on the arm's bending motions. On the other hand, in the absence of damping ($\nu = 0$), $\dot{\mathscr{E}}(t) > 0$ if $\dot{\ell}(t)w_t(t,\ell(t)) > 0$. Evidently, when the arm extends with constant velocity, the total energy increases as long as the arm tip velocity is nonzero. Thus, the extending motion of the arm has a destabilizing effect on the bending motions. Figure 2 shows typical motions of the arm tip during the extension or contraction of the arm with constant velocities and zero damping. These numerical results are computed using (15) with N=3. Note that the growth or decay of the arm-tip vibration amplitude is induced by the arm motion only. More detailed numerical results are given in [1].

Now, we consider the feedback stabilization problem associated with the robot arm. Here, it is of interest to find an appropriate control law for damping the arm-tip vibrations so that the end effector at the arm-tip can perform its tasks as quickly as possible.

In order to formulate a physically meaningful problem, we introduce the following additional equation from Newton's law describing the translational motion of the robot arm:

$$M\,\delta\ddot{\ell}(t) + c\,\delta\dot{\ell}(t) = -f(t), \tag{19}$$

where $\delta\ell(t) = \ell_d - \ell(t)$ with ℓ_d being the desired position of the arm tip; M is the total effective mass of the arm and the actuator; c is a friction coefficient and f is the control force. A simple control law can be derived by considering the time rate-of-change of the total energy $\mathscr{E}(t)$ given by

$$\dot{\mathscr{E}}(t) = \frac{d}{dt}\left\{\frac{1}{2}\int_0^{\ell(t)} (\rho w_t^2 + EI(w_{xx})^2)\,dx + \frac{1}{2}M(\delta\dot{\ell}(t))^2 + k_1(\delta\ell(t))^2\right\}$$

$$= \delta\dot{\ell}(t)\left\{-\frac{1}{2}\rho(\ell(t))w_t^2(t,\ell(t)) + k_1\delta\ell(t) - f(t)\right\} - c(\delta\dot{\ell}(t))^2$$

$$-\int_0^{\ell(t)} \nu w_t\,dx - \delta\ddot{\ell}(t)\int_0^{\ell(t)} \rho w_t(w_x - (\ell(t)-x)w_{xx})\,dx, \tag{20}$$

where k_1 is a positive constant. If the effect of the axial force induced by the arm acceleration on the bending motion is small (i.e. the magnitudes of the last two terms in (1) are small as compared to the remaining terms), then an effective control law is given by

$$f(t) = -\frac{1}{2}\rho(\ell(t))w_t^2(t,\ell(t)) + k_1\delta\ell(t) + k_2\delta\dot{\ell}(t), \tag{21}$$

where k_1 is a positive constant. Note that this control law depends only on $\delta\ell(t), \delta\dot{\ell}(t)$ and the arm-tip velocity. Hence it can be readily implemented physically. It is apparent from (21) that during arm extension, the control force f tends to slow down the extension when the arm-tip velocity is nonzero.

Substituting (22) into (20) gives

$$M\ddot{\ell}(t) + (c+k_2)\dot{\ell}(t) + k_1\ell(t) = -\frac{1}{2}\rho(\ell(t))w_t^2(t,\ell(t)) + k_1\ell_d. \tag{22}$$

Consequently, the equations of motion for the feedback-controlled robot-arm are given by (1) with boundary conditions (2) and initial conditions (3), along with (22) for the free boundary $\{\ell(t)\}$. Thus, we have a nonlinear free-boundary problem involving an elastic beam. A detailed study of this problem will be given elsewhere.

3. ATTITUDE CONTROL OF A SPACE STATION WITH FLEXIBLE STRUCTURAL COMPONENTS

Consider a space station consisting of a rigid core and two large identical flexible solar panels as shown in Fig.3. It is of interest to derive implementable control laws for automatic attitude regulation with respect to certain fixed reference frames. Since the space station consists of both rigid and flexible components, its dynamic behavior associated with large angle maneuvers is described by coupled nonlinear partial and ordinary differential equations. The determination

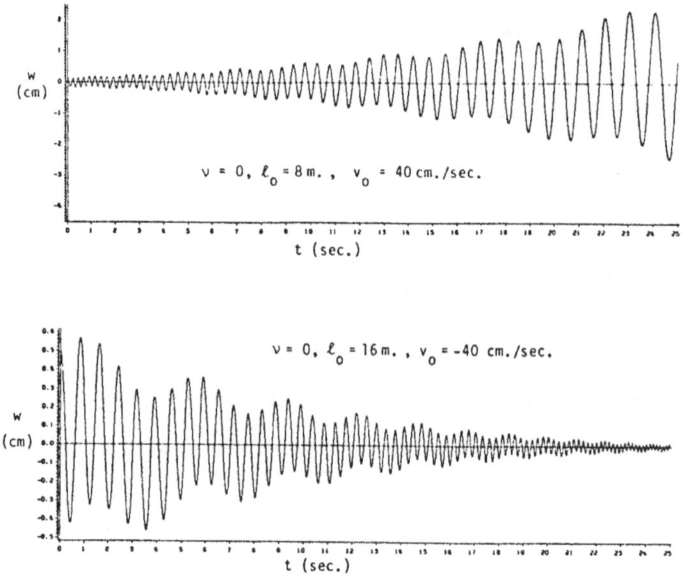

Fig.2 Arm tip displacement as functions of time for $\ell(t) = \ell_o + v_o t$. Arm parameters: $EI = 2.7872 \times 10^{11}\, g.cm^2$; $\rho = 0.4\, g.sec^2/cm^2$.

Initial conditions:

Top: $\alpha_1(0) = -2$ cm, $\dot{\alpha}_1(0) = 0.1$ cm/sec.
$\alpha_i(0) = 0$ cm, $\dot{\alpha}_i(0) = 0$ cm/sec, $i = 2,3$;

Bottom: $\alpha_1(0) = -10$ cm, $\dot{\alpha}_1(0) = 0.5$ cm/sec.
$\alpha_i(0) = 0$ cm, $\dot{\alpha}_i(0) = 0$ cm/sec, $i = 2,3$.

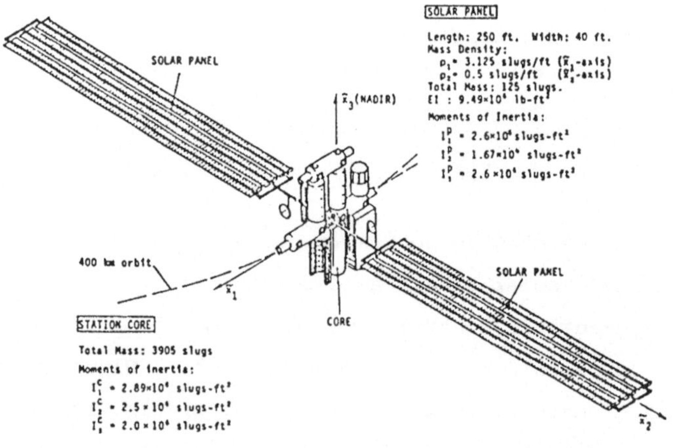

Fig.3 Sketch of a space station with flexible solar panels.

of appropriate control laws using such an infinite dimensional model is a formidable task. Moreover, the resulting control laws are most likely to be very complex. Therefore, it is desirable to derive simple effective control laws which can be implemented physically. The basic approach taken here is to use a rigid-body model to derive an attitude control law which globally stabilizes its equilibrium state. Then the effect of the elastic deformations of the solar panels on the stability of equilibrium of the resulting feedback system is determined using a nonlinear infinite dimensional model.

3.1 Mathematical Model

Let X be a Cartesian inertial coordinate system with its origin at the Earth's center with orthonormal basis $B_0 = \{\underline{e}_1^0, \underline{e}_2^0, \underline{e}_3^0\}$. We set the origin of a moving coordinate system $\tilde{X}$ at the center of mass of the space station with undeformed solar panels. Let $B_1 = \{\underline{e}_1^1, \underline{e}_2^1, \underline{e}_3^1\}$ be an orthonormal basis for $\tilde{X}$. The basis vectors in B_0 and B_1 are related by a linear transformation C defined by $\underline{e}_i^1 = C\underline{e}_i^0$, i=1,2,3, whose representation with respect to basis B_0 is given by the direction cosine matrix:

$$C(\underline{q}) = \begin{bmatrix} q_1^2 - q_2^2 - q_3^2 + q_4^2 & 2(q_1q_2 + q_3q_4) & 2(q_1q_3 - q_2q_4) \\ 2(q_1q_2 - q_3q_4) & -q_1^2 + q_2^2 - q_3^2 + q_4^2 & 2(q_2q_3 + q_1q_4) \\ 2(q_1q_3 + q_2q_4) & 2(q_2q_3 - q_1q_4) & -q_1^2 - q_2^2 + q_3^2 + q_4^2 \end{bmatrix}, \quad (23)$$

where $\underline{q} = (q_1, q_2, q_3, q_4)$ denotes the Euler quaternion with q_i being the Euler symmetric parameters [2] defined by

$$q_i = \varepsilon_i \sin(\phi/2), \quad i=1,2,3; \quad q_4 = \cos(\phi/2), \quad (24)$$

where ϕ is the principal angle, and the ε_i's are the components of the principal vector of rotation $\underline{\ell}$ defined by

$$\underline{\ell} = \varepsilon_1\underline{e}_1^0 + \varepsilon_2\underline{e}_2^0 + \varepsilon_3\underline{e}_3^0 = \varepsilon_1\underline{e}_1^1 + \varepsilon_2\underline{e}_2^1 + \varepsilon_3\underline{e}_3^1. \quad (25)$$

The time rate-of-change of $\underline{q}$ is related to the angular velocity $\underline{\omega}$ of the moving coordinate system $\tilde{X}$ relative to the inertial frame X by

$$d\underline{q}/dt = \Omega([\underline{\omega}]_1)\underline{q}, \quad (26)$$

where $[\underline{\omega}]_1 = (\omega_1^1, \omega_2^1, \omega_3^1)^T$ denotes the representation of $\underline{\omega}$ with respect to B_1, and

$$\Omega([\underline{\omega}]_1) = \frac{1}{2}\begin{bmatrix} 0 & \omega_3^1 & -\omega_2^1 & \omega_1^1 \\ -\omega_3^1 & 0 & \omega_1^1 & \omega_2^1 \\ \omega_2^1 & -\omega_1^1 & 0 & \omega_3^1 \\ -\omega_1^1 & -\omega_2^1 & -\omega_3^1 & 0 \end{bmatrix}. \tag{27}$$

Let $\Gamma(t) \subset \mathbb{R}^3$ be the spatial domain occupied by the space station at time t. The position of a material point in $\Gamma(t)$ can be specified by a vector $\underline{x}$ in the inertial system X, or by a vector $\underline{\tilde{x}}$ in the moving coordinate system $\tilde{X}$. Their representations are related by

$$[\underline{\tilde{x}}]_1 = C(\underline{q})[\underline{x} - \underline{r}(t)]_0 \quad \text{or} \quad [\underline{x}]_0 = [\underline{r}(t)]_0 + C(\underline{q})^T[\underline{\tilde{x}}]_1, \tag{28}$$

where $\underline{r}(t)$ specifies the position of the center of mass of the space station at time t.

For simplicity, we consider the case where the space station is in a circular orbit with radius r_o about the Earth's center with angular speed $\omega_o = \sqrt{\mu_1/r_o^3}$. Let the basis vectors $\underline{e}_i$, i=1,2,3 of the moving coordinate system $\tilde{X}$ be along the principal axes of the space station with undeformed solar panels such that the elastic deformations of the solar panels are restricted to bending in the $(\tilde{x}_2,\tilde{x}_3)$-plane only, and the effect of axial forces induced by the system motion is negligible. Then it can be shown [3] that a simplified model for the deforming motion of the solar panels is given by

$$\rho_2\tilde{w}_{i,tt} = -EI\tilde{w}_{i,xxxx} - \nu_i\tilde{w}_{i,t} - \rho_2 s_i(\omega_2^1\omega_3^1 + \dot{\omega}_1^1), \quad i=1,2; \; x \in I_p =]\ell_o,\ell_1[, \tag{29}$$

where $s_1 = -s_2 = x$, $\tilde{w}_i = \tilde{w}_i(t,x)$ denotes the $\tilde{x}_3$-component of displacement of the i-th panel about its equilibrium at time t and a point x along the x-axis; ρ_j is the mass density per unit length along the $\tilde{x}_j$-axis; EI is the flexural rigidity per unit length of the panel. Both ρ_j and EI are positive constants. The boundary conditions for (29) are

$$\tilde{w}_i(t,\ell_o) = 0, \quad \tilde{w}_{i,x}(t,\ell_o) = 0, \quad \tilde{w}_{i,xx}(t,\ell_1) = 0, \quad \tilde{w}_{i,xxx}(t,\ell_1) = 0, \quad i=1,2. \tag{30}$$

The rotational motion of the space station can be derived by considering the rate of change of angular momentum about its center of mass [3]. The resulting equation has the following representation with respect to basis B_1:

$$[\Pi(t)]_1[\dot{\underline{\omega}}]_1 + [\underline{\omega}]_1 \times ([\Pi(t)]_1[\underline{\omega}]_1) = \int_{\Gamma(t)} [\tilde{\underline{x}} \times \underline{f}]_1 \rho \, dV$$

$$- \int_{\Gamma(t)} [\tilde{\underline{x}}]_1 \times \{D[\tilde{\underline{v}}]_1/Dt + 2[\underline{\omega}]_1 \times [\tilde{\underline{v}}]_1\} \rho \, dV + [\underline{\tau}_c]_1 + [\underline{\tau}_d]_1, \quad (31)$$

where ρ is the mass density of the space station; $\underline{\tau}_d$ is an external disturbance torque; $\underline{f}$ is the extrinsic body force per unit mass; $D[\tilde{\underline{v}}]_1/Dt = [\tilde{\underline{v}}_t + \tilde{\underline{v}} \cdot \nabla \tilde{\underline{v}}]_1$; $\Pi(t)$ is the inertia tensor of the space station at time t and is given by

$$\Pi(t) = [\text{trace } \mathcal{G}(t)] I - \mathcal{G}(t), \quad (32)$$

where I is the identity transformation and $\mathcal{G}(t)$ is the Euler tensor defined by

$$\mathcal{G}(t) = \int_{\Gamma(t)} \tilde{\underline{x}} \otimes \tilde{\underline{x}} \, \rho \, dV \quad (33)$$

and $\otimes$ denotes the tensor product.

The matrix representation of $\Pi(t)$ with respect to basis B_1 can be written in the form:

$$[\Pi(t)]_1 = [\Pi_o]_1 + [\Delta\Pi(t)]_1, \quad (34)$$

where Π_o and $\Delta\Pi(t)$ are the inertia tensor and the perturbed inertia tensor of the undeformed and deformed space station respectively. They are given explicitly by

$$[\Pi_o]_1 = \int_{\Gamma_o} \{\|\tilde{\underline{x}}\|^2 I_3 - \text{diag}[\tilde{x}_1^2, \tilde{x}_2^2, \tilde{x}_3^2]\} \, \rho(\tilde{\underline{x}}) \, dV$$

$$= \text{diag}[I_1^c + 2\rho_2(\ell_1 - \ell_o)^3/3,\ I_2^c + \rho_1 \ell_1^3/2,\ I_3^c + 2\rho_2(\ell_1 - \ell_o)^3/3], \quad (35)$$

$$[\Delta\Pi(t)]_1 = \int_{\Gamma(t) \setminus \Gamma_o} H(\tilde{\underline{x}}) \, \rho(\tilde{\underline{x}}) \, dV - \int_{\Gamma_o \setminus \Gamma(t)} H(\tilde{\underline{x}}) \rho(\tilde{\underline{x}}) \, dV$$

$$\simeq \int_{\ell_o}^{\ell_1} \rho_2 \{\tilde{w}_1^2(t,x) + \tilde{w}_2^2(t,x)\} \, dx \begin{bmatrix} 1 & 0 & 0 \\ 0 & 1 & 0 \\ 0 & 0 & 0 \end{bmatrix}$$

$$- \int_{\ell_o}^{\ell_1} \rho_2 \{\tilde{w}_1(t,x) - \tilde{w}_2(t,x)\} x \, dx \begin{bmatrix} 0 & 0 & 0 \\ 0 & 0 & 1 \\ 0 & 1 & 0 \end{bmatrix}, \quad (36)$$

where $H(\tilde{x}) = \|\tilde{x}\|^2 I_3 - [\tilde{x}]_1 [\tilde{x}]_1^T$; $\Gamma(t)\setminus\Gamma_o$ denotes the difference between $\Gamma(t)$ and the spatial domain Γ_o of the undeformed space station, and I_i^c is the moment of inertia of the station core about the $\tilde{x}_i$-axis. Note that $[\Delta\Pi(t)]_1$ given in (36) is valid only for sufficiently small displacements $\tilde{w}_1$ and $\tilde{w}_2$, and $\Pi(t)$ is positive if

$$\Pi_{o22} > \rho_2(\ell_1^3 - \ell_o^3)/6 \quad \text{and} \quad \Pi_{o33} > \int_{\ell_o}^{\ell_1} \rho_2 \,(|\tilde{w}_1| + |\tilde{w}_2|)\, x\, dx, \tag{37}$$

where Π_{oii} is the i-th diagonal element of $[\Pi_o]_1$. The spatial domain variation due to elastic panel deformation is neglected in the approximation.

Thus, a simplified mathematical model of the space station consists of the nonlinear ordinary differential equation (26) for the quaternion $\underline{q}$, and two partial differential equations (29) for the solar panel displacements $\tilde{w}_i$ which are nonlinearly coupled with the integrodifferential equation (32) for the angular velocity $\underline{\omega}$. Note that if we retain the exact expression for the inertia tensor $\Pi(t)$ given by (31) and consider large deformations for the solar panels, the resulting mathematical model would involve a distributed system whose spatial domain $\Gamma(t)$ depends on the system motion.

3.2 Attitude Control Law

To derive an attitude control law based on rigid-body dynamics, we consider (26) along with the following equation corresponding to (31) in the absence of elastic deformations and the moment induced by the extrinsic body force $\underline{f}$:

$$[\Pi_o]_1[\dot{\underline{\omega}}]_1 + [\underline{\omega}]_1 \times ([\Pi_o]_1[\underline{\omega}]_1) = [\underline{\tau}_c]_1 + [\underline{\tau}_d]_1. \tag{38}$$

Let the desired attitude of the space station be specified by the quaternion $\underline{q}^d = (q_1^d, q_2^d, q_3^d, q_4^d)$. We introduce a nonlinear transformation $\underline{z} = \underline{z}(\underline{q})$ defined by

$$\underline{z}(\underline{q}) = (q_1/q_4, q_2/q_4, q_3/q_4). \tag{39}$$

Let $\underline{z}^d = \underline{z}(\underline{q}^d)$, $\Delta\underline{z} = \underline{z}^d - \underline{z}$, and $\Delta\underline{\omega} = \underline{\omega} - \underline{\omega}^d$, where $\underline{z}^d$ and $\underline{\omega}^d$ are the desired $\underline{z}$ and $\underline{\omega}$ respectively. It can be verified by using (26) and (38) that the equations for $\Delta\underline{z}$ and $\Delta\underline{\omega}$ are given by

$$\Delta\dot{\underline{z}} = -\{([\underline{\omega}]^T(\underline{z}^d - \Delta\underline{z})I_3 - B([\underline{\omega}]_1))(\underline{z}^d - \Delta\underline{z}) + [\underline{\omega}]_1/2, \tag{40}$$

$$[\Pi_o]_1[\Delta\dot{\underline{\omega}}]_1 + [\Delta\underline{\omega}]_1 \times \{[\Pi_o]_1([\Delta\underline{\omega}]_1 + [\underline{\omega}^d]_1)\} + [\underline{\omega}^d]_1 \times \{[\Pi_o]_1([\Delta\underline{\omega}]_1 + [\underline{\omega}^d]_1)\} = [\underline{\tau}_c]_1 + [\underline{\tau}_d]_1, \tag{41}$$

where

$$B([\underline{\omega}]_1) = \begin{bmatrix} 0 & -\omega_3^1 & \omega_2^1 \\ \omega_3^1 & 0 & -\omega_1^1 \\ -\omega_2^1 & \omega_1^1 & 0 \end{bmatrix}. \tag{42}$$

By considering the time rate-of-change of the positive definite function $V_1(\Delta\underline{z},\Delta\underline{\omega}) = k_p\|\Delta\underline{z}\|^2 + [\Delta\underline{\omega}]_1^T[\Pi_o]_1[\Delta\underline{\omega}]_1$ on $\mathbb{R}^6$ along the trajectories of (40) and (41), with k_p being a specified positive constant, we can establish the following result [3] :

THEOREM 1: In the absence of the external disturbance torque ($\underline{\tau}_d(t) \equiv 0$ for all t), the control law

$$\begin{aligned}[\underline{\tau}_c]_1 &= [\underline{\omega}^d]_1 \times ([\Pi_o]_1([\underline{\omega}^d]_1 + [\Delta\underline{\omega}]_1)) \\ &+ k_p[(\underline{z}^d - \Delta\underline{z})(\underline{z}^d - \Delta\underline{z})^T + I_3](\Delta\underline{z})/2 + k_p\gamma_1[\Pi_o]_1[\Delta\underline{\omega}]_1/2 \\ &- (\mathbb{K}_1 + \gamma_2 I_3)[\Pi_o]_1[\Delta\underline{\omega}]_1/2\end{aligned} \tag{43}$$

with

$$\gamma_1 = \frac{(\underline{z}^d)^T B([\underline{\omega}]_1)(\Delta\underline{z}) + [\underline{\omega}^d]_1^T[(\underline{z}^d - \Delta\underline{z})(\underline{z}^d - \Delta\underline{z})^T + I_3](\Delta\underline{z})}{[\Delta\underline{\omega}]_1^T[\Pi_o]_1[\Delta\underline{\omega}]_1},$$

$$\gamma_2 = \frac{(\Delta\underline{z})^T\mathbb{K}_2(\Delta\underline{z})}{[\Delta\underline{\omega}]_1^T[\Pi_o]_1[\Delta\underline{\omega}]_1} \tag{44}$$

and $\mathbb{K}_1$, $\mathbb{K}_2$ being arbitrary positive definite real symmetric matrices, exponentially stabilizes the zero state $(\Delta\underline{z},\Delta\underline{\omega}) = (\underline{0},\underline{0})$ of system (40)-(42) for any given $(\underline{\omega}^d,\underline{z}^d) \in \mathbb{R}^6$.

Remark: By making use of LaSalle's Invariance Principle, it is possible to establish the global asymptotic stability of the zero state when $\mathbb{K}_1$ is positive definite and $\mathbb{K}_2 = 0$.

The next result pertains to system (40)-(42) in the presence of external disturbance $\underline{\tau}_d$, and with $\mathbb{I}_o$ replaced by $\mathbb{I}_o + \Delta\mathbb{I}(t)$, where $\Delta\mathbb{I}(t)$ is a specified time-dependent inertia tensor perturbation.

THEOREM 2: Assume that $\mathbb{I}(t)$ is C_1 and uniformly positive definite on the time interval $\mathbb{R}^+ = [0, \infty[$, and there exist positive constants δ_1, δ_2, $\check{\delta}_3$ and $\hat{\delta}_3$ such that

$$
\begin{aligned}
&\lambda_{min}(\mathbb{K}_1) - 2\|\Delta\mathbb{I}(t)\|\,\|\underline{\omega}^d\| - \|\Delta\dot{\mathbb{I}}(t)\| \geq \delta_1, \\
&\|\mathbb{I}(t)\|\,\|\underline{\omega}^d\|^2 + \|\underline{\tau}_d(t)\| \leq \delta_2, \\
&\check{\delta}_3\|\Delta\underline{\omega}\|^2 \leq [\Delta\underline{\omega}]_1^T[\mathbb{I}(t)]_1[\Delta\underline{\omega}]_1 \leq \hat{\delta}_3\|\Delta\underline{\omega}\|^2
\end{aligned}
\tag{45}
$$

for all $t \geq 0$, the all the trajectories of system (40)-(42) with control law (43)-(44) are attracted to the set $S = \{(\Delta\underline{z}, \Delta\underline{\omega}) \in \mathbb{R}^6 : k_p\|\Delta\underline{z}\|^2 + \check{\delta}_3\|\Delta\underline{\omega}\|^2 < 2\,\delta_2^2/(\delta_1\beta)\}$ as $t \to \infty$, where $\beta = \min\{\lambda_{min}(\mathbb{K}_2)/k_p,\ \delta_1/(2\hat{\delta}_3)\}$, and $\lambda_{min}(\mathbb{K}_i)$ denotes the minimum eigenvalue of $\mathbb{K}_i$.

The above result can be established by estimating the time rate-of-change of the uniformly positive definite function $V_2(t, \Delta\underline{z}, \Delta\underline{\omega}) = k_p\|\Delta\underline{z}\|^2 + [\Delta\underline{\omega}]_1^T[\mathbb{I}(t)]_1[\Delta\underline{\omega}]_1$ on $\mathbb{R}^6$ along the system trajectories for $t > 0$. This result shows that the asymptotic behavior of the feedback-controlled system corresponding to the rigid-body model is robust with respect to slowly time-varying inertial perturbations and external disturbance torque $\underline{\tau}_d$.

Now, we consider the asymptotic behavior of the feedback-controlled space station with elastic deformations modelled by (26),(29)-(31) with control law (43), and $\underline{\tau}_d(t) \equiv 0$ for all t. Here, we may take the system state at any time $t > 0$ as $(\Delta\underline{z}(t), \Delta\underline{\omega}(t), \tilde{w}_1(t,\cdot), \tilde{w}_2(t,\cdot), \tilde{w}_{1t}(t,\cdot), \tilde{w}_{2t}(t,\cdot))$ with $(\Delta\underline{z}(t), \Delta\underline{\omega}(t)) \in \mathbb{R}^6$, $\tilde{w}_i(t,\cdot) \in V(I_p)$, $\tilde{w}_{it}(t,\cdot) \in L^2(I_p)$, i=1,2, where $V(I_p)$ is the subspace of functions in the Sobolev space $H^2(I_p)$ satisfying boundary conditions (30).

By estimating the time rate-of-change of the functional V_3 given by

$$
\begin{aligned}
V_3(\Delta\underline{z}, \underline{\omega}, \tilde{w}_1, \tilde{w}_2, \tilde{w}_{1t}, \tilde{w}_{2t}) &= k_p\|\Delta\underline{z}\|^2 + [\underline{\omega}]_1^T[\mathbb{I}(t)]_1[\underline{\omega}]_1 \\
&+ \int_{\ell_o}^{\ell_1} \{\rho[\tilde{w}_{1t}^2 + \tilde{w}_{2t}^2 + 2x\omega_1^1(\tilde{w}_{1t} - \tilde{w}_{2t})] + EI(\tilde{w}_{1xx}^2 + \tilde{w}_{2xx}^2)\}\,dx
\end{aligned}
\tag{46}
$$

along the system trajectories, we can establish the following result [3]:

THEOREM 3: Let $\underline{\omega}^d = \underline{0}$. If the initial data satisfy (37) and

$$\lambda_{min}(\mathbb{K}_1) > \int_{\ell_o}^{\ell_1} \rho_2 \sum_{i=1}^{2} (\tilde{w}_i^2(0,x) + \tilde{w}_{it}^2(0,x))\,dx\,, \tag{47}$$

then the corresponding trajectories of the feedback-controlled system (26),(29)-(31) and (43) are bounded in the norm of $H^2(I_p)$ and tend to the zero state as $t \to \infty$.

The localized nature of this result is partially attributed to the fact that both the solar panel equations (29) and the inertia tensor perturbation (36) are invalid for large deformations.

Figure 4 shows typical simulation results of the feedback-controlled system for the case where $k_p = 10^5$, $\mathbb{K}_1 = I_3$, $\underline{q}^d = [1/\sqrt{2},0,0,1/\sqrt{2}]^T$,

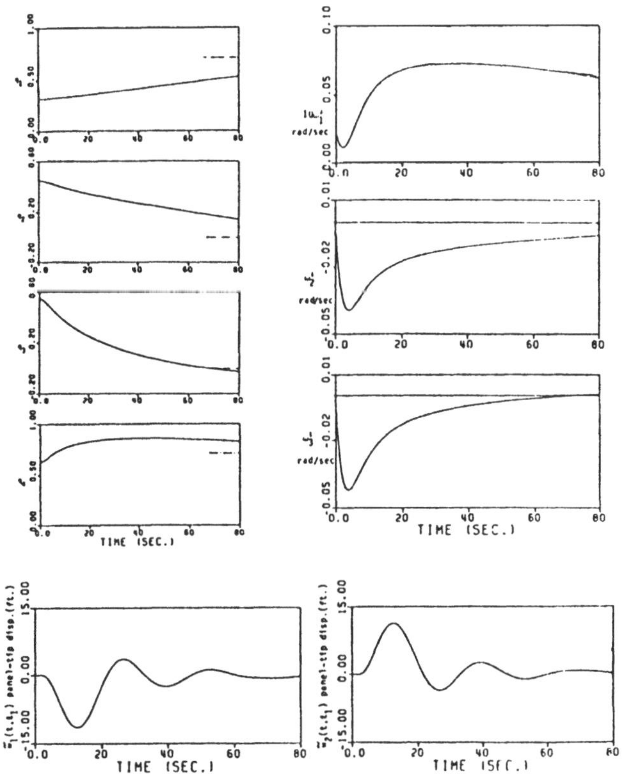

Fig.4 System response in the presence of elastic deformation of the solar panels and with damping coefficient $\nu_i = 0.05$ slugs/(ft sec), i=1,2.

$[\underline{\omega}(0)]_1 = [0.002,0,-0.002]^T$ rad/sec. and $\underline{q}(0) = [\sqrt{0.1},\sqrt{0.2},\sqrt{0.3},\sqrt{0.4}]^T$. The beam equations (29) are solved numerically with zero initial displacements and velocities using a 20-point spatial finite-difference approximation. Although the maximum magnitude of the panel-tip displacement induced by the slewing motion is approximately 12 feet, the attitude response behaves essentially the same as that of a rigid-body model.

4. ABSTRACT CONTROL PROBLEM

Now, we consider an abstract control problem motivated by the concrete problems discussed in Sections 2 and 3. Here, the evolution of the spatial domain of the distributed system with time is governed by an ordinary differential equation depending on the control. It is of interest to control the dynamical behavior of the distributed system by manipulating its spatial domain. In what follows, the equations of motion for the spatial domain and the distributed system will be discussed separately.

4.1 Spatial Domain Motion

Let I_T denote the time interval $]0,T[, T\to\infty$; and Ω_o be a specified nonempty bounded open connected subset of $\mathbb{R}^n$. Let $\underline{x}_u = \underline{x}_u(t,\underline{x}_o)$ denote the solution of

$$d\underline{x}/dt = \underline{h}(t,\underline{x},\underline{u}), \quad \underline{x}(0) = \underline{x}_o \in \mathbb{R}^n, \tag{48}$$

where the control $\underline{u} \in \mathcal{U}_{ad}(I_T)$-- the set of all admissible controls defined on $\bar{I}_T =]0,T[$; $\underline{h}$ is a C_1 function of its arguments defined on $\bar{I}_T \times \mathbb{R}^n \times \mathbb{R}^m$. The spatial domain of the distributed system at time t corresponding to a specified $\underline{u}(\cdot)$ is given by $\Omega_u(t) = \underline{x}_u(t,\Omega_o)$. We assume that $\Omega_u(t)$ remains bounded for each $t \in \bar{I}_T$ and all $\underline{u} \in \mathcal{U}_{ad}(\bar{I}_T)$. This assumption can be satisfied by imposing suitable restriction on $\underline{h}$ (eg. There exists a positive constant C such that the inner product $\langle \underline{x},\underline{h}(t,\underline{x},\underline{u})\rangle_{\mathbb{R}^n} \leq C\|\underline{x}\|^2$ for all $t \in \bar{I}_T$ and $(\underline{x},\underline{u}) \in \mathbb{R}^{n+m}$).

An important case of physical interest (eg. an incompressible body) involves isochoric motions in which the volume of $\Omega_u(t)$ remains constant for all t. In this case, div $\underline{h}(t,\underline{x},\underline{u}(t))$ = trace $J_h(t,\underline{x},\underline{u}(t))=0$ for all $t \in \bar{I}_T$ and $\underline{x} \in \mathbb{R}^n$, where J_h denotes the Jacobian of $\underline{h}$ with respect to $\underline{x}$.

When (48) is a linear system with $\underline{h}(t,\underline{x},\underline{u}) = F(t)\underline{x} + G(t)\underline{u}$ with $F(t)$ and $G(t)$ being linear transformations from $\mathbb{R}^n$ and $\mathbb{R}^m$ to $\mathbb{R}^n$ respectively, we have

$$\Omega_u(t) = \Phi(t)\Omega_o + \underline{g}_u(t), \tag{49}$$

where $\Phi(t)$ is the state transition operator and

$$\underline{g}_u(t) = \int_0^t \Phi(t)\Phi(\tau)^{-1}G(\tau)\underline{u}(\tau)\,d\tau. \tag{50}$$

Evidently, the effect of the control $\underline{u}$ is a translation of the set $\Phi(t)\Omega_o$ by $\underline{g}_u(t)$.

4.2 Distributed System Motion

Here, we adapt Lions' formalism for evolution equations[4] to our case. As before, the time interval $I_T =]0,T[$, $T < \infty$, is specified. We are given real Hilbert spaces $V_u(t)$ and $H_u(t)$ consisting of functions defined on $\Omega_u(t)$ such that $V_u(t) \subset H_u(t) \subset V'_u(t)$ (the dual of $V_u(t)$) with $V_u(t)$ dense in $H_u(t)$, $H_u(t)$ dense in $V'_u(t)$, and with continuous corresponding injections.

Let $L^2(I_T; V_u(\circ))$ denote the Hilbert space of measurable functions $t \to w(t)$ such that

$$\int_0^T \|w(t)\|^2_{V_u(t)}\,dt < \infty$$

endowed with the inner product

$$\int_0^T (w(t), \widetilde{w}(t))_{V_u(t)}\,dt.$$

For each fixed $t \to I_T$, we consider a family of bilinear forms on $V_u(t)$: $\phi(t), \psi(t) \to a(t; \phi(t), \psi(t))$ having the coercive property:

(P1) There exists a positive number α_o, independent of t, such that $a(t; \phi, \phi) \geqslant \alpha_o \|\phi\|^2_{V_u(t)}$ for all $\phi \in V_u(t)$ and $t \in I_T$.

Let $A(t)$ be an operator from $\mathcal{L}(V_u(t), V'_u(t))$ defined by

$$a(t; \phi, \psi) = (A(t)\phi, \psi)_{V'_u(t) \times V_u(t)}, \qquad A(t)\phi \in V'_u(t), \tag{51}$$

where $(\circ, \cdot)_{V'_u(t) \times V_u(t)}$ denotes the duality pairing between $V_u(t)$ and $V'_u(t)$. We introduce the space

$$W_u(I_T) = \{w: w \in L^2(I_T; V_u(\cdot)), \, dw/dt \in L^2(I_T; V'_u(\cdot))\} \tag{52}$$

with the norm

$$\|w\|_{W_u(I_T)} = \left\{ \int_0^T \|w(t)\|^2_{V_u(t)} + \left\| \frac{dw(t)}{dt} \right\|^2_{V_u'(t)} dt \right\}^{\frac{1}{2}}. \tag{53}$$

Now, we consider a distributed system described by the following first-order evolution equation:

$$dy/dt + A(t)y = f(t), \qquad y(0) = y_o \in H_u(0), \tag{54}$$

where f is a specified input in $L^2(I_T; V_u'(\cdot))$, and the spatial domain is given by $\underline{x}_u(t,\Omega_o)$.

For this system, we have the following theorem for the existence and uniqueness of solution to the evolution problem of finding a $y \in W_u(I_T)$ satisfying (54) for a given $\underline{u} \in \mathcal{U}_{ad}(I_T)$:

THEOREM 4: Assume that $\Omega_u(t) = \underline{x}_u(t;\Omega_o)$ is bounded on $\overline{I}_T$. If

$$\sup_{t \in \overline{I}_T} \ \sup_{\underline{x} \in \Omega_u(t)} |\mathrm{trace}\ J_h(t,\underline{x},\underline{u}(t))| < \alpha_o, \tag{55}$$

where α_o is defined in property (P-1), then the evolution problem for (54) has a unique solution.

The proof relies on Galerkin approximations of the form:

$$y_m(t) = \sum_{i=1}^{m} g_{im}(t)\phi_i(t) \tag{56}$$

with $y_m(0) = \sum_{i=1}^{m} \xi_{im}\phi_i(0) \to y_o \in H_u(0)$ as $m \to \infty$, where $\{\phi_1(t), \phi_2(t), \ldots\}$ is a countable basis for $V_u(t)$ such that $d\phi_i(t)/dt \in V_u'(t)$. The ordinary differential equation for $\underline{g}_m(t) = (g_{1m}(t),\ldots,g_{mm}(t))$ has the form:

$$\begin{aligned} &W_m(t)\frac{d\underline{g}_m(t)}{dt} + [\widetilde{W}_m(t) + A_m(t)]\underline{g}_m(t) = \underline{f}_m(t), \\ &\underline{g}_m(0) = (\xi_{1m}(0),\ldots,\xi_{mm}(0)), \end{aligned} \tag{57}$$

where $\underline{f}_m(t) = ((f(t),\phi_1(t))_{V_u'(t)\times V_u(t)}, \ldots, (f(t),\phi_m(t))_{V_u'(t)\times V_u(t)})$, and $W_m(t)$, $\widetilde{W}_m(t)$ and $A_m(t)$ are $m \times m$ matrices given by

$$\begin{aligned} &\widetilde{W}_m(t) = [(dw_i(t)/dt, w_j(t))_{V_u'(t)\times V_u(t)}], \\ &W_m(t) = [(w_i(t),w_j(t))_{V_u'(t)\times V_u(t)}], \quad A_m(t) = [a(t;w_i(t),w_j(t))]. \end{aligned} \tag{58}$$

The details of the proof will be given in a forthcoming paper.

We may also consider a second-order evolution equation of the form $d^2y/dt^2 + A(t)y = f(t)$. However, the condition for existence and uniqueness of solution to the corresponding evolution problem is more involved.

4.3 Control Problems

An optimal control problem associated with the foregoing distributed system (54) and (48) can be stated as follows:

Given $\mathcal{U}_{ad}(I_T)$ and Ω_o along with (48), (54) and the cost functional

$$J(u) = \int_0^T \int_{\Omega_u(t)} |y_u(t,\underline{x}) - y_d(\underline{x})|^2 \, dx\, dt, \tag{59}$$

where y_d is the desired state of the system. We seek a $\underline{u}^* \in \mathcal{U}_{ad}(I_T)$ such that $J(\underline{u}^*) = \inf_{\underline{u} \in \mathcal{U}_{ad}} J(\underline{u})$.

For the special case where $\Omega_u(t)$ is given by (49) and (50), and $\mathcal{U}_{ad}(I_T)$ consists of measurable functions with their values in a compact subset of $\mathbb{R}^m$, it follows from Lyapunov's theorem on the range of a vector measure [5] that the set $\{\underline{g}_u(t) \in \mathbb{R}^n : \underline{u} \in \mathcal{U}_{ad}(I_t)\}$ is compact and convex, where $\underline{g}_u(t)$ is defined in (50). The cost functional J can be rewritten as:

$$J(\underline{u}) = \int_0^T \int_{\Omega_o} |y_u(t, \Phi(t)\underline{x}_o + \underline{g}_u(t)) - y_d(\Phi(t)\underline{x}_o + \underline{g}_u(t))|^2 \cdot \left\{ \exp \int_0^t \text{trace } F(s)\, ds \right\} d\underline{x}_o\, dt. \tag{60}$$

Thus, a necessary condition for which an optimal control $\underline{u}^*$ must satisfy can be written in the form of a variational inequality

$$J'(\underline{g}_{u^*}) \cdot (\underline{g}_u - \underline{g}_{u^*}) \geqslant 0 \quad \text{for all } \underline{g}_u \in \Gamma(I_T), \tag{61}$$

where $\Gamma(I_T) = \{\underline{g}_u(\cdot) : \underline{u} \in \mathcal{U}_{ad}(I_T)\}$. Having found a $\underline{g}_{u^*}$ satisfying (61), the corresponding $\underline{u}^*$ can be determined from the equation:

$$g_{u^*}(t) = \int_0^t \Phi(t)\Phi(\tau)^{-1} G(\tau)\underline{u}(\tau)\, d\tau. \tag{62}$$

Finally, we mention briefly the following feedback stabilization problem associated with (48) and (54):

Let $\{\underline{x}^i, i=1,\ldots,K\}$ be a set of specified points on $\partial\Omega_o$ (the boundary of Ω_o), and $\{\underline{x}_d^i, i=1,\ldots,K\}$ be the corresponding desired boundary points of the spatial domain. Let y_d be the desired state of the distributed system. We wish to find a control law depending on the instantaneous deviations $\delta y(t) = y_d - y(t)$ and $\delta\underline{x}^i(t) = \underline{x}_d^i - \underline{x}_u(t;\underline{x}_o^i)$, $i=1,\ldots,K$ such that $\underline{x}_u(t;\underline{x}_o^i) \in \partial\underline{x}_u(t,\Omega_o)$, $i=1,\ldots,K$; and $\underline{x}_u(t,\Omega_o)$, $\|y_d - y(t)\|_{V_u(t)}$ remain bounded for all $t > 0$, and $\sum_{i=1}^{K} \|\underline{x}_d^i - \underline{x}_u(t,\underline{x}_o^i)\|_{\mathbb{R}^n}$ and $\|y_d - y(t)\|_{V_u(t)}$ tend to zero as $t \to \infty$. This problem is nontrivial. We observe that if such a stabilizing control law can be found, the evolution problem associated with the feedback-controlled system involves a free boundary as in the case of the feedback-controlled robot arm discussed in Sec. 2.

ACKNOWLEDGMENT

This work was supported by the National Science Foundation Grant ECS 85-09145.

REFERENCES

[1] Wang, P.K.C., and Wei, Jin-Duo. "Vibrations in a Moving Flexible Robot Arm." University of Calfornia at Los Angeles, Engineering Report No. UCLA-ENG-86-20, March, 1986.

[2] Wertz, J.R., ed. Spacecraft Attitude Determination and Control. Boston: D. Reidel, 1980.

[3] Wang, P.K.C., "A Robust Nonlinear Attitude Control Law for Space Stattions with Flexible Structural Components." In Dynamics and Control of Large Structures, Proceedings of the Fifth VPI & SU/ AIAA Symposium Held in Blacksburg, Virgina. ed. L. Meirovitch. June, 1985.

[4] Lions, J.L. Optimal Control of Systems Governed by Partial Differential Equations. (Translated by S.K. Mitter), New York: Springer-Verlag, 1971.

[5] Hermes, H., and LaSalle, J.P., Functional Analysis and Time-Optimal Control. New York: Academic, 1969.

NULL CONTROLLABILITY AND EXACT CONTROLLABILITY FOR PARABOLIC EQUATIONS

N. Weck

Universität Gesamthochschule Essen
The Federal Republic of Germany

ABSTRACT

Which functions can be final states of a system described by a homogeneous parabolic differential equation? For inhomogeneous boundary conditions this is the problem of characterizing the set R of states reachable by boundary control. We construct more spaces of reachable states. In particular we can show that this question is "independent" of the eigenfunctions and eigenvalues of the corresponding elliptic problem.

O. INTRODUCTION

We consider a bounded region $\Omega \subset \mathbb{R}^n$, a uniformly strongly elliptic operator L of order $2m$ and a system B of boundary operators. Let us assume that L, B generate a selfadjoint positive operator L . In the mixed initial boundary value problem

$$\begin{aligned} (\partial_t + L_x)\, w(t,x) &= o \\ Bw(t,\xi) &= u(t,\xi) \\ w(o,\cdot) &= y \end{aligned} \qquad \text{(IBVP)}$$

$$t \in [o,T] \quad , \quad x \in \Omega \ , \quad \xi \in \partial\Omega$$

we regard u as a boundary control i.e. we want to solve the

EXACT CONTROLLABILITY PROBLEM

Given $y, z \in L^2(\Omega)$ *find* $u \in U$ *(Banach space of boundary controls) such that* $w(T,\cdot) = z$ *for the solution* w *of* (IBVP) .

We shall write this symbolically as

$$y \overset{o \qquad T}{\underset{u}{\longmapsto}} z$$

and define the reachable set

$$\mathcal{R}(y,T) \;:=\; \{\, z \mid y \overset{o \qquad T}{\underset{u}{\longmapsto}} z \quad \text{for some } u \,\}$$

A special case of (P) is the

NULL CONTROLLABILITY PROBLEM

> *Given* $y \in L^2(\Omega)$ *find* $u_o \in U$ *such that*
> $w(T,\cdot) = o$ *for the corresponding* $\qquad (P_o)$
> *solution* w *of* (IBVP) .

We want to show that (as in the group case) it suffices to solve (P_o) in order to solve (P). This is meant in the following sense: It is possible to prove theorems concerning (P_o) which by elementary means yield all the results for (P) which have been obtained so far by moment methods (and at the same time generalize these considerably).

1. NULL CONTROLLABILITY

Without going into details we want to recall two methods which yield the following basic result:

$$(P_o) \;\textit{ is solvable for all }\; y \;\textit{ and all }\; T > o \tag{1}$$

1) Russell's method [4] : Investigate a corresponding hyperbolic problem and use a Fourier transform technique to translate the results to the parabolic problem. This method strongly relies on "local decay of energy"-results from scattering theory and is confined to second order equations, but can handle some cases of
   - i) x-dependent coefficients ([7],[2])
   - ii) t-dependent coefficients ([8])
   - iii) control only on part of $\partial\Omega$ ([5]).

2) Littman s method [3] : Construct a fundamental solution for the parabolic operator with support in $[o,T) \times \mathbb{R}^n$. Littman's construction uses the Radon transform to reduce the general case to the case $n=1$ and therefore (up to now) only works with constant coefficients in the general case, but can handle higher order operators as well.

Using (1) it can be seen [1] that

$$R := R(T,y)$$

is in fact independent of y and T and hence a vector space.

2. EXACT CONTROLLABILITY

D.L. Russell has shown that

$$\mathcal{D} := \{ z \in L^2(\Omega) \mid \sum_{k=1}^{\infty} |\langle z,\phi_k\rangle|^2 \exp(2\beta\lambda_k^{1/2}) < \infty \} \tag{2}$$

is a subspace of R . (Here λ_k and ϕ_k are the eigenvalues resp. eigenfunctions of L and thus $\mathcal{D}$ is the domain of definition of $\exp(\beta L^{1/2})$.) Using the observation that (1) gives the freedom to construct trajectories running into z by disregarding y and T the present author constructed another subspace of R , namely [9]

$$X := \{ z \mid L^i z \in L^2(\Omega) \text{ for all } i \text{ and } \sum_{i=o}^{\infty} \lambda_1^{-i} \, ||L^i z|| < \infty \}$$

Although $\mathcal{D}$ is optimal if one insists on characterizing reachability by growth properties of the Fourier coefficients $\langle z,\phi_k\rangle$ (cf. [4]) it can be shown that the sum $X + \mathcal{D}$ is direct. Treating suitable finite dimensional eigenspaces separately we can prove the following generalization of [9],[1o] $(L_\mu := L-\mu)$:

THEOREM 1 *If* (1) *holds, then for arbitrary* $\mu \in \mathbb{C}$ *and arbitrary* $d \in \mathbb{R}^+$

$$X_\mu^d := \{ z \mid L_\mu^i z \in L^2(\Omega) \text{ for all } i \text{ and } \sum_{i=o}^{\infty} d^{-i} ||L_\mu^i z|| < \infty \} \subset R$$

To simplify notation let us fix d:=1 (for instance) and define

$$p_\mu(z) := \sum_{i=o}^{\infty} ||L_\mu^i z||$$

for $z \in X_\mu := X_\mu^1$. Then we can prove a quantitative version of Theorem 1 :

THEOREM 1´ *For any* $\mu \in \mathbb{C}$ *, any* $z \in X_\mu$ *and any* $T \in \mathbb{R}^+$ *there exist* y,u *such that*

i) $y \overset{o \qquad T}{\underset{u}{\longmapsto}} z$

ii) $||y||\,,\ ||u|| \leq C\,(1 + |\mu|)\,\exp((\mathrm{Re}\,\mu)T)\;p_\mu(z)$

where C *is independent of* μ,z,T *while* μ *resp.* T *vary in* $\mathrm{Re}\,\mu \geq \rho_o$ *resp.* $(o,T_{max}]$.

REMARK In fact the cases $-\mathrm{Re}\,\mu$ large and/or T large are even easier to handle; they have been excluded for the sake of simplicity,only.

Looking on Russell´s space $\mathcal{D}$ from the viewpoint of Theorem 1 we see that the elements of $\mathcal{D}$ are rapidly converging series over very special functions from some X_{μ_k} , namely

i) $\mu_k = \lambda_k$ (eigenvalues)

ii) $z_k := \langle z,\phi_k\rangle\phi_k$ satisfies homogeneous boundary conditions

iii) z_k satisfies the differential equation $L_{\mu_k} z_k = o$ and thus belongs trivially to X_{μ_k} , in fact (since $L_{\mu_k}^i z_k = o$ for $i > o$) we have $p_{\mu_k}(z_k) = ||z_k||$

All of these restrictions can be removed and we can show

THEOREM 2 *In cases where Russell's method is applicable there is an* $\alpha \in \mathbb{R}^+$ *such that any series*

$$z = \sum z_k$$

belongs to $\mathcal{R}$ *provided*

i) $z_k \in X_{\mu_k}$ *for some complex sequence* (μ_k)

ii) $\sum_{k=1}^{\infty} p_{\mu_k}(z_k)\ (1 + |\mu_k|)\ \exp(\alpha(\mathrm{Re}\ \mu_k)^{1/2}) < \infty$

To make a comparison with (2) we have to assume something about the behavior of the complex sequence (μ_k) :

COROLLARY 1 *If* $\mathrm{Re}\ \mu_k$ *is bounded then in Theorem 1 ii) may be replaced by*

ii') $\sum_{k=1}^{\infty} p_{\mu_k}(z_k)\ (1 + |\mu_k|) < \infty$

Condition ii') looks very mild when compared to (1). But since the eigenvalues λ_k are real and $\lambda_k \to \infty$ the following Corollary is better suited for a comparison:

COROLLARY 2 *If* $\mathrm{Re}\ \mu_k \geq K \ln|\mu_k|$ *(which is trivially satisfied for* μ_k *(real)* $\to \infty$*) then in Theorem 1 ii) may be replaced by*

ii'') $\sum_{k=1}^{\infty} p_{\mu_k}(z_k)\ \exp(\alpha(\mathrm{Re}\ \mu_k)^{1/2}) < \infty$

For higher order operators we have

THEOREM 3 *In cases where Littman's method is applicable any series* $z = \Sigma\, z_k$ *belongs to* R *provided*

i) $z_k \in X_{\mu_k}$ *for some complex sequence* (μ_k)

ii) $\sum_{k=1}^{\infty} p_{\mu_k}(z_k)\ (1 + |\mu_k|)\ \exp([\ln \rho_k]^{\frac{1+\varepsilon}{2m}}\ \rho_k^{1/2m}) < \infty$

for some $\varepsilon > 0$ *(* $\rho_k := \mathrm{Re}\ \mu_k$ *;* $2m$ *: order of* L *)*

Of course, if something is known about the behavior of the complex sequence (μ_k) then similar conclusions as in Corollaries 1 and 2 can be drawn from Theorem 3. It is interesting to note that for higher order operators (where no comparable results from moment methods are available) the convergence of the series z_k need not be as rapid as in the case of order 2 .

Theorems 2 and 3 can be proved using the following elementary technique [1o] : Choose times T_k individually for the z_k and (applying Theorem 1´) pick y_k, u_k such that

$$y_k \;\xmapsto[u'_k]{T-T_k \qquad T}\; z_k \tag{3}$$

Now use (1) to steer $-y_k$ into zero:

$$-y_k \;\xmapsto[u''_k]{T-T_k \qquad T}\; o \tag{4}$$

Add the two processes and adjoin the trivial process

$$o \;\xmapsto[o]{o \qquad T-T_k}\; o$$

Then z has been reached from zero by the control

$$u_k := E_o(u'_k + u''_k) \qquad (\, E_o \text{ : extension by zero }\,)$$

Trying to sum over k we realize that we have to estimate the u_k . Theorem 1´ furnishes such an estimate for process (3) whereas (4) will be investigated in the following section.

3. NULL-CONTROLLABILITY WITH AN ORDER

DEFINITION 1 *The system under consideration is "null-controllable with order* $\rho(\cdot)$ *" iff for all* $T > o$ *and all* y *there exists* u *such that*

i) $y \;\xmapsto[u]{o \qquad T}\; o$

ii) $||u|| \leq K \exp(\rho(T))\, ||y||$

where K *is independent of* T *and* y .

The sketch of proof for Theorems 2 and 3 can now be completed using

THEOREM 2′ (Th. Seidman [7]) *In cases where Russell′s method is applicable we have null-controllability with order* γ/T .

THEOREM 3′ *In cases where Littman′s method is applicable we have null-controllability with order*

$$\rho(T) := |\ln T|^{\frac{1+\varepsilon}{2m-1}} \; T^{\frac{-1}{2m-1}}$$

Theorem 3′ is proved by choosing the cut-off function ϕ used in [3] from an almost optimal Gevrey class and by doing a careful study of the asymptotic behavior of the constructed fundamental solution when T goes to zero. The extra logarithmic terms in Theorems 3 and 3′ could only be removed for a ϕ in an even better Gevrey class which would make ϕ quasi-analytic and hence useless as a cut-off.

REFERENCES

[1] Fattorini, H.O. 1978. Reachable states in boundary control of the heat equation are independent of time. *Proc. Roy. Soc. Edinburgh* 81: 71-77.

[2] Lagnese, J. 1983. Decay of solutions of wave equations in a bounded region with boundary dissipation. *J. Diff. Eq.* 5o: 163-182.

[3] Littman, W. 1978. Boundary control theory for hyperbolic and parabolic partial differential equations with constant coefficients. *Ann. Sc. Norm. Sup. Pisa* 37: 567-58o.

[4] Russell, D.L. 1973. A unified boundary controllability theory for hyperbolic and parabolic partial differential equations. *Studies in Appl. Math.* 52: 189-211.

[5] - 1974. Boundary value controllability of wave and heat processes in star-complemented regions. Proc. Conference on Differential Games and Control Theory (Kingston, R.I., 1973), Marcel Dekker, New york.

[6] - 1978. Controllability and stabilizability theory for linear partial differential equations: recent progress and open questions. *SIAM Rev.* 2o: 639-739.

[7] Seidman, Th. 1977. Observation and prediction for the heat equation. IV: Patch observability and controllability. *SIAM J. Control and Opt.* 15: 412-427.

[8] - 1984. Two results on exact boundary control for parabolic equations. *Appl. Math. Opt.* 11:145-152.

[9] Weck, N.1984. More states reachable by boundary control of the heat equation. *SIAM J. Control and Opt. 22:* 699-71o.

[1o] - 1984. On exact boundary controllability for parabolic equations. In *Optimal Control of Partial Differential Equations (Conference Held at the Mathematisches Forschungsinstitut Oberwolfach, December 5-11, 1982)*, ed. K.-H. Hoffmann and W.Krabs. Birkhäuser,Basel.

VARIABLE STRUCTURE CONTROL FOR SOME EVOLUTION EQUATIONS†

T. Zolezzi

Dipartimento di Matematica
Universita di Genova, Italy

We consider control systems given by

$$y \quad + A(y) \quad B\,u, \quad 0 \leq t \leq T; \quad y\,(0) \;=\; y_o \tag{1}$$

with control objective given by a *sliding manifold*

$$s[y(t)] \;=\; 0, \qquad 0 \leq t \leq T. \tag{2}$$

The problem is to obtain admissible states y which satisfy (2).

Here T is fixed. We are given a real reflexive Banach space V, Hilbert spaces H, W, Z such that

$$V \subset H \subset V^*$$

with continuous and dense imbedding. All spaces are assumed to be separable. The state variable $y(t) \in V$, the control variable $u \in U$ is a given closed convex subset of W. We assume that

$$s : V \text{ or } H \to Z$$

is Frechet differentiable;

$$B : W \to V^*$$

is a bounded linear mapping;

$$A : V \to V^*$$

is monotone, semicontinuous, and

$$\langle A(x), x\rangle \geq c_1 \| x \|^p; \; \| A(x) \| \leq c_2 + c_3 \| x \|^{p-1}$$

---

† Work partially supported by M.P.I.

for some $p > 1$, $c_1 > 0$. Admissible controls are further constrained by

$$\int_0^T \|u\|^{p'} dt \leq c_4 < +\infty ,$$

where p' is the conjugate exponent to p. Solutions y to (1) are meant in the sense of [7].

We want to extend to the present infinite-dimensional setting some known results of variable structure control theory for ordinary differential equations, see [1], [2], [3], [4], [5], [6].

Our approach is the following. We first fix bases (possible finite) $v_1,\ldots,v_n,\ldots,$ in V; $w_1,\ldots,w_n,\ldots,$ in W; $z_1,\ldots,z_n,\ldots,$ in Z. Then we consider for all m a suitable closed convex $\cup_m \subset sp\ (w_1,\ldots,w_m)$. Next we solve the following finite-dimensional variable structure control problem, obtained from the Faedo-Galerkin approximations to (1), (2):

$$\langle y, v_j\rangle + \langle A(y), v_j\rangle = \langle Bu, v_j\rangle , \quad j = 1,\ldots,n,$$

$$\begin{cases} y(0) \longrightarrow y_0 \text{ in } H, \\ s_m(y) = 0, \quad 0 \leq t \leq T; \quad m \leq n, \end{cases} \qquad (3)$$

where

$$s_m(y) \quad = \quad \text{vector of components} \quad (s(y, z_j)), \quad j = 1,\ldots,m,$$

and

$$y(t) \in sp\ (v_1,\ldots,v_n); \quad u \in \cup_m.$$

It is well known that solutions to (3) involve using discontinuous feedback (see [1]).

Existence of solutions to (3) may be obtained under suitable conditions from viability theory (see [9]), in such a way that there is equivalence between almost everywhere solutions and Filippov solutions to (3) (see [3]).

Then we take limits in $L^p(0,T,V)$ for the approximate states in (3), under suitable boundedness and continuity assumptions. In this way we get solutions to (1), (2). Examples of operators A to which the above theory applies are the following:

(i) bounded linear operators A;

$$\text{(ii)} \quad A(y) = -\sum_{i.j=1}^{N} {}^{\partial}/_{\partial x_i} [a_{ij}(x,y)\, {}^{\partial y}/_{\partial x_j} + b_i(x,y)],$$

$$y \in H_o^1(D) = V,$$

where a_{ij} and b_i are Caratheodory functions of at most linear growth with respect to y, and the matrix (a_{ij}) is uniformly positive definite for $x \in D$ a bounded open set in R^N and all y.

A key problem is to properly relate the behavior of the control system (1) *near* the sliding manifold (2) to that obtained by keeping the state variable *on*such a manifold. The mere existence of solutions to (1), (2) is not sufficient for solving mathematically the original variable structure control problem (see [1]).

Generalizing the definition of *approximability* given for the first time in [3] and [4], we isolate a relevant property of (1), (2) which rules out ambiguous systems (see [1] for terminology).

Sufficient conditions are obtained to fulfill such a property.

Some results in temperature control obtained in [8] by variable structure methods are generalized by the present results, which may be considered a first step towards a mathematically well-defined approach to variable structure distributed control problems.

Detailed results will be presented elsewhere.

REFERENCES

[1] V.I. Utkin. *Sliding Modes and Their Application in Variable Structure Systems*. Moscow: Mir, 1978. (English translation)

[2] ———. "Variable Structure Systems. Present and Future." In *Autom. Remote Control* 44 (1984): 1105-20. (English translation)

[3] G. Bartolini, and T. Zolezzi. "Control of Nonlinear Variable Structure Systems." *J. Math. Anal. Appl*. To appear.

[4] ———. "Variable Structure Systems Nonlinear in the Control Law." *IEEE Trans. Autom. Control* AC-30 (1985): 6818-24.

[5] ———. "Nonlinear Control of Variable Structure Systems." *Lecture Notes in Control Information Sciences* 62, pt. 1 (1984): 542-9.

[6] ———. "Dynamic Output Feedback for Observed Variable Structure Systems." *Systems Control Letters.*

[7] J.L. Lions. *Quelques methodes de resolution des problemes aux limites non lineaire*. Paris: Dunod-Gauthier-Villars, 1969.

[8] Yu.V. Orlov, and V.I. Utkin. "Use of Sliding Modes in Distributed Systems Control Problems (in Russian)" *Automatika i Telemekhanika* 9 (1982): 36-46.

[9] J.P. Aubin, and A. Cellina. *Differential Inclusions*. Berlin Heidelberg New York Tokyo: Springer-Verlag, 1984.

Lecture Notes in Control and Information Sciences

Edited by M. Thoma

Vol. 43: Stochastic Differential Systems
Proceedings of the 2nd Bad Honnef Conference
of the SFB 72 of the DFG at the University of Bonn
June 28 – July 2, 1982
Edited by M. Kohlmann and N. Christopeit
XII, 377 pages. 1982.

Vol. 44: Analysis and Optimization of Systems
Proceedings of the Fifth International
Conference on Analysis and Optimization of Systems
Versailles. December 14–17, 1982
Edited by A. Bensoussan and J.L. Lions
XV, 987 pages, 1982

Vol. 45: M. Arató
Linear Stochastic Systems
with Constant Coefficients
A Statistical Approach
IX, 309 pages. 1982

Vol. 46: Time-Scale Modeling of Dynamic Networks
with Applications to Power Systems
Edited by J.H. Chow
X, 218 pages. 1982

Vol. 47: P.A. Ioannou, P.V. Kokotovic
Adaptive Systems with Reduced Models
V, 162 pages. 1983

Vol. 48: Yaakov Yavin
Feedback Strategies for Partially
Observable Stochastic Systems
VI, 233 pages, 1983

Vol. 49: Theory and Application of Random Fields
Proceedings of the IFIP-WG 7/1
Working Conference
held under the joint auspices of the
Indian Statistical Institute
Bangalore, India, January 1982
Edited by G. Kallianpur
VI. 290 pages. 1983

Vol. 50: M. Papageorgiou
Applications of Automatic Control Concepts
to Traffic Flow Modeling and Control
IX, 186 pages. 1983

Vol. 51: Z. Nahorski, H.F. Ravn, R.V.V. Vidal
Optimization of Discrete Time Systems
The Upper Boundary Approach
V, 137 pages 1983

Vol. 52: A.L. Dontchev
Perturbations, Approximations and Sensitivity Analysis
of Optimal Control Systems
IV, 158 pages. 1983

Vol. 53: Liu Chen Hui
General Decoupling Theory of Multivariable
Process Control Systems
XI, 474 pages. 1983

Vol. 54: Control Theory for Distributed
Parameter Systems and Applications
Edited by F. Kappel, K. Kunisch,
W. Schappacher
VII, 245 pages. 1983.

Vol. 55: Ganti Prasada Rao
Piecewise Constant Orthogonal Functions
and Their Application to Systems and Control
VII, 254 pages. 1983.

Vol. 56: Dines Chandra Saha, Ganti Prasada Rao
Identification of Continuous
Dynamical Systems
The Poisson Moment Functional
(PMF) Approach
IX, 158 pages. 1983.

Vol. 57: T. Söderström, P.G. Stoica
Instrumental Variable Methods
for System Identification
VII, 243 pages. 1983.

Vol. 58: Mathematical Theory of
Networks and Systems
Proceedings of the MTNS-83 International
Symposium
Beer Sheva, Israel, June 20–24, 1983
Edited by P.A. Fuhrmann
X, 906 pages. 1984

Vol. 59: System Modelling and Optimization
Proceedings of the 11th IFIP Conference
Copenhagen, Denmark, July 25-29, 1983
Edited by P. Thoft-Christensen
IX, 892 pages. 1984

Vol. 60: Modelling and Performance
Evaluation Methodology
Proceedings of the International Seminar
Paris, France, January 24–26, 1983
Edited by F. Bacelli and G. Fayolle
VII, 655 pages. 1984

Vol. 61: Filtering and Control of Random
Processes
Proceedings of the E.N.S.T.-C.N.E.T. Colloquium
Paris, France, February 23–24, 1983
Edited by H. Korezlioglu, G. Mazziotto, and
J. Szpirglas
V, 325 pages. 1984

Lecture Notes in Control and Information Sciences

Edited by M. Thoma and A. Wyner

Vol. 62: Analysis and Optimization of Systems
Proceedings of the Sixth International Conference on Analysis and Optimization of Systems
Nice, June 19–22, 1984
Edited by A. Bensoussan, J. L. Lions
XIX, 591 pages. 1984.

Vol. 63: Analysis and Optimization of Systems
Proceedings of the Sixth International Conference on Analysis and Optimization of Systems
Nice, June 19–22, 1984
Edited by A. Bensoussan, J. L. Lions
XIX, 700 pages. 1984.

Vol. 64: Arunabha Bagchi
Stackelberg Differential Games in Economic Models
VIII, 203 pages, 1984

Vol. 65: Yaakov Yavin
Numerical Studies in Nonlinear Filtering
VIII, 273 pages, 1985.

Vol. 66: Systems and Optimization
Proceedings of the Twente Workshop
Enschede, The Netherlands, April 16–18, 1984
Edited by A. Bagchi, H. Th. Jongen
X, 206 pages, 1985.

Vol. 67: Real Time Control of Large Scale Systems
Proceedings of the First European Workshop
University of Patras, Greece, Juli 9–12, 1984
Edited by G. Schmidt, M. Singh, A. Titli, S. Tzafestas
XI, 650 pages, 1985.

Vol. 68: T. Kaczorek
Two-Dimensional Linear Systems
IX, 397 pages, 1985.

Vol. 69: Stochastic Differential Systems – Filtering and Control
Proceedings of the IFIP-WG 7/1 Working Conference
Marseille-Luminy, France, March 12-17, 1984
Edited by M. Metivier, E. Pardoux
X, 310 pages, 1985.

Vol. 70: Uncertainty and Control
Proceedings of a DFVLR International Colloquium
Bonn, Germany, March, 1985
Edited by J. Ackermann
IV, 236 pages, 1985.

Vol. 71: N. Baba
New Topics in Learning Automata Theory and Applications
VII, 231 pages, 1985.

Vol. 72: A. Isidori
Nonlinear Control Systems: An Introduction
VI, 297 pages, 1985.

Vol. 73: J. Zarzycki
Nonlinear Prediction Ladder-Filters for Higher-Order Stochastic Sequences
V, 132 pages, 1985.

Vol. 74: K. Ichikawa
Control System Design based on Exact Model Matching Techniques
VII, 129 pages, 1985.

Vol. 75: Distributed Parameter Systems
Proceedings of the 2nd International Conference, Vorau, Austria 1984
Edited by F. Kappel, K. Kunisch, W. Schappacher
VIII, 460 pages, 1985.

Vol. 76: Stochastic Programming
Edited by F. Archetti, G. Di Pillo, M. Lucertini
V, 285 pages, 1986.

Vol. 77: Detection of Abrupt Changes in Signals and Dynamical Systems
Edited by M. Basseville, A. Benveniste
X, 373 pages, 1986.

Vol. 78: Stochastic Differential Systems
Proceedings of the 3rd Bad Honnef Conference, June 3–7, 1985
Edited by N. Christopeit, K. Helmes, M. Kohlmann
V, 372 pages, 1986.

Vol. 79: Signal Processing for Control
Edited by K. Godfrey, P. Jones
XVIII, 413 pages, 1986.

Vol. 80: Artificial Intelligence and Man-Machine Systems
Edited by H. Winter
IV, 211 pages, 1986.

Lecture Notes in Control and Information Sciences

Edited by M. Thoma and A. Wyner

Vol. 81: Stochastic Optimization
Proceedings of the International Conference, Kiew, 1984
Edited by I. Arkin, A. Shiraev, R. Wets
X, 754 pages, 1986.

Vol. 82: Analysis and Algorithms of Optimization Problems
Edited by K. Malanowski, K. Mizukami
VIII, 240 pages, 1986.

Vol. 83: Analysis and Optimization of Systems
Proceedings of the Seventh International Conference of Analysis and Optimization of Systems
Antiba, June 26-27, 1986
Edited by A. Bensoussan, J. L. Lions
XVI, 901 pages, 1986.

Vol. 84: System Modelling and Optimization
Proceedings of the 12th IFIP Conference
Budapest, Hungary, September 2–6, 1985
Edited by A. Prékopa, J. Szelezsán, B. Strazicky
XII, 1046 pages, 1986.

Vol. 85: Stochastic Processes in Underwater Acoustics
Edited by Charles R. Baker
V, 205 pages, 1986.

Vol. 86: Time Series and Linear Systems
Edited by Sergio Bittanti
XVII, 243 pages, 1986.

Vol. 87: Recent Advances in System Modelling and Optimization
Proceedings of the IFIP-WG 7/1 Working Conference
Santiago, Chile, August 27-31, 1984
Edited by L. Contesse, R. Correa, A. Weintraub
IV, 199 pages, 1987.

Vol. 88: Bruce A. Francis
A Course in H_∞ Control Theory
XI, 156 pages, 1987.

Vol. 89: G. K. H. Pang/A. G. J. McFarlane
An Expert System Approach to Computer-Aided Design of Multivariable Systems
XII, 223 pages, 1987.

Vol. 90: Singular Perturbations and Asymptotic Analysis in Control Systems
Edited by P. Kokotovic, A. Bensoussan, G. Blankenship
VI, 419 pages, 1987.

Vol. 91 Stochastic Modelling and Filtering
Proceedings of the IFIP-WG 7/1 Working Conference
Rome, Italy, Decembre 10-14, 1984
Edited by A. Germani
IV, 209 pages, 1987.

Vol. 92: L. T. Grujić, A. A. Martynyuk, M. Ribbens-Pavella
Large-Scale Systems Stability Under Structural and Singular Perturbations
XV, 366 pages, 1987.

Vol. 93: K. Malanowski
Stability of Solutions to Convex Problems of Optimization
IX, 137 pages, 1987.

Vol. 94: H. Krishna
Computational Complexity of Bilinear Forms
Algebraic Coding Theory and Applications to Digital Communication Systems
XVIII, 166 pages, 1987.

Vol. 95: Optimal Control
Proceedings of the Conference on Optimal Control and Variational Calculus
Oberwolfach, West-Germany, June 15-21, 19
Edited by R. Bulirsch, A. Miele, J. Stoer and K. H. Well
XII, 321 pages, 1987.

Vol. 96: H. J. Engelbert/W. Schmidt
Stochastic Differential Systems
Proceedings of the IFIP-WG 7/1 Working Conference
Eisenach, GDR, April 6-13, 1986
XII, 381 pages, 1987.

Vol. 97: I. Lasiecka/R. Triggiani (Eds.)
Control Problems for Systems Described by Partial Differential Equations and Applications
Proceedings of the IFIP-WG 7.2 Working Conference
Gainesville, Florida, February 3-6, 1986
VIII, 400 pages, 1987.